Advances in Material Research and Technology

Series Editor

Shadia Jamil Ikhmayies, Physics Department, Isra University, Amman, Jordan

This Series covers the advances and developments in a wide range of materials such as energy materials, optoelectronic materials, minerals, composites, alloys and compounds, polymers, green materials, semiconductors, polymers, glasses, nanomaterials, magnetic materials, superconducting materials, high temperature materials, environmental materials, Piezoelectric Materials, ceramics, and fibers.

Shadia Jamil Ikhmayies
Editor

Advanced Ceramics

Springer

Editor
Shadia Jamil Ikhmayies
Department of Physics, School of Science
University of Jordan
Amman, Jordan

ISSN 2662-4761 ISSN 2662-477X (electronic)
Advances in Material Research and Technology
ISBN 978-3-031-43917-9 ISBN 978-3-031-43918-6 (eBook)
https://doi.org/10.1007/978-3-031-43918-6

This Springer imprint is published by the registered company Springer Nature Switzerland AG
The registered company address is: Gewerbestrasse 11, 6330 Cham, Switzerland

Paper in this product is recyclable.

Preface

Advanced ceramics were defined by the 1993 Versailles Project on Advanced Materials and Standards (VAMAS), which described an advanced ceramic as "an inorganic, nonmetallic (ceramic), basically crystalline material of rigorously controlled composition and manufactured with detailed regulation from highly refined and/or characterized raw materials giving precisely specified attributes." Advanced ceramics are either based on oxides, non-oxides, or combinations of the two. Typical oxides used are alumina (Al_2O_3) and zirconia (ZrO_2), while non-oxides are most often carbides, nitrides, borides, and silicides. Due to their remarkable high-temperature strength, hardness, and electrical properties, advanced ceramics are used to replace metals, polymers, and refractory materials in a variety of applications. By changing the chemical composition or manufacturing processes, their properties can be adapted to meet a wide range of challenging conditions. Nowadays, advanced ceramic materials are well established in many uses and industries including metals production and processing, aerospace, electronics, automotive, and personnel protection. In medicine, advanced ceramics are playing an increasingly important role, with alumina and zirconia being used as dental implants, and as bone substitutes in orthopedic operations such as hip and knee replacement.

This book introduces the reader to advanced ceramics through comprehensive reviews as well as research findings which include types, classifications, chemical and physical properties, requirements of different types of ceramics, synthesis methods, production, processing methods characterization techniques, properties, and applications of different types of ceramics. Therefore, it is suitable as text or a reference for graduate and undergraduate students in ceramic-related disciplines and is suitable as a reference for researchers in advanced ceramic research, and professionals in ceramic industries. In addition, it is a reference for researchers in perovskite ceramics, perovskite solar cells, nanoscale rare-earth-based mixed-metal oxides for solar photocatalytic applications, uses of advanced ceramics for self-healing coatings and treatment of wastewater. The book includes eight chapters which are described briefly in the following paragraphs.

Chapter one, “Advanced Ceramics: Stages of Development” by Mahmoud F. Zawrah et al., sheds light on the stages of development of advanced ceramics. The authors present the types and classifications of advanced ceramics, advanced processing techniques, properties, and sintering as well as new forms of applications. Examples of these kinds of advanced ceramics, e.g., alumina, zirconia, Mg–Al spinels, silicon carbide, silicon nitride, ceramic composites, thin films, etc., with their specific applications, are also presented in this chapter. Chapter two, “Recent Advances in Perovskite Ceramics: Synthesis, Processing, and Applications” by José Luis Clabel Huamán et al., presents the recent advances in perovskite ceramics. In this chapter, the authors reported the synthesis, processing, characterization, and advanced applications of perovskite ceramics. Chapter three, “Advanced Perovskite Solar Cells” by Yuqin Tian et al., introduces the most common perovskite thin films and their preparation methods, organic–inorganic perovskite solar cells, etc., focusing on the development status of perovskite solar cells, and the main factors affecting their stability. The current problems and development prospects in the research and application of perovskite solar cells are also presented, which will lay a solid foundation for a deeper understanding of perovskite solar cells and the preparation of new and efficient perovskite ones.

Chapter four, “Advanced Ceramics (Self-healing Ceramic Coatings)” by Ali Shanaghi et al., describes different types of ceramics-based self-healing coatings including titania, zirconia, titanium–alumina, and zirconia–alumina incorporated with Benzotriazole (BTA) as an inhibitor. Fabrication processes are also described in this chapter. Chapter five, “Utilization of Advanced Ceramics Towards Treatment of Wastewater” by Deepti et al., focuses on various advanced ceramic materials that have been used for wastewater treatment purposes. The chapter first presents a detailed insight into ceramic adsorbents, resins, and aerogels, as well as ceramic membranes, which have been extensively used for wastewater treatment. Then discusses advanced ceramics to modify membrane-based technology to handle wastewater treatment. The future and perspective of these advanced ceramic materials and their modifications to ensure better efficacy toward environmental remediation purposes are highlighted. Chapter six, “Nanoscale Rare-Earth-Based Mixed-Metal Oxides for Solar Photocatalytic Applications” by Sahar Zinatloo-Ajabshir and Seyyed Javad Heydari-Baygi, summarizes the recent advances in production techniques and photocatalytic uses of $Ln_2B_2O_7$ (Ln: lanthanide, B=Zr, Ce, and Sn) nanostructures, where various methods for the synthesis of lanthanide zircons, lanthanide stannates, and lanthanides for ceramic production have been presented. The authors discussed the advantages and disadvantages of each manufacturing method and reviewed the applications of solar photocatalysis for mixed-metal oxide nanostructures based on rare earths.

Chapter seven, “Electrophoretic Deposition of Hydroxyapatite Incorporated Composite Coatings on Metallic Substrates: A Review of the Fundamentals” by Sandeep Singh et al., presents a review of the fundamental and technical aspects of the electrophoretic deposition (EPD) method to prepare hydroxyapatite (HA) coatings. The factors influencing the deposition process, suspension preparation, control of suspension mechanism, and deposition of composite coatings obtained by EPD are all discussed. The authors also give a comprehensive analysis of the kinetics involved in EPD of hydroxyapatite (HA)-reinforced coatings, and the different factors which can influence the surface morphology, corrosion behavior, and in vitro bioactivity assessment. Chapter eight, "Geopolymers Prepared by Microwave Treatments" by Yuta Watanabe and Takaomi Kobayashi, highlights the use of microwaves as efficient auxiliary processing for geopolymer production. The chapter also describes the feasibility of synthesizing the porous morphology and dense-structured geopolymers without micro-sized pores using microwave energy.

Amman, Jordan Shadia Jamil Ikhmayies

Contents

Advanced Ceramics: Stages of Development

Mahmoud F. Zawrah, Mohammed A. Taha, and Rasha A. Youness

Abstract Ceramics are inorganic nonmetallic materials (oxides, carbides, nitrides, etc.) processed after sintering of natural or synthetic precursors at high temperature. They can also be applied at or resist high firing temperatures. They are highly crystalline (most of the advanced and traditional ceramics), semi-crystalline (vitrified ceramics such as earthenware, stoneware, and porcelain), or completely amorphous (glasses). The composition/structure relationship, method of processing, raw materials, and applications determine the properties of ceramics and whether the ceramics are traditional or advanced ones. The first man-made ceramics were pottery objects and figurines from clays after firing. Several stages have been considered in the development of ceramic industry until reaching the production of advanced ceramics. The second stage for development of ceramic includes the production of glazed-colored ceramics, ceramic arts, and building products. Recently, new categories of advanced ceramics have been developed for electronics, biomedical, semiconductors, energy, and optical and structural applications. In the present chapter, we are going to shed light on the stages of development for advanced ceramics. Types and classifications, advanced processing techniques, properties, sintering as well as new forms of applications will be presented in the current chapter. Examples of these kinds of advanced ceramics, e.g., alumina, zirconia, Mg–Al spinels, silicon carbide, silicon nitride, ceramic composites, thin films, etc., with their specific applications will be also presented.

M. F. Zawrah (✉)
Refractories, Ceramics and Building Material Department-Center of Excellence for Advanced Sciences, Advanced materials and Nano group, National Research Centre, El Buhouth St., Dokki, Cairo 12622, Egypt
e-mail: mzawrah@hotmail.com

Pharos University in Alexandria, Canal Mahmoudiah Street, Smouha, Alexandria, Egypt

M. A. Taha
Solid State Physics Department, National Research Centre, El Buhouth St., Dokki, Cairo 12622, Egypt

R. A. Youness
Spectroscopy Department, National Research Centre, El Buhouth St., Dokki, Cairo 12622, Egypt

S. J. Ikhmayies (ed.), *Advanced Ceramics*, Advances in Material Research and Technology, https://doi.org/10.1007/978-3-031-43918-6_1

Keywords Advanced ceramics · Classification · Processing · Sintering · Properties

1 Historical Background of Ceramics

Ceramic, as a word, comes from a Greek word called kéramos, "potter's clay". It is one of the most ancient products and its industry has been discovered since thousands of years. Its discovery returns to the discovery of clays and clay firing. The first ceramic artifact was produced in 28,000 BCE (before the Common Era), i.e., through the Paleolithic age. It was a statue of a woman, called the Venus of Dolní Věstonice, from Brno, Czech Republic. After thousands of years (18,000–17,000 BCE), the first pottery was detected in China and then in Japan and Russia in 14,000 BCE. The application of ceramic has been raised intensively though the Neolithic period (9,000 BCE) and the ceramic-clay products became common as vessels for water and food, art stuffs, tiles and bricks, and their application moved to the Middle East and Europe. The first products were withered under the sun or sintered at $\leq$1,000 °C in simple furnaces dug in the ground. In 7,000 BCE, sharp tools prepared from naturally occurring volcanic glass have been used. It has been stated by the Roman historian Pliny that the first artificial glass was unintentionally prepared by Phoenician merchants in 5,000 BCE during cooking on the fired sodium-containing rocks while relaxing on the beach. The heating led to fusion of rocks and formed glass. After that, some glass objects have been produced in Mesopotamia and Egypt dating to 3,500 BCE. Also, during that period the wheel-forming tool for the production of ceramic artifacts and pottery has been started to be applied. A new development, i.e., glazing of pottery, has been introduced in Mesopotamia in 3000 BCE. On the other hand, the Egyptians began to build firms for glassware industries in 1500 BCE. At the same time, the Greeks reached the maximum development for vases industry. In 600 CE (Common Era), the first kiln working at 1350°C was made in China to produce porcelain from kaolinite. This industry continued in Europe and the Middle East during 1600 CE. The first blast furnace made from natural or synthetic refractory materials working at 1500 °C was produced in Europe in 1500 CE for melting of metals, glasses, building materials, and porcelain. With the progression of industries, new materials with new properties have appeared. The electrical insulator ceramic was invented in 1850. After 1939, the application of advanced ceramics in different areas such as electronics–optoelectronics, biomedical, energy, automotive, and aerospace has been increased. With the emergence of nanotechnology, new applications such as transparent ceramics, ductile ceramics, hyperelastic-bones, microscopic capacitors, and ceramic coatings are developed. From the aforementioned stages of ceramic development, it is expected that the worldwide market will reach 1.1 trillion dollars against 800 billion dollars in 2018 [1].

2 Introductory Remarks on Ceramics

Ceramics are inorganic solid materials composed of metal/metalloid and nonmetal elements with ionic or covalent bonds, e.g., metal oxides, carbides, and nitrides. They can be prepared at and withstand high temperature. The nature of chemical bonds intensely affects the properties of ceramic materials such as high melting point, and thermomechanical and optical properties [2, 3]. Most of the ceramics are insulators for electricity due to the deficiency of free electron as in the case of metals [4]. The most common conventional raw materials are kaolin and the advanced ones are alumina, spinels, silicon carbide, tungsten carbide, silicon nitride, etc. These materials can be applied as wear-resistant materials, cutting tools, biomedical, electronics, and structural application. Most of the ceramic materials are crystalline, while the amorphous ones are being glasses. The properties of ceramics depend mainly on the structure and chemical composition of ceramics. The microstructure, density, grain size distribution, amount of porosity, and liquid phase contents are important factors affecting the overall properties of ceramic materials. The microstructure comprises the main grains, secondary phases, grain boundaries, closed and open pores, microcracks, and structural defects. These parameters in addition to the processing conditions are totally affecting mechanical, optical, thermal, electrical, and magnetic properties.

3 Classification of Ceramics

Ceramics are classified into two main categories including traditional (conventional) and advanced ones. The first group involves whitewares, structural and heavy clay products, cement, and refractories and they are derived from natural sources [5]. Also, it includes glass and glass-ceramics [4] which can be produced from pure chemicals or natural resources. Some types of glass and glass-ceramics belong to the category of advanced ceramics depending on their advanced applications. On the other hand, the advanced ceramics are fabricated by sophisticated chemical processing methods from synthetic or raw materials giving the desirable product with high purity and improved physical, mechanical, electrical, magnetic, superconducting as well as optical properties. The following scheme (Fig. 1) represents the classification of ceramic materials.

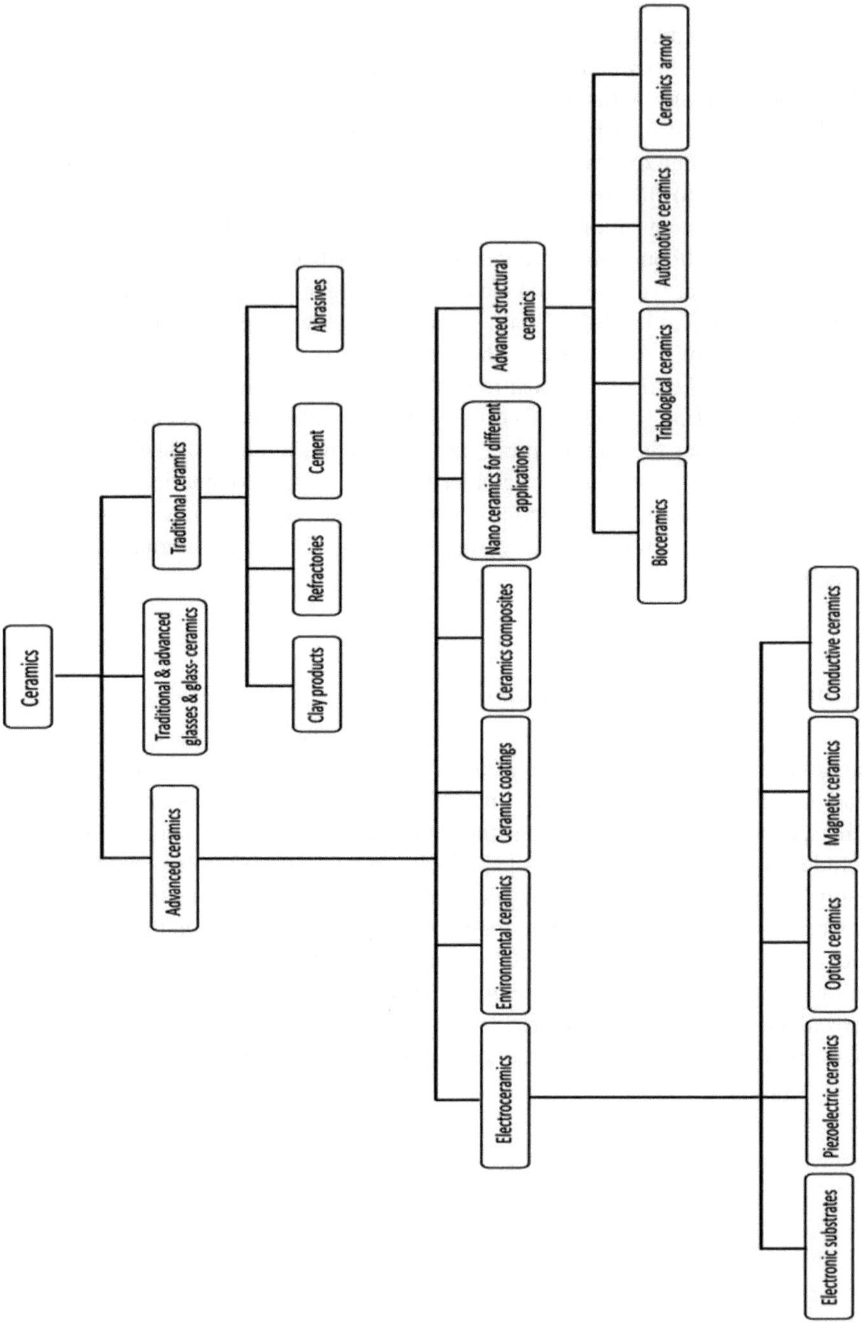

Fig. 1 Classification scheme of ceramic materials

3.1 Traditional Ceramics

3.1.1 Clay Products

Clay is an interesting widespread natural material widely applied in the globe [6]. It is the most known raw material utilized worldwide owing to its astonishing properties as abundance, feasibility, environmental friendly, and suitability for numerous practical and creative purposes containing applications like agriculture, engineering, geology, environment, pharmaceuticals, etc. [7, 8]. Economically, it is a low price raw material as compared with other pure and natural materials [9]. It is worth mentioning that the clay is formed through elongated time by regular chemical weathering of silicate rocks using diluted carbonic acid and solvents. Thus, the clay is categorized into two kinds; the main one is formed as a residual deposit in the soil and remains at the place of formation. The second one is that translated from its original place by water erosion and deposited in a new sedimentary deposit. It is crystalline with fine particles of pseudo-hexagonal shapes and few micrometer sizes [10]. It is mainly composed of hydrous-aluminum silicates with minor amounts of iron (Fe), magnesium (Mg), alkali metals, alkaline earths, etc. [11]. On the other side, their atomic structure is mainly composed of two basic units, i.e., octahedral- and tetrahedral-sheets. The first unit consists of closely packed oxygen atoms and hydroxyl groups in which Al, Mg, and Fe atoms are ordered in octahedral symmetry. On the contrary, silicon atoms are equidistant from four oxygen atoms or hydroxyls, arranged in a tetrahedron pattern [9, 12].

After giving a brief introduction on clay, now we are going to provide some examples of traditional clay products. They are classified as unvitrified or vitrified products. The vitrified one is fired at a higher temperature and formed of fused clay and other compounds having glassy feature, e.g., whiteware. Unvitrified products are fired at lower temperature and form porous ware such as pottery and earthenware. The following are some clay products:

(a) Pottery is a category of traditional ceramic products which has high porosity and is densified at low temperature $\leq$900 °C. It is not glazed but can undergo surface modification and decoration. Recently, they can be applied in flowerpots, roof tiles, bricks, and art ware.
(b) Earthenware is a known porous (10–25%), and opaque product which is densified at 900–1200 °C. Its raw materials are ball clay (25%), China clay (25%), flint/sand (35%), and feldspar (15%). It can be applied to bricks, tiles, and tin enameled majolica.
(c) Whitewares have many types of clay products such as hard porcelain for artware, tableware, semi-vitreous tableware, hotel china, vitreous sanitary ware, vitreous tile ware, electrical porcelain, and dental porcelain. Porcelain is one of the most interesting products in this category. Its raw materials are flint/quartz sand (13%, SiO_2), feldspars ($KAlSi_3O_8$–$NaAlSi_3O_8$–$CaAl_2Si_2O_8$, 12%), China clay (63% kaolinite, $Al_2Si_2O_5(OH)_4$), and ball clay (8.7%). It is produced by multiple steps of cycle heating at around 1400°C. The main phases of the fabricated porcelain

are quartz, mullite ($Al_6Si_2O_{13}$), and glass; which are equivalent to about 60% SiO_2, 32% Al_2O_3, 4% K_2O, and 2% Na_2O oxides with minor amounts of iron, titanium, calcium, and magnesium oxides. The role of clay is to offer small particles, and good plasticity for easier shaping, while feldspar gives fluxes and forms liquid phases that increase the strength. On the other side, flint can be considered as filler that forms liquid phase and controls the thermal expansion of densified bodies.

The second interesting example of this category is sanitary ware. It includes several products such as bidets, wall urinals, water closets, washbasins, and cisterns. The main raw materials are ball clay (23–25%), china clay (25–27%), flint or quartz sand (20–30%), and feldspar or nepheline syenite (20–30%). It is formed by slip casting process with controlled rheology and then glazed by a spray gun. The body is coated by several layers of engobe, and glaze.

The third class of these ceramic products is the wall tiles. This type of ceramic has low shrinkage during firing (about 1%). The main composition of wall tile is ball clay (22.5–25%), china clay (22.5–25%), flint/quartz sand (40%), and limestone ($CaCO_3$) (10–15%). It is shaped by pressing in a die cavity and sintered at about 1100 °C. After fabrication, it is subjected to glazing.

3.1.2 Cement

There are several kinds of cement-binding materials as lime, gypsum plaster, natural cement, high alumina cement, pozzolana, masonry cement, magnesium oxychloride cement, calcium aluminate cement, Portland cement, and geopolymers. In spite of the pozzolana cement has been invented (in 1824) before Portland cement, the most famous one is the Portland cement. The pozzolana cement is formed by the reaction of lime water with silica to form bomorite gel $(CaO)_3(SiO_2)_2(H_2O)_3$. On the other hand, the Portland cement could be prepared by firing the limestone with clay at 1450 °C to produce clinker. The produced clinker is milled with about 5% gypsum to produce the Portland cement. The addition of gypsum is to control the setting. The chemical composition of Portland cement is 64–67% CaO, 5–5.5% Al_2O_3, 3% Fe_2O_3, 22% SiO_2, 1.4%MgO, and 2.1 % SO_3, in addition to minor quantities of alkali. The clinker consists of four main phases, namely, 50–70% alite (tricalcium silicate Ca_3SiO_5), 15–30% belite (dicalcium silicate Ca_2SiO_4), 5–10% aluminate (tricalcium aluminate $Ca_3Al_2O_6$), and 5–15% ferrite (tetracalcium aluminoferrite Ca_2AlFeO_5). Upon the reaction of cement with water, it hardens and forms hydrated gel due to the reaction of water with all cement phases. At the early ages of reaction, the strength is developed owing to the fast reaction of alite, while after a long time, the strength is developed due to the reaction of alite and belite with water [13].

Calcium aluminate cement (CAC) is fabricated by firing a mix of limestone and bauxite or any materials that contain alumina at 1450–1600 °C. The formed fused clinker is crushed and ground until getting very fine powder [14].

3.1.3 Refractories

Refractory ceramics are materials that can be fabricated and applied at high temperatures. They are mainly applied for lining and repairing the industrial furnaces, e.g., steel and metallurgical furnaces, cement kilns and glass melting furnaces. According to the type of industry in which they are applied, they can be classified into acidic, basic and neutral refractories. The acidic ones are based on SiO_2 and silica-containing refractories as fireclay series, sillimanite, and andalusite; these types resist the acidic slags. The second ones are basic refractories which depend on MgO as magnesite, dolomite, chrome–magnesite, and magnesite–chrome; they are resisting the basic slags. The third class is neutral refractories which are relatively inert in acidic and basic slags. This type includes carbon, chromite, forsterite, and spinels. The refractories can be applied in the forms of bricks of standard shapes and sizes, tubes, castbles, sheet, cloth, tape, and formable shapes. The recommended refractory properties are low thermal expansion, high thermal shock resistance, and low thermal conductivity, high refractoriness, good cold and hot strength, as well as excellent slag resistance, low porosity, and high bulk density [15–18].

One of the most interesting systems in refractory ceramics is MgO-Al_2O_3. It exhibits highly refractory oxide materials, namely, periclase (MgO), spinel $MgAl_2O_4$, and corundum (α-Al_2O_3). The melting points of these phases are about 2800, 2100, and 2050 °C, respectively. This system includes periclase, spinel-bonded magnesia, spinel, spinel–alumina, and corundum. All of these materials are highly refractory and have a wide range of applications for industrial furnaces lining or for advanced applications as structural, electronic, and biomaterials [15–18].

The known stages for the production of refractory are crushing and grinding of raw materials, mixing, molding/casting, drying, and firing. The raw materials include clay, bauxite, chamote, sand, magnetite, dolomite, etc. Generally, the refractory can be applied in the form of shaped refractories, e.g., bricks, or unshaped refractories as castables/concrete, and mortars [15–18].

3.1.4 Abrasive Material

An abrasive material is a hard and tough grain material that can be produced naturally and synthetic, and utilized as a medium for cutting, grinding, and polishing processes of materials to make a smooth and refractive surface. The natural ones are impure natural materials such as emery, corundum, quartz, flint, garnet, diamond, tripoli, diatomaceous earth, sandstone, pumice, and natural-sharpening stones [19]. The synthetic ones are pure materials since we can control the method of preparation. They are like SiC, Al_2O_3, glass, steel wool, and shot, as well as grit. The abrasive grains must be harder and tougher than the medium in which they are applied; moreover, they must resist thermal and mechanical shock as well as friability. The abrasive processes include grinding, honing, lapping, and polishing.

3.1.5 Glasses and Glass-Ceramics

Historically, glass has been discovered thousands of years ago. It is dissimilar to ceramic, which is mostly crystalline material; glass is an amorphous material fabricated after fusion of its precursors at relatively high temperature, casting, and then annealing at lower temperature. It has attractive properties as optical transparency, hardness and easy for shaping and can be applied for conventional and advanced applications. For example, they can be applied for window glass, light bulbs, TV/computer display tubes, optics, fiber optics, cookware, and biomedical applications. The commercial glass includes soda-lime, lead, borosilicate, aluminosilicate, and high-silica glasses. The soda-lime is the most common one; it is composed mainly of $Na_2O{\cdot}CaO{\cdot}6SiO_2$ in the presence of few amounts of alumina, borate, and arsenic oxide or antimony oxide. The alumina is added to improve the corrosion and crystallization resistances, while borate enhances the workability and arsenic oxide or antimony oxide helps to eliminate the bubbles. This kind of glass is inexpensive and is applied in bottles, electric bulbs, and glazing. Borosilicate glass (Pyrex glass) has good corrosion and thermal shock resistance and could be applied in automobile headlamps, cookware, and laboratory apparatus Alumino-silicate glass can be applied when high chemical durability, strength, and resistance to devitrification are needed, e.g., cookware, fibers, and seals. Lead glass has a high refractive index, so it has a high luster and is named crystal glass. It is applied for high-quality artware and tableware, radiation shielding, lamp envelopes, seals, and optical ware. High-silica glass has high resistance to chemical attack. Fused-silica glass is a special one having >99% silica content. It is used at high temperature cycling as crucibles. Fused-quartz is a high expensive one and is applied in special optical applications.

Comparing to the other types of materials, glass is characterized by larger optical properties which are affected by the imperfection, impurities, stresses inside glass, and surface roughness.

The addition of nucleating agent and thermal treatment of the glass lead to the formation of crystalline materials called glass-ceramics. It is worth to mention that the glass-ceramics are regularly not wholly crystallized; the crystallization percentage ranges between 50 and 95 vol%, and the remaining is residual glass [20]. To obtain glass-ceramic, many parameters should be controlled, e.g., appropriate heat treatment process for the right glass composition since some glasses are highly stable and cannot simply crystallize [21]. Nowadays, there is an emerging mandate for developing the glass-ceramics to be engaged in various applications. These include construction instead of stone [22], optical applications since they have translucency/transparency properties [23–25], and military applications (aircraft and missiles). That is glass-ceramics have excellent properties such as low thermal expansion coefficient, excellent abrasion resistance, high mechanical strength, and high radar wave transparency [26, 27]. Moreover, glass-ceramics have various biomedical application [22, 28].

3.1.6 Ceramics from Waste Streams

A huge amount of solid industrial wastes are produced as by-products from many industries. These wastes might cause environmental pollution for air, soil, and water resources. Recently, a great interest has been directed to use these wastes in the production of traditional and advanced ceramic products [29]. These wastes include silica fume, waste clays, grog, granite sludge, aluminum dross, blast furnace slag, slags from different metals industries, cement dust, glass culets, etc. For example, silica fume is a by-product produced after the production of silicon metal and ferrosilicon alloy in a huge amount. It consists of very fine and active spherical particles, and has been applied in many industrial ceramic products such as cordierite, forsterite, mullite, building concrete, refractory castables, and many other ceramic products. It has an efficient effect on improving the mechanical properties of building concrete and refractory castables. Also, silica fume has been used for the production of foamed and non-foamed geopolymers [30]. Moreover, it has been utilized for the production of advanced ceramics, nanosilicon carbide, and silicon carbide/silicon nitride composites by mechanical alloying and thermal activation method [31–34]. The use of various slags, and tailings in the production of conventional ceramics is also advantageous for both circular economy and virgin resources' preservation [35, 36]

3.2 Advanced Ceramics

These types of ceramics have received a great attention in the recent technologies and they play interesting roles in various industrial applications such as energy and energy storage, transport, life sciences, electronics, and communication. This is due to their outstanding properties which cannot be originated in the conventional materials, such as mechanical, chemical, physical, electrical, optical, and magnetic properties [37–39]. Their name varies according to the society; the Japanese say "fine ceramics", American society says "advanced ceramics" or "technical ceramics", while the European society says "technical ceramics". One of the most important things is that these materials are cheaper and more cost-effective than metallic ones [37]. Advanced ceramics can be classified into structural ceramics, electroceramics, environmental-related ceramics, ceramic coatings, ceramic composites, and nanoceramics for different applications.

3.2.1 Structural (Engineering) Ceramics

One of the most interesting types of advanced ceramics is the structural ceramic. It includes different types such as bioceramics, tribological ceramics, automotive ceramics, and ceramic armors. The demand for structural ceramics comes from their applications to protect and serve human beings; thus, they exposed a higher progressive rate. These types of ceramics include alumina, partially/fully stabilized zirconia, silicon carbide, silicon nitride and boron nitride, ceramic composites, etc. [40]. The following are some examples of structural ceramic materials.

(A) **Alumina (Al_2O_3)**

Alumina is the most common structural ceramic, since it has several attractive properties like high melting point, high wearing resistance, high chemical durability, low density, good hardness, as well as good strength and availability. It is worth to mention that alumina has several crystal metastable polymorphs in addition to the stable form (corundum or alpha-alumina). The metastable forms are classified into two main classes; a face-centered cubic (fcc), or a hexagonal close-packed (hcp) arrangement of oxygen anion. The type of each polymorph depends on the distribution of cations inside it. Fcc packing of oxygen contains γ, η (cubic), θ (monoclinic), and δ (tetragonal or orthorhombic), while the hcp forms include α (trigonal), κ (orthorhombic), and $\varkappa$ (hexagonal). Some other monoclinic alumina forms have been recently recognized. The metastable forms of alumina (e.g., γ-alumina) have small particle size, and high surface area, thus they can be applied as adsorbents, catalysts or catalyst carriers, coatings, and soft abrasives. Alumina occurs in nature as alpha-alumina which forms gemstones ruby and sapphire. They contain some impurities of titanium (Ti), iron (Fe), and chromium (Cr), and have blue and red colors and therefore, they can be used in jewelry industry. On the opposite side, leucosapphire is a synthetic and colorless form that has high-strength optical components and is commonly used for cabin windows and active laser components. Polycrystalline Al_2O_3 can be applied for various uses like electrical insulation, wear resistance, mechanical ingredients, biomedical applications, etc. However, it has relatively low fracture toughness along with low fracture strength which leads to the rapid spreading of cracks and consequently it has a negative effect on its performance for some applications [41]. However, this serious drawback can be overcome by preparing composites containing metals to enhance the fracture toughness and strength [42–49].

(B) **Zirconia (ZrO_2)**

Owing to its superior mechanical properties along with its high ionic and thermal conductivity, ZrO_2 has many potential applications [50]. Generally, it has high strength, toughness, hardness, and wear resistance, good frictional behavior, none magnetic, electrical insulation, low thermal conductivity, corrosion resistance in acids and alkalis, modulus of elasticity similar to steel, and coefficient of thermal expansion similar to iron. It is chemically inert and has three polymorphs, i.e., monoclinic, tetragonal, and cubic. The monoclinic transforms into a tetragonal at 1173 °C, the tetragonal transforms into a cubic phase at 2370 °C, these transformations are reversible. The transformation of tetragonal to monoclinic is accompanied by 3–5 expansion and shape deformation. Zirconia can be partially or fully stabilized by magnesia or CaO and Y_2O_3. The main applications of zirconia are pumping components, blade edges, wear resistance parts, nozzles, dies, telecommunications, biomedical, refractories, refractory fibers, thermal barrier coatings, electrolytes, oxygen sensors, fuel cells, furnaces elements, and gemstones [51].

(C) **Silicon Carbide (SiC)**

SiC is one of the most attractive structural ceramics since it has remarkable characteristics such as good chemical inertness, high hardness, low density, and coefficient of thermal expansion, as well as excellent thermal and chemical stability [52]. It is found in nature as a mineral moissanite which is rare thus most of the silicon carbides are synthetic. One-dimensional (1D) nanostructured materials (e.g., nanowires and nanotubes) are highly necessitated for different potential applications. 1D SiC nanostructure has highly desired physical and electronic properties as a wide band gap semiconductor of band gap energy ranging from 2.2 to 3.4 eV. This variation in band gap is owing to their polytypes with a high electronic mobility and thermal conductivity. Since SiC has a high melting point (2873 °C), it can be applied at high temperature. Due to its chemical inertness, it can resist the harsh environmental conditions. Based upon these outstanding characteristics, SiC is extensively used in different industrial fields such as electronics, heating elements, and structural materials. Another promising usage for such high-strength material is enhancing the thermal and mechanical properties of other ceramics like alumina and mullite via composite fabrication [52]. It could be fabricated industrially by the Acheson method by silica sand with coke (carbon) at high temperature in electric furnaces. The sintering of SiC needs high temperature and inert atmosphere.

(D) **Other Non-oxide Ceramics**

TiC, ZrC, TaC, Si_3N_4, TiN, BN, AlN, TiB_2, ZrB_2, B_4C, and HfB_2 are also interesting types of non-oxide ceramics having serious properties and applications when applied separately or in composite forms [53–61]. For example, Si_3N_4 is one of the main structural ceramics; it has high mechanical properties and excellent wear resistance. This is due to the developed microstructure that arises after liquid phase sintering in which high aspect ratio grains and intergranular glass phase lead to excellent fracture toughness and high strength. Si_3N_4 was fabricated for the first time in 1955 for use as thermocouple tubes, molten metal crucibles, and also rocket nozzles. It was prepared by nitridition of silicon powder compacts, and attention initiated to rise for possible use in gas turbines. There are reaction-bonded and sintered silicon nitride types. Non-oxide ceramics as carbides and nitrides of group IV transition metals have an interesting mix of ionic, covalent, and metallic bonding. This arrangement of bonding makes the materials having significant valuable properties such as high hardness (about 25 GPa), very high melting points (>2750 °C), good thermal, and electrical conductivity ($\geq$10 W m^{-1} K^{-1} and about 200 $\times$ 10^4 $ohme^{-1}$ m^{-1}, respectively). For example, TiN could be extensively applied for biomedical and engineering applications since it has excellent hardness and high melting point, high electrical conductivity, and low thermal conductivity. TiN and its composites (as nc-TiN/a-BN/a-TiB_2) have commonly been utilized as super hard coatings to enhance the wear resistance, thermal stability, and oxidation resistance of mechanical devices such as cutting tools and fuel claddings. Most of the TiN composites have an exciting color range so they could be applied as super hard decorative coatings. TiB_2 is one of the high-temperature ceramics which has astonishing characteristics similar to that of TiN; mostly with the same applications, but with some limitations owing to the difficulties of getting full sintered bodies even after applying high pressure and temperature. The high sintering temperature increases the grain growth, which leads to decreasing the fracture toughness. Its high melting point, low self-diffusion coefficient, strong covalent bonding, and presence of oxygen-rich layer on the surface of the TiB_2 powder are the main causes of weak sinterability. One of the most interesting high-temperature non-ceramic oxides for modern structural applications is ZrB_2. It crystalizes in a hexagonal-system with c/a crystal-parameter-ratio of 1.11, this makes it having isotropic properties compared to other classes of high temperature ceramics. It has also high thermal stability (melting point >3240 °C), relatively low density (6.1 g/cm^3), and low coefficient of thermal expansion (6.7×10^{-6} K^{-1}). These properties make it suitable for an extensive range of applications as aerospace and high-temperature industries. The main problem which faces ZrB_2 is the poor sinterability; it comes from intrinsic strong covalent atomic bonds leading to very low self-diffusion. Thus, diffusion-promoting parameters like time, temperature, and pressure are usually required to increase its sintering, but they have some drawbacks as grain coarsening. This grain growth is typically occurred when the grain-boundary diffusion rate is considerably higher than the bulk diffusion rate. This means the diffusing atoms incline to move at grain boundaries rather than in bulk of the crystals. This performance might become leading in materials with stronger atomic bonds. So, the addition of diffusion promoters may improve the overall diffusion through the sintering process. On the other hand, ZrC and ZrN are high-temperature ceramics having interesting applications in exciting high-temperature environments as nuclear applications in generation IV reactors. These materials have required properties as high thermal and electrical conductivities as well as high hardness and melting point. Due to the high melting point and oxidation of these non-oxide ceramic materials, their consolidation methods are a vast challenge. There are many sintering approaches that have been utilized for the preparation of non-oxide bulk ceramics and coatings such as pressureless sintering (PS), hot-pressing (HP), and spark plasma sintering (SPS), chemical vapor deposition, physical vapor deposition, etc.

Bioceramics

A large number of scientists worldwide have exerted much effort on the opportunity to repair and regenerate the diseased/damaged human tissues and organs. This approach is the so-called "tissue engineering" which gained more developments after discovering the biomaterials [62]. This approach depends on the utilization of appropriate biomaterial for proliferation of cells and growth factors to regulate the cellular functions [63, 64]. The selection of the proper biomaterial for accomplishing this purpose is not easy since it should meet numerous requirements. Based on this principle, the main requirements can be presented briefly in the following points:

(i) **Porosity**

Owing to the porous structure of trabecular and cortical bone, which is essential for blood movement, perfect scaffold should have a relatively high porosity with suitable inter-connectivity. The porosity permits the passage of substantial nutrients, osteogenic cells, and molecules that are required for mineralization and vascularization throughout the structure.

(b) **Biocompatibility**

In 2008, biocompatibility has been defined as the ability of the material to facilitate natural cellular and molecular activity within a scaffold without inducing systemic toxicity [65]. Meanwhile, it also involves the osteoconductivity, osteoinductivity, and blood vessels formation around its implantation site.

(iii) **Biodegradability**

One of the most important requirements of optimum scaffold is that it should possess controlled biodegradation rate over time when it is inserted into the human body, taking in mind that the biodegradation rate should be comparable to that of tissue regeneration. It should be noted that this rate is highly correlated to the kind of chosen material and its properties for specific application.

(d) **Mechanical properties**

It is widely accepted that the clinical success of scaffold is based on the matching of its mechanical properties with that of living bones. However, the stiffness of the scaffold should not hinder in-vivo bone growth to help the structural integrity to physically support itself [66].

Recently, tissue engineering approach has been effectively developed for treating several organs like skin [67], bone [68], and cartilage [69]. This wide application for such an approach is accomplished through the great development of biomaterials which are substantial for tissue regeneration [70]. Therefore, it is very useful for the readers of this chapter to get an overview of the biomaterials and their classification.

Generally, a material that is utilized for a medical application is called "biomaterial" [71]. Biomaterials are divided into polymers, composites, metals, and bioceramics. Keeping in mind that they should have many desirable properties such as biocompatibility, stability, and having mechanical properties comparable to human bones [72]. Bioceramics are considered as a large subset of ceramics which are specific for replacing damaged/diseased parts of the human body. They involve crystalline, partially crystalline, or amorphous materials, and they can be divided, according to their behavior when they implanted in the body into three categories as follows [73].

a. **Nearly bioinert ceramics**: Materials that exhibit no reactions within the body, but they are biocompatible and accordingly they are not suffering from body-rejection problems. The most famous examples of such a category are Al_2O_3 and ZrO_2 [49].

b. **Biodegradable materials**: They are able to rapidly dissolve in the body under physiological conditions. Tricalcium phosphate (TCP) and porous magnesium alloys are the common examples of such a type of biomaterials [74, 75].
c. **Bioactive materials**: They show a favorable reaction when they are inserted into the human body through the formation of hydroxyapatite (HA)-like layer on their surfaces which is responsible for the formation of strong bond with the surrounding bone tissues. This category includes bioactive glass [76–78], bioactive glass-ceramic [79], hydroxyapatite [80–84], fluorapatite [85], and silicates compounds [86].

It is substantial to stress that, beside the aforementioned desirable properties of these specific ceramics, they also possess other excellent characteristics such as their high corrosion and wear resistances, thermal stability, high strength, hydrophilicity, and excellent chemical durability [6, 39, 87]. In this regard, in 1970, traditional stainless steel head of hip has been replaced by the sintered Al_2O_3, where the latter material possesses low density which is very beneficial for surgical operations. A great evolution for using ceramics in orthopaedic and dental applications has been approved after the approval of the US Food and Drug Administration in 2003 for ceramic-on-ceramic hip joint replacements. In spite of these amazing properties for such materials, they also have some drawbacks which need to be solved like their low fracture toughness compared to natural bone [88].

Since the bioactive ceramics are the most interesting categories for dental and orthopedic applications, we have to briefly discuss some important information about these promising ceramics.

Bioactive glasses (BGs): Highly bioactive glass compositions form the desirable HA or carbonated hydroxyapatite (CHA)-like layer on their surfaces when inserted in the body within few days or even within few hours, and can also stimulate the proliferation and differentiation of osteoblasts through the release of the leached ions from BGs. Therefore, the past four decades have already witnessed the development of different systems of BGs including silicate, phosphate, borate borophosphate, etc. All these types exhibit biocompatibility, bioactivity, and good bioresorbability. Additionally, some metal oxides like silver oxide (Ag_2O), gold oxide (Au_2O_3), cupric oxide (CuO), zinc oxide (ZnO), and titanium oxide (TiO_2) can be added to glass composition to give the prepared glasses good and even excellent antibacterial behavior against different species. In spite of the brilliant biological properties of BGs, their clinical applications are restricted due to their inappropriate mechanical properties compared to those of human bones. Accordingly, the crystallization of these glasses into their corresponding glass-ceramics or incorporation of another phase, which possesses suitable mechanical properties with the aim of producing biocomposites having the desired mechanical and biological characteristics have been developed. Based on these outstanding properties, this bioceramic type receives great attention from researchers around the world [76, 77].

Bioactive Glass-Ceramics: They are characterized by the abovementioned amazing biological properties for BGs along with better mechanical properties due to that crystallization process is responsible for significant shrinkage in their microstructure giving good physical and mechanical properties [77]. Unfortunately, crystallization process may lead to severe reduction in their bioactivity behavior depending on the starting glass composition, the used sintering conditions, and the resulting phases [79].

Hydroxyapatite (HA; $Ca_{10}(PO_4)_6(OH)_2$): It is a very promising alternative for damaged/diseased bone where its structure and composition are very similar to those of human hard tissues [89]. Moreover, when HA is in nanoscale range, it exhibits better bioactivity, because of its high surface-to-volume ratio. As a rule, there are two kinds of HA; namely biological and synthetic ones. The biological one is characterized by the presence of many substitutions such as fluorine (F), sodium (Na), zinc (Zn), magnesium (Mg), and carbonates (CO_3) into its crystal structure giving it superior biological characteristics. Notably, synthetic HA can be fabricated by several methods such as sol–gel, hydrothermal, solid-state reactions, mechanochemical synthesis and wet precipitation methods. For a more comprehensive view, it should be noticed that these methods require some cautions during preparation because the characteristics of the final product are strongly correlated to the various conditions. In spite of the aforementioned characteristics, its major limitation for use in clinical applications is its poor mechanical properties and accordingly, its applications are restricted in bone augmentation and middle-ear implants [82].

Fluorapatite (FA): It is commonly utilized as a biomaterial because of its superior biological properties such as high durability, good hardness values, and reduced solubility. It is possible to say that the latter property is very beneficial for biomedical applications where high solubility rate can lead to considerable weakness in the chemical bond between the biomaterial and fresh grown hard tissues. The major reason behind these excellent characteristics is the existence of F^- ions in the center of Ca^{2+} triangle and its crystal structure. Additionally, the presence of F^- ions is favorable for the prevention of tooth decay, and inhibiting bacterial activity. In addition, through bone formation, F^- ions work to promote the mineralization as well as the crystallization of calcium phosphate. Another benefit of FA, compared to HA, is that it shows better cell adhesion, good adsorption of the proteins, and improved effectiveness for alkaline phosphatase [86].

Calcium silicates (CS): They are usually employed for bone regeneration purposes and they include several promising types like wollastonite (β-$CaSiO_3$), pseudo-wollastonite (α-$CaSiO_3$), and dicalcium silicate (Ca_2SiO_4). All these aforementioned types show an excellent bioactivity behavior and positively affect the mineralization process of hard tissues. On the opposite side, like the abovementioned biomaterials, the major limitations of their broad uses in biomedical applications are their poor mechanical properties and high biodegradation rate [90]. Based on this fact, some other materials encompass Al_2O_3, graphene oxide, carbon nanotubes, and polymers can be added to CS ceramic to give the desirable mechanical properties and control their biodegradation rate.

Amazingly, some bioactive ceramics like HA or FA can be also employed as delivery systems for proteins, genes, vitamins, anticancer, and anti-inflammatory drugs [91–95].

Tribological and Automotive Ceramics

Nowadays, ceramics having high thermal stability, and excellent wear resistance are of great interest for tribological and automotive industry. Some ceramics such as SiC, Si_3N_4, Al_2O_3, AlN, TiN, MgO, and their composites with each other or with metals can be also used for such applications. Generally, aluminum matrix composites (AMCs) have excellent properties as strength, stiffness, and wear resistance. Accordingly, they can be broadly utilized in different applications when mixed with the aforementioned ceramics with the aim of improving their physical, mechanical, and tribological properties. The incorporation of ceramic particles in the alloys leads to increasing the strength and decreasing the ductility of alloys [96].

Ceramic Armors

In the past years, numerous kinds of metals and alloys have been utilized in the production of armors. Recently, ceramics and their composites have been developed to be used as armors due to their excellent mechanical properties. Particularly, boron carbide represents the strongest and lightest ceramic kind utilized in armors. The ceramic plates are hard enough to scatter the bullets considering that the armor pushes out on the bullet with the same force with which the bullet pushes in. Consequently, the bullet is banned from penetrating the body. In spite of the effectiveness of ceramics used for this purpose, the protective value is severely reduced owing to the successive impacts; this still is a major drawback. Hence, one talented strategy to overcome this disadvantage is to reduce the size of ceramic tiles to the smallest possible size, while the matrix substances have a thickness of at least 25 mm [97].

3.3 Ceramic Composites for Different Applications

1. **Aircraft industry**

One of the most significant applications of ceramic composites is the aircraft industry. The famous examples of such composites are fiberglass, carbon fiber, and fiber-reinforced matrix systems [98].

2. **Heat-shield systems**

These systems are crucial for space vehicles since the ambient temperature is elevated throughout the re-entry phase which subsequently, leads to the occurrence of thermal shock and heavy loads. Moreover, they can also be used in the fabrication of brake system components due to their ability to be subjected to excessive thermal shock with high mechanical resistance [99].

3. **Biomedical applications**

Despite the great advantages of ceramics, their biomedical applications are restricted due to their brittleness. Therefore, one of the most promising solutions to counter this problem is the preparation of ceramic-containing composites. Thanks to this solution, the uses of these precious materials were widely expanded to [100]:

A. **Dentistry**
   - Dental restorative materials.
   - Fixed/removable dental prostheses.

B. **Oral and maxillofacial surgery**
   - Dental implants.
   - Temporomandibular joint prostheses.
   - Cranial bone repair.

C. **Tissue Engineering**

D. **Orthopedic**
   - Bone graft.
   - Bone fracture internal fixation devices.
   - Joint prostheses.
   - Artificial tendons.
   - Artificial ligaments.
   - Artificial cartilage.

3.4 Electroceramics

Generally, ceramic materials are attractive materials for optical, electrical, and magnetic applications. The electroceramics with low electric-resistivity are valuable in making integrated circuits. On the contrary, those with high electric resistivity can be employed in the manufacturing of capacitors. Some electroceramics display piezo-electricity property and thus, can be needed in the manufacturing of transducers for microphones. It is significant to note that these ceramics are usually based on (Pb, Ti, and Zr)O_3, and accordingly, they are abbreviated as "PZT" compositions. Moreover, other types of electro-ceramics have good magnetic properties and accordingly, are favorable for transformer cores/permanent magnets. Other types of electroceramics are attractive for optical properties, i.e., luminescence, and lasing. Bearing in mind that the optical properties of these types are changed as a result of applying electric fields and thus used as modulators, demodulators, and switches [101, 102].

3.4.1 Conductive Ceramics

It is well established that the several types of ceramics display high electrical resistivity where they resist the flow of electric current. Based on this fact, they are usually used as electric insulators. Remarkably, some ceramics show excellent electric conductivity. As we know, electric conductivity can be divided into electronic and ionic. The first kind takes place because of the passage of free electrons via the substance. Since the atoms are well bonded with each other due to the ionic bonds, free electrons are forbidden to pass. In some circumstances, the inclusion of impurities can work as donors/acceptors of electrons. Conversely, the second type of electric conductivity depends on the transport of ions, through vacancies in the crystal structure of the material, from one site to another. It is important to underline that ions hopping is very little at normal ambient temperatures, while, at higher temperatures, vacancies become mobile and special ceramics show the so-called "fast ionic conduction". Noteworthy, these ceramics are very desirable in some applications such as gas sensors and batteries [103, 104].

3.4.2 Thermoelectric Ceramics

These ceramics are continuously engaged as semiconductors. For such applications, the ceramic should have a high Seebeck coefficient which is also well known as thermoelectric power or thermoelectric sensitivity of the material. It is the magnitude of produced thermoelectric-voltage due to the temperature difference across that material. For metals, it should have high electrical conductivity with relatively poor Seebeck coefficient. The thermoelectric ceramics can be widely applied in devices that are working at high temperature as combustion engines, gas turbines, power plants including nuclear power plants, furnaces, heaters, or associated with solar cells/solar heaters. However, the energy source for such conversion devices depends on the difference between the temperature inside the chamber and that of external environment. The application of thermoelectric ceramics in these devices makes them cost-effective devices along with their long lifetime, and the produced waste heat can be easily turned into electricity. This means that these devices are deemed as environmentally friendly and economical [105].

3.4.3 Ceramic Coatings

Recently, high-technology applications need the development of materials with superior surface properties to be used in different industries like mining, forestry, construction, cutting tools, electronics, and engineering ceramics for automotive and aerospace industries. In spite of the importance of such industries for economy, they are continuously threatened by many factors such as the price, the erosion, and corrosion of the used raw materials, etc. Thus, one of the best solutions to face the wear, erosion, and corrosion problems is to coat the used materials with super hard ceramics

[106, 107]. In this regard, advanced ceramics can be utilized as coatings for different materials, especially metals and bulk ceramics to be suitable for different technological applications. For example, ceramics coatings on metallic materials received noticeable development in the 1970s, and since then, this field gained more interest for coating of many alloys which need excellent performance in different applications like adiabatic engines [108], and cutting tools. It is important to stress that ceramics coatings exhibit superior properties as they work to improve the thermomechanical performance of metallic substrate [109, 110]. Meanwhile, these coatings are desirable for protecting metallic substrate against corrosion and oxidation. Additionally, they act also to minimize wear damage and decrease the base metal temperature in thermal barrier coatings [111]. There are several approaches for surface modification of metals and ceramics. These include surface modification of ceramic by laser technique [112], and surface coating of metals using different tools such as chemical vapor deposition, physical vapor deposition, and so on [113, 114].

3.5 Environmental-Related Ceramics

After the oil crisis in the 1970s, there is a high demand for the development of advanced energy-efficient engines. The traditional engines are different from the advanced ones; they are working at high temperatures. This means that the ceramics are the most suitable materials that can be applied at higher temperature while keeping their characteristics without noted damage in their functionality. These ceramics include silicon carbide (SiC), silicon nitride (Si_3N_4), and partially stabilized zirconia (PSZ). It is worth to mention that these kinds of engines are cost-effective ones with excellent performance and consequently they are highly recommended for power-generation purposes as stationary gas turbine parts [115]. Furthermore, these ceramics are playing an effective role in preserving the environment from pollution, since they are broadly used in manufacturing of hot gas filter elements that are able to operate at elevated temperatures [38]. Due to their high surface area, outstanding mechanical properties, and their resistances against thermal shock, these filters are very talented for catalyst support.

It is well known that the heat exchanger produced from ceramic can be applied at high temperature like 1500 °C for waste recovery purposes. Thus, it can save 50% of the fuel compared to the metallic one that can save only 20–30% [103]. SiC is characterized by high thermal-conductivity, wear and corrosion resistances, and thus, it is considered as the one of most promising ceramics for such application.

4 Ceramics Processing

There are three essential stages for ceramic processing:

(a) Powder preparation;
(b) Shaping or forming of the needed bodies;
(c) Sintering of the shaped bodies at suitable temperature.

4.1 Powder Preparation Process

For conventional and advanced ceramics, the powder may be prepared by grinding and calcination (solid state route) of natural or synthetic materials like oxides, hydroxides, carbonates, or other salts [116]. Other approaches as chemical precipitation (for single oxides) or co-precipitation (for complex oxides) [60, 117–121], sol–gel [122], gel-combustion [123] and hydrothermal reactions [124] are also accessible for accomplishing this purpose. However, in spite of these altered methods, solid-state route is quite the most promising among the others due to its cost-effectiveness. Unfortunately, this method may give inhomogeneous powder. On the other side, the other wet-chemical methods offer an outstanding homogeneity, but they are fitting only for laboratory studies since they are sophisticated, expensive and give the resulting product in low quantities. Interestingly, wet chemical methods can be combined with solid state where the main materials are solids, while the dopants or additives are added via chemical processes [125].

4.2 Forming (Shaping) Processes

These processes are utilized to convert ceramic powders into the required shapes and bodies. Briefly, powders are firstly well mixed and converted into wet powder by water or suitable mixing agent as polyvinyl alcohol, starch, etc., or into slurry by suitable slurry forming techniques as slip-casting or gel-casting, etc. The wetted powder is compacted in the suitable dies by different techniques. It is worth mentioning that the current forming processes are axial and isostatic dry pressing, hot pressing, porous molding, tape molding, extrusion or injection molding. Of course, the obtained bodies after forming have low mechanical properties, and high apparent porosity. However, the main purposes of forming processes are to accomplish high packing of ceramic particles and removing most of the pores to avoid the possible microstructure defects, which affect the characteristics of the resulting product. This issue depends on several parameters, and should be considered during preparation. These factors are particle-size and particle-size distributions, homogeneity, agglomeration, and the flow of powders during the shaping process. Furthermore, some other advanced routes are available for getting green bodies by thermal treatment prior to sintering process such as hot pressing, deposition of molten particles, and deposition of vapor phase (CVD) [126]. In these methods, obtaining green pieces with the preferable shapes is followed by drying with the aim of removing the organic-parts. In all cases, the shaped products should be dried to remove the solvents by evaporation. Unfortunately, this process can generate defects inside the bodies, and consequently cracking the green sample due to the fast formation of solvent-gases and contractions of the sample. So, the drying process should be completed under a controlled atmosphere and temperature conditions [127].

4.3 Sintering Process

It is well known that the key-parameter for production of final ceramic parts is the sintering process, which is responsible for the solid-state reaction between the grains and elimination of pores scattered in the sintered body. This stage involves applying high temperature on the shaped ceramic bodies, keeping in mind that it should be lower than its melting point. Several phenomena are occurring during this process, which stimulates the consolidation and densification of the shaped part. Also, many reactions can occur and lead to considerable changes in the chemical composition and/or phase composition of the sintered product. Hence; the ultimate properties of ceramics are mostly affected by this process [128]. It is important to highlight that the exact adjustment of essential temperature and time is significantly associated with the required connectivity, porosity, and densification. The main properties of sintered ceramics depend mainly on temperature, time, initial density, particle size, particle size distribution, and the whole process [129].

It is well confirmed that the properties of polycrystalline ceramics are strongly affected by their microstructure. Consequently, in order to improve the properties and the reliability of ceramics, the density, grain size, and microstructure homogeneity should be exactly controlled [130]. Unfortunately, the production of highly dense ceramics having nanometric or submicrometric grain size is difficult through conventional sintering. Instead, other manufacturing routes are available for the preparation of these ceramics such as colloidal powder processing with controlled distribution of particle sizes [131] with or without using sintering additives [132], pressure-assisted sintering [133], spark plasma sintering (SPS) [134] and pulsed electric current sintering (PECS) [135]. However, these methods are not cost-effective due to their expensiveness and complexity [136].

4.3.1 Mechanisms of Sintering

Solid State Sintering

In this type of sintering, a relatively high temperature is required to facilitate the diffusion process. Since the diffusion is a matter of transportation mechanism, it induces both densification and grain growth. According to this aspect, the sintering environments that permit the occurrence of densification without simultaneously encouragement of grain growth are most suitable for microstructural refinement with the production of highly dense ceramics and nanometric grains [137]. As reported in Ref. [126], solid-state sintering can occur through three sequential stages:

a. The powder particles should be compacted to attain the proper connection with each other.
b. The particles begin to form necks between each other when the sintering temperature reaches 2/3 of their melting point, and then they are intensely bonded together after the formation of necks. Furthermore, increasing the contact areas between the grains tends to the successive increase in the density of sintered bodies and decrease in the total void volume (porosity). Remarkably, the particles having small diameter possess high surface area and high surface free energy which are considered as the driving force of sintering process. Accordingly, the surface area is thermodynamically reduced by bonding the particles together, which is responsible for considerable reduction in the energy.
c. The third stage is characterized by full bonding of grains with the completion of solid state reaction. Moreover, the residual porosity becomes closed-off pores which consequently leads to improving the microstructure and mechanical properties of sintered bodies. It should be noticed that the particle size of the original powder greatly affects the pores sizes and their distribution. Examples of materials which have been sintered by this mechanism were published in our recent works [45, 138, 139].

Liquid Phase Sintering

This kind of sintering is highly appropriate for numerous industrial fields like steels, cemented carbides, heavy alloys, bronzes, and silicon nitride systems. This mechanism is intensely dependent upon the presence of liquid phase during all or part of the sintering process to improve the material's densification. There are two different processes:

a. Normal liquid phase sintering in which the original ceramic compressed in situ forms liquid phase, and is accompanied by the formation of one or more components (multi-phase formation) after heat treatment.
b. In this type of liquid phase sintering, infiltration of original ceramic bodies by liquid formed outside the ceramic bodies occurred throughout the early sintering period.

Liquid phase sintering proceeds via three steps as follows:

1. Liquid flow and pore/particle rearrangement.
2. Accommodation or dissolution and re-precipitation.
3. Coalescence and final densification.

It is worth mentioning that these steps typically occur with their approximate sequence and might be overlapped for some systems. The development in liquid phase sintering is responsible for dropping the densification-kinetics. The early densification is enhanced by raising the liquid amount up to closely 35 vol% [140].

Activated Sintering

This kind of sintering deals with the possibility of depressing the sintering temperature by adding various activating agents or sintering aids as certain transition metals and oxides. For example, it is well known that SiC has exciting features owing to strong and stiff Si-C bonds. This makes significant difficulties in its sintering due to low self-diffusion coefficient [141]. Accordingly, the presence of some additives is crucial for facilitating the sintering process. The most utilized aids for this purpose are Al_2O_3 and ZrO_2 [142] which are responsible for reducing the needed temperature for the accomplishment of the sintering process. On the other side, hardness and thermal stability may be negatively affected by these aids [143]. On the contrary, the occurrence of boron and carbon may overwhelm this problem by taking into consideration that sintering should be occurring at elevated temperatures to let the sintered bodies have high hardness values and relative density similar to theoretical one [144]. It is significant to emphasize that the existence of numerous polymorphic structures in SiC, i.e., β and α is considered as another difficulty for obtaining dense structure. Although, β is the major phase that exists during SiC synthesis, it can easily convert to α one [145, 147].

Reaction Sintering

It is a modern sintering type, discovered with production of new materials. Reaction (chemical) sintering is specific for the materials that cannot be compacted under the traditional mechanisms as a result of physical processes, but it occurs due to the chemical reactions. The existence of reagents in the starting materials or their entrance from outside as a result of gaseous medium existence in furnace space leads to the creation of gaseous, or liquid product throughout sintering process. In this process two or more components react together with the possibility of dimension and density changes during densification [146]. In this regard, cold isotactic pressing (CIP) is very useful for accomplishing high densification way during the final sintering process, where it helps to promote the densities of green bodies by 5–10% [145]. On the opposite side, pressure-less sintering process such as vacuum or microwave leads to the enlargement of grains sizes, while pressure sintering such as hot-pressing (HP), hot isotactic pressing (HIP) and spark plasma sintering (SPS) help in suppressing the grain growth [146] and promote the reaction sintering as well as reducing the sintering time and temperature.

4.3.2 Factors Affecting Sintering Process

There are many factors affecting sintering process. Some of these factors are specific to the materials, while the others belong to the process itself. The followings are the brief description of such factors.

Process Factors

A. Temperature: it is a very significant parameter for controlling both the rate and magnitude of the sintering-associated changes.

b. Time: is another effective factor for improving the densification with less effectiveness than temperature.

c. Atmosphere: is crucial for obtaining the desirable results.

Materials Factors

A. Particle size: It is well known that reducing particle size contributes to considerable enhancement of sintering process.

b. Particle shape: Reduced sphericity and increasing macro or micro-surface roughness are responsible for close contacts between particles, and considerable increase in the internal surface area encourages the sintering process.

Particle structure: A fine grain structure within the original particles induces sintering process because of its desirable effect on material transport mechanisms.

Particle composition: It is well known that the presence of additions or impurities within the metals can positively or negatively affect the sintering kinetics. The distribution and the reaction of impurities greatly identify whether this effect is harmful or beneficial. For example, the oxidation is frequently harmful, while dispersed phases within the matrix strongly encourage the sintering by prohibiting the motion of grain boundaries. On the other hand, at high sintering temperatures, the expected reaction between the impurities and either base metal or alloying elements can be highly unfavorable.

Green density: High driving force for sintering can be easily achieved through increasing the amount of internal surface area which arises from the decreased grain density [147].

4.4 Simulation of Sintering Process

It is well known that the sintering process is the key-parameter for ceramic fabrication, and it significantly affects the value, yield, and price of ceramic products. However, it is important to note that the quality of the sintering process depends on the temperature distribution and the experience of researchers. Therefore, sintering of large ceramic bodies leads to a huge economic loss due to the enormous waste of resources. Moreover, from the environmental-protection point of view, and the resources management, the traditional sintering is not suitable for industrial progress. Based on these important facts, the use of computers is beneficial for improving this important process. In this sense, computer-aided sintering (CAS) helps in smart ceramics sintering. CAS technology is a breakthrough for good quality coupled with lower energy consumption [148–151].

Recently, neural network has received a great interest after the revolution of integrated circuits and computer technology. Nowadays, this network is applied in several fields like intelligence control, knowledge processing, robot technology, computer vision, sonar signal processing, etc. In this context, this network has been also applied in the ceramic industries. Neural network is classified into two classes; namely feed network and feedback network. However, Backpropagation (BP) neural network algorithm is the most common algorithm of the first class. Using the neural network, some researchers have successfully built a BP network model to determine some factors such as thermal insulation time, heating rate for different stages, and average of high temperature section. The use of this model was helpful in identifying 20 samples with an accuracy of 90%. It is important to say that the application of the neural network in ceramic industries is very successful for identifying and predicting the materials' properties as well as their defects, etc. So, we hope that scientists will expand their interest toward the application of the neural network for various ceramic materials [148].

5 Novel Processing Ceramic Techniques

5.1 Magnetic Slip Casting

Generally, slip casting is a method used to cast a colloidal ceramic-suspension into a porous mold, dried and consolidated by sintering. This method is very attractive due to the fact that it is a water-based process, which provides a high flexibility of slurry compositions. Therefore, it can be used in the case of slurries made of ceramic powders, metal/ceramic mixtures, etc. Unfortunately, it is not desirable for the industrial scale and therefore must be combined with other technologies. For instance, during casting, pressing/centrifuging can be used. Also, the manufacturing processes can be coupled with magnetic fields to enhance properties of dense bodies through texturing. It is well established that the attractive/repulsive forces of paramagnetic/diamagnetic materials are generated by magnetic fields. In liquid suspensions, magnetic fields are able to manipulate the location and orientation of the particles. By taking advantage of their remoteness as well as tenability, magnetic fields are preferred over other processing techniques. Moreover, when it is generated by passing an electric current through coil or electromagnet, it can be turned on and off. In magnetic slip casting, a magnetic field is applied through the casting process to orient the preferential crystalline axis of the ceramic grains taking into account that most ceramic materials are diamagnetic. Therefore, one can expect that when an external magnetic field is applied to them, an internal magnetic one is induced in the other direction. Thus, a repulsive force is created. It is worth to mention that the strength of the applied external magnetic field must be extremely high to generate such an induced internal magnetic field as the diamagnetic materials are characterized by their low magnetic susceptibility. In order to reduce the strength of the external magnetic field, the magnetic sensitivity of the diamagnetic materials can be increased via transforming the diamagnetic ceramic powder by magnetic elements. Particularly, the crystalline/shape anisotropy of the starting ceramic powder controls the orientation of the ceramic powders [152].

5.2 Laser Surface Texturing

In 1969, LASER (Light Amplification by Stimulated Emission of Radiation) has been discovered by Maiman and since then, this technology is applied in various important industrial and medical fields. In short, if the laser beam hits a material, it is partially reflected, partially transmitted, and partially absorbed. Hence, the electrons in this material are excited, which leads to an increase in thermal energy. The laser beam can be a continuous wave or a pulsed wave. The first type acts to transform uniform energy to the material's surface. On the contrary, the second one allows achieving a higher peak power. Indeed, the pulse duration is short to convey the energy to the desired area before that conveyed by the previous pulse has dissipated.

Laser surface texturing relies on utilizing the laser beam to produce ablation in which the material is selectively removed from the surface. In this respect, a pulsed or high-intensity continuous wave laser is suitable for this purpose. The ablation process can be performed through three mechanisms. They are photochemical (bond-breaking), photothermal (temperature increase), and photophysical (ejection of material from the surface due to a laser-induced shockwave [153, 154].

Laser surface treatment of ceramic materials has also been paid great attention since it is an effective and economical process for improving the surface properties. Re-melting, sealing, or alloying by laser can change the ceramic material into functional materials. It has been reported that tribological properties such as wear resistance, and friction can be significantly improved by laser treatment. Zawrah et al. reported on surface alloying of alumina ceramics using IR-CO_2 laser by adding HfO_2 or HfO_2 with 5 or 10 mol.% SiO_2, CeO_2 and cordierite. They found that the tendency of crack formation and grain growth was significantly decreased after laser treatment. The treated layers exhibited no open porosity, and contained 20–40 vol% second phases homogeneously distributed around the grain boundaries of the alumina matrix. Due to alloying, average grain size, hardness, and elastic modulus of the composite layers have been reduced, but resistance to cracking under the mechanical action of sliding diamond (scratch test) and abrasive wear was significantly enhanced compared with the unmodified dense alumina ceramic [155].

5.3 Cold Sintering

Generally, all bulk ceramic materials are fabricated using green shaping followed by sintering at temperatures usually higher than 1000 °C, so this process is considered as a highly energy-intensive process and high-energy dissipation approach. Most of the energy used for firing dissipates irreversibly through the furnace insulation and lost during cooling. The demand to decrease the consumption of energy and CO_2 produced during firing of ceramics urged the scientists to develop consolidation techniques operating at lower temperatures, ideally not more than 300 °C. Cold sintering is a remarkably low-temperature process that uses a transient transport water phase and an applied uniaxial force to support the densification of powder compacts. The liquid phase can accelerate plasticity, grain boundary/surface diffusion, and mass transport. By utilizing this method, many ceramic powders can be transformed into highly dense monoliths at temperature far lower than the melting point. The existence of water can promote the atomic diffusion at the particle surface without raising the temperature. Using cold sintering process, the researchers succeeded to reduce the consumed energy for consolidation of $BaTiO_3$ powders from 2800 to 30 M kg^{-1}. Other researchers conducted an interesting experiment to compare between hot and cold sintering of 1 Kg of ZnO. In the case of hot sintering, the consolidation required 291 MJ kg^{-1}, while cold sintering required 162 MJ kg^{-1} [156].

5.4 Flash Sintering

As previously discussed, the main disadvantage of conventional sintering process is the high energy consumption along with the release of carbon dioxide. In response to this drawback, the European Commission approved in 2007 the heat consumption required for production of different types of ceramics [157]. The greatest energy needed for sintering process is estimated by some mega-joules per kilogram of resulting ceramic product. Flash sintering (FS) has been discovered in 2010 by Cologna et al. [158]. This technique is based on applying an electric field to the sample powders. Under coupling the electric field and furnace temperature, the densification of ceramic material occurred in a very short time. This time ranges between a few seconds to a few minutes. Actually, the success of Cologna's experiment was based on the previous study by Gosh et al. [159] in 2009. In that work, the effect of applying a weak DC-electric field for 3 mol.% yttria stabilized zirconia (3YSZ) was described. The application of a weak electric current evidently reduced the rate of grain growth.

5.5 3D Printed Ceramics

Three-dimensional (3D) printing is a modern manufacturing and an additive technology. In this technique, the physical parts are fabricated in an additive manner, separated point by point, line by line, or layer by layer are digitally sliced into 2D cross sections using 3D machine. In this context, this technique is a promising one for the flexible preparation of highly fine and sophisticated structures that are extremely difficult to obtain through conventional preparation methods such as casting and machining. Besides this feature, 3D printing technology provides high throughput since many materials can be constructed in one process. Due to these important benefits, 3D technology has attracted the attention of researchers in science and engineering [158, 160]. This technique was first produced by Marcus et al. [161] and Sachs et al. [162] in the 1990s. Depending on the shape of the raw materials pretreated before printing, this technology can be divided into: slurry-based, powder-based, and bulk solid-based methods. In slurry-based technologies utilized for ceramic/polymer mixtures, the viscosities are ranging from low-viscosity (~mPa·s) with a low ceramic loading (up to 30 vol.%) as inks to high-viscosity(~Pa s) with a much greater ceramic loading (up to 60 vol.%) as pastes.

5.6 Porous Ceramics

Advanced porous ceramics "ceramic foams or ceramic filters or ceramic membranes" are used for many applications such as structural, environmental, biological and biomedical ones. Since these types of ceramics possess various advantages over metals and polymers, they gained huge interests from scientists to use them in the above-mentioned applications. Notably, the functionality of these types of ceramics closely depends on the factors related to pore structure such as pore size, shape, distribution, and connectivity. For example, the pores can behave as insulators at elevated temperatures, capture impurities in a filter, expedite tissue ingrowth in a bio-scaffold, or provide the architecture for reinforcement in a ceramic–metal composite. Meanwhile, their properties can be manipulated to suit a specific application through the modification and refinement of processing techniques to control their microstructure. Figure 2 shows the applications of porous ceramics according to their pore sizes [163]. Several processing techniques are responsible for integrating pores into the structure like sintering of coarse powders, and partial sintering of green powders. Other processing techniques are well known for the production of porous ceramics as replica method, the sacrificial phase technique, direct foaming method, paste extrusion, and most recently developed rapid prototyping technique [163].

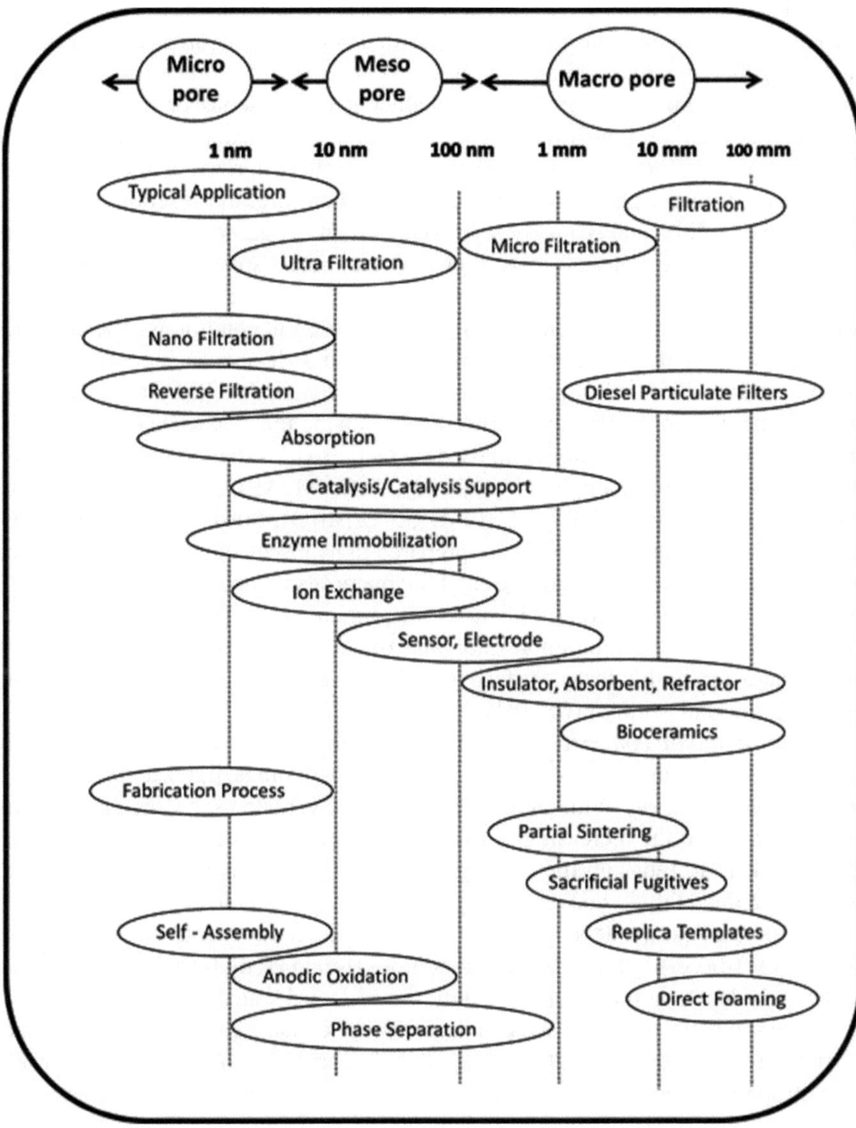

Fig. 2 Schematic of classification of porous ceramic: Redrawn after Ohji and Fukushima [163]

5.7 Functional Ceramics for Catalytic Applications

The ceramics and bioceramics can also play an important role as catalysts of some chemical reactions. About 90% of the new processes undertaken in chemical industry are catalysis-dependent. The catalysts are divided into two categories, heterogeneous and homogeneous. Homogenous catalysis takes place if the catalyst is statically dispersed in the mixture of reactants, whether it is a gaseous or liquid solution. This category includes transition metals, enzymes, and inorganic acids/bases. On the other hand, the second class, i.e., heterogeneous, is characterized by the presence of catalyst in a different phase from the reactants. Therefore, the reaction happens at the surface of solid (catalyst) particle in contact with the gaseous or liquid solution.

Since the ceramic materials resist high temperatures, display chemical-durability and wear resistance, they have unlimited potential as catalytic materials for usage in a variety of reactor structures, and can operate at different conditions. The ceramics catalysts are applied as (i) direct use as catalysts and (ii) use as support materials (substrates) to anchor and disperse a variety of active metals. Ceramic catalysts and catalyst supports are the biggest market sector outside the electronic ceramics. The ceramic catalysts are typically composed of metal oxides that contain at least two different cations. For examples; V_2O_5/SiO_2, $CeO_2/MgAl_2O_4$, CeO_2/ZrO_2, V_2O_5/TiO_2, CeO_2/Al_2O_3, $BaO/SrTiO_3$, Sr/La_2O_3, SrO/Nd_2O_3, Sm_2O_5/MgO, N_2O/MgO, La_2O_3/CaO. When these materials are prepared in nanoscale, they exhibit improved properties. Zeolites are also ceramic catalytic materials composed of aluminum silicates that have cavities or pores with molecular dimensions as part of their crystalline structure. Zeolites are unique among inorganic oxides because of their uniform microporous character, high internal surface area, high thermal stability, and their ability to develop internal acidity [165–171].

Ceramic catalytic converters: Ceramics are applied also in construction of many parts of modern car industry as windows, mirrors, spark plugs, lights and electronics, catalytic converters, etc. With increasing worldwide vehicle usage, the air pollution is increased due to the increasing of exhausts produced by the cars. This has urged the automakers to redesign the car engines and fuel systems using the incorporation of ceramic catalytic converter. Engine exhaust gases consist of a complicated mixture, its composition depends on the type of engine (two- or four-stroke, spark- or compression (diesel)-ignited), driving conditions (urban vs. highway), vehicle speed, acceleration/deceleration, etc. The catalytic converter is located in the downstream of the engine's exhaust manifold and can deal with the exhaust gases by chemical reaction to treat the pollutants as CO, NOx, and hydrocarbons into CO_2, nitrogen, water vapor through the reduction and oxidation actions. A typical catalytic converter unit consists of either a ceramic honeycomb monolithic structure (cordierite) or ceramic beads coated with precious metals (Pt, Pd, and Rh) [165].

Solid oxide fuel cells: In the near future the fuel cell devices will efficiently contribute to power generation. They are classified according to the electrolyte and the fuel used to generate electricity. They include a barrier to select and allow ions to pass as well as pushing the electrons to flow through an exterior circuit to generate electricity. The most important fuel cell that was fabricated from ceramic material as ZrO_2 is the solid oxide fuel cell (SOFC). The basic SOFC unit contains three oxide ceramic units: a fuel electrode; an air electrode; and a solid electrolyte. It operates as follows: the oxygen in supplied air can accept electrons from the external circuit to form O^{2-} ions at the cathode then move via electrolyte to the fuel electrode (anode) and create H_2O and/or CO_2 (depending on the fuel used), liberating electrons to the external circuit. There are several advanced ceramic materials which show greater electrical properties. These materials are known as electroceramics which act as fast ion conductors and are capable to improve a high flux of ions in the solid state. Ceramics which can quickly generate oxygen ions (O^{2-}) are interesting candidates for SOFC developments. Recently, many ceramic materials have been utilized as potential SOFC constituents. The anode can consist of nickel/yttria-stabilized zirconia (Ni/YSZ) cermet (i.e., ceramic/metal composite). The cathode can be formed from a mixed conducting perovskite, lanthanum manganite ($LaMnO_3$). Yttria-stabilized zirconia (YSZ) is used for the oxygen ion conducting electrolyte. To improve the performance of SOFC, it is important to develop materials with elevated electrical conductivity and catalytic performance of the anode. Improved SOFC performance must contain greater energy-conversion productivity, lesser chemical and acoustical pollution, fuel flexibility, cogeneration capability, and a rapid load response [163–171].

6 Properties of Ceramics

It is well established that ceramics possess promising physical, chemical, and mechanical characteristics which enable them to be valid for different industrial purposes. Therefore, we briefly mention these characteristics as follows [147]:

- Low density.
- Low thermal expansion coefficient.
- High hardness value.
- High temperature as well as corrosion resistance.
- High wear resistance.
- High melting temperature.
- Stability of mechanical properties at elevated temperatures.

In spite of these interesting properties, their major drawbacks can be summarized as follows:

- Brittleness.
- Highly demanding processing.
- Joining different ceramic components with each other or with other materials is so difficult.

Tables 1, 2, and 3 illustrate the properties of some advanced oxide and none oxide ceramics [172].

Table 1 Properties of some advanced oxide ceramics (alumina, cordierite, zirconia and mullite) [172]

Property	Ceramic type/value of property				Unit	Test
Physical	Alumina	Cordierite	Zirconia	Mullite		
Chemical formula	Al_2O_3	$2MgO–2Al_2O_3–5SiO_2$	ZrO_2	$3Al_2O_3–2SiO_2$		
Density, ρ	3.7–3.97	2.60	6.04	2.80	g cm^{-3}	ASTM C20
Color	Ivory/white	Tan	White	Tan		
Crystal structure	Hexagonal	Orthorhombic	Tetragonal	orthorhombic		
Water absorption	0.0	0.02–3.2	0.0	0.0	% at R.T.*	ASTM C373
Hardness	9	7	6.5	8	Mohs	
Hardness	2000	–	1600	1450	Knoop (kg mm^{-2})	Knoop 100 g
Mechanical0						
Compressive strength	2070–2620	350	2500	551	MPa at R.T.	ASTM C773
Tensile strength	260–300	25.5	248	103.5	MPa at R.T	ACMA Test #4
Modulus of elasticity (Young's Modulus)	393	70	207	150	GPa	ASTM C848
Flexural strength (MOR)	310–379	117	900	170	MPa at R.T	ASTM F417
Poisson's ratio, υ	0.27	0.21	0.32	0.25		ASTM C818
Fracture toughness, K_{Ic}	4.5	–	13.0	2.0	MPa.$m.^{1/2}$	Notched beam test
Thermal						
Max. use temperature (in air)	1750	1371	500	1700	°C	No load cond
Thermal shock resistance	200	500	280–360	300	ΔT (°C)	Quenching
Thermal conductivity	35	3.0	2.7	3.5	W/m–K at R.T	ASTM C408
Coefficient of linear thermal expansion, α_l	8.4	1.7	11.0	5.3	µm/m °C l (~25 °C through ±1000 °C)	ASTM C372

(continued)

Table 1 (continued)

Property	Ceramic type/value of property				Unit	Test
Physical	Alumina	Cordierite	Zirconia	Mullite		
Specific heat, c_p	0.21	0.35	0.10	0.23	cal/g °C at R.T	ASTM C351
Electrical						
Dielectric constant	9.6	4.7	26 @ 100 kHz	6.0	1 MHz at R.T.	ASTM D150
Dielectric strength	15	5.11	9.0	9.8	kV/mm	ASTM D116
Electrical resistivity	$>10^{14}$	$>10^{14}$	$>10^{4}$	10^{-13}	Ω cm at R.T.	ASTM D1829

* R.T: Room temperature

Table 2 Properties of some advanced none oxide ceramics (silicon carbide, silicon nitride, boron nitride and boron carbide) [172]

Property	Ceramic type/Value of property				Unit	**Test**
Physical	Silicon carbide	Silicon nitride	Boron nitride	Boron carbide		
Chemical formula	α-SiC	Si_3N_4	BN	B_4C		
Density, ρ	3.21	3.31	2.28	3.25	g cm^{-3}	ASTM C20
Color	Dark gray	Dark gray	White	White/tan/gray		
Crystal structure	Hexagonal	Hexagonal (α & β)	Hexagonal	Hexagonal		
Water absorption	0.0	0.0	0.0–1.0	Ng	% at R.T.	ASTM C373
Hardness	9–10	9	2	36	Mohs	
Hardness	2800	2200	25–205	Ng	Knoop (kg mm^{-2})	Knoop 100 g
Mechanical						
Compressive strength	1725–2500	689–2760	23.5	2.9	MPa at R.T.	ASTM C773
Tensile strength	310	360–434	2.41 (1000 °C)	155	MPa at R.T.	ACMA Test #4
Modulus of elasticity (Young's Modulus)	476	317	675	445	GPa	ASTM C848
Flexural strength (MOR)	324	679–896	51.8	375	MPa at R.T	ASTM F417
Poisson's ratio, υ	0.19	0.23	0.05	0.19		ASTM C818
Fracture toughness, K_{Ic}	4.0	5.0–8.0	2.6	Ng	MPa.$m^{1/2}$	Notched Beam Test
Thermal						
Max. use temperature (in air)	1400	1500	985	2450	°C	No load cond
Thermal shock resistance	350–500	750	>1500	Ng	ΔT (°C)	Quenching
Thermal conductivity	41	27	20	28	W/m K at R.T.	ASTM C408
Coefficient of linear thermal expansion, α_l	5.12	3.4	1.0–2.0	945	μm/m °C l (~25 °C through ±1000 °C)	ASTM C372
Specific heat, c_p	0.15	0.17	0.19	0.25	cal/g °C at R.T.	ASTM C351

(continued)

Table 2 (continued)

Property	Ceramic type/Value of property				Unit	**Test**
Physical	Silicon carbide	Silicon nitride	Boron nitride	Boron carbide		
Electrical						
Dielectric constant	10.2	7.0	4.08	Ng	1 MHz at R.T.	ASTM D150
Dielectric strength	–	s	374	Ng	kV/mm	ASTM D116
Electrical resistivity	10^{18}	10^{13}	$>10^{13}$	Ng	Ω.cm at R.T.	ASTM D1829

Table 3 Properties of some other important advanced ceramics [145]

Property	Ceramic type/value of property				Unit	Test
Physical	Sapphire	Steatite L-5	Graphite	AlN		
Chemical formula	α-Al_2O_3	$H_2Mg_3(SiO_3)_4$	C	AlN		
Density, ρ	3.97	2.71	2.28	3.25	g cm^{-3}	ASTM C20
Color	White/transparent	Buff	Black	White/tan/gray		
Crystal structure	Trigonal	Hexagonal	Hexagonal	hexagonal		
Water absorption	0.0	0.0–0.2	0.1–1.5	0.0	% at R.T	ASTM C373
Hardness	9	7.5	0.1–1.5	5	Mohs	
Hardness	2200	–	–	1170	Knoop (kg mm^{-2})	Knoop 100 g
Mechanical						
Compressive strength	2000	621	96	2068	MPa at R.T.	ASTM C773
Tensile strength	250–400	62	4.8	–	MPa at R.T.	ACMA Test #4
Modulus of elasticity (Young's Modulus)	250–400	138	4.8	308	GPa	ASTM C848
Flexural strength (MOR)	760–1035	140	50	428	MPa at R.T.	ASTM F417
Poisson's ratio, υ	0.29	–	–	0.25		ASTM C818
Fracture toughness, K_{Ic}	1.89	–	–	3.5	MP a.m.$^{1/2}$	Notched Beam Test
Thermal						
Max. use temperature (in air)	~2000	1425	3650	1600	°C	No load cond
Thermal shock resistance	200	190	200–250	400	ΔT (°C)	Quenching
Thermal conductivity	40	2.9	24	82.3–170	W/m K at R.T.	ASTM C408
Coefficient of linear thermal expansion, α_l	7.9–8.8	7.0	8.39	4.6–5.7	μm/m °C l (~25 °C through ±1000 °C)	ASTM C372

(continued)

Table 3 (continued)

Property	Ceramic type/value of property				Unit	Test
Physical	Sapphire	Steatite L-5	Graphite	AlN		
Specific heat, c_p	0.18	0.22	0.16	0.25	cal/g °C at R.T.	ASTM C351
Electrical						
Dielectric constant	9.3–11.4	6.3	–	8.0–9.1	1 MHz at R.T.	ASTM D150
Dielectric strength	15–50	9.3	–	15	kV/mm	ASTM D116
Electrical resistivity	10^{17}	10^4	7×10^{-3}	$>10^{14}$	Ω cm at R.T.	ASTM D1829

7 Conclusion

Nowadays, advanced ceramics are considered as promising materials for different applications such as engineering, biomedical, electronics, energy, and environmental issues. Most of the advanced ceramics depend on pure oxides or non-oxides prepared by new chemical, physical, and mechanical routes. Through their preparation, different parameters such as temperature, concentration of reactants, time of reaction, and atmosphere should be controlled. To convert the prepared ceramic powders into consolidated bodies, sintering processes should be conducted under suitable conditions such as temperature, time, and atmosphere. The sintering process includes solid-state sintering, liquid phase sintering, activated sintering, and reaction sintering. The prepared ceramic powder and sintered ceramic bodies can be investigated by different tools for their physical, chemical, mechanical, morphological, magnetic, electrical, optical, and biological properties according to the needed applications. This chapter dealt only with the most common types of advanced ceramics and their sintering without details on the aforementioned properties.

References

1. https://ceramics.org/about/what-are-engineered-ceramics-and-glass/brief-history-of-ceramics-and-glass. Accessed January 1, 2020
2. D. Eliche-Quesada, Luis Pérez-Villarejo, Pedro José Sánchez-Soto, Introduction to Ceramic Materials: Synthesis, Characterization, Applications, and Recycling, (IntechOpen, 2019), pp. 1–6. https://doi.org/10.5772/intechopen.84710
3. W. Freiman, in *Introduction to Ceramic and Glasses*, Engineered materials handbook, ceramic and glasses, vol. 4 (ASM International, 1991), pp. 41–122
4. R Nathan Katz, Overview of ceramic materials, design and application, in *Mechanical Engineers' Handbook*, ed. by Myer Kutz (Wiley, 2006), pp. 433–449. https://doi.org/10.1002/0471777447.ch12
5. F. Bergaya, G. Lagaly, in *General Introduction: Clays, Clay Minerals and Clay Science*, ed. by F. Bergaya, B.K.G. Theng, G. Lagaly, Handbook of clay science, vol. 1 (Elsevier, Amsterdam, 2000), pp. 1–18
6. W.R. Matizamhuka, Advanced ceramics-the new frontier in modern-day technology: part I. J. S. Afr. I. Min. Metall. **118**, 757–764 (2018). https://doi.org/10.17159/2411-9717/2018/v118n7a9
7. I. Anastopoulos, A. Bhatnagar, E. Lima, Adsorption of rare earth metals: A review of recent literature. J. Mol. Liq. **221**, 954–962 (2016). https://doi.org/10.1016/j.molliq.2016.06.076
8. L. Su, X. Zeng, H. He, Q. Tao, S. Komarneni, Preparation of functionalized kaolinite/epoxy resin nanocomposites with enhanced thermal properties. Appl. Clay Sci. **148**, 103–108 (2017). https://doi.org/10.1016/j.clay.2017.08.017
9. A. Alshameri, X. Wei, H. Wang, Y. Fuguo, X. Chen, H. He, C. Yan, F. Xu, in *A review of the role of natural clay minerals as effective adsorbents and an alternative source of minerals*, ed. by K.S. Essa, Minerals, (IntechOpen 2019), p. 49. Doi: https://doi.org/10.5772/intechopen.
10. A. Bouquillon, in *History of Ceramics*, ed. by P. Boch, J-C. Nièpce, *Ceramic Materials: Processes, Properties and Applications* (ISTE Ltd, 2007), pp. 30–53
11. F. Bergaya, G. Lagaly, ed. *Developments in Clay Science*, Handbook of clay science, (Amsterdam: Elsevier, 2006), p. 1224

12. F. Brigatti, E. Galan, B. Theng, in *Structures and Mineralogy of Clay Minerals, Developments In Clay Science* (Amsterdam: Elsevier, 2006), pp. 19–86
13. L. Black, P. Purnell, J. Hill, Adv. Appl. Ceram. **109**(5), 253 (2010)
14. C. Parr, R. Roesky, C. Wohrmeyer, C N Refractories **5**, 6 (2001)
15. M.A. Serry, S.M. Hammad, M.F.M. Zawrah, Phase composition and microstructure of refractory $MgAl_2O_4$ spinel grains. Brit. Ceram. Trans. **97**(6), 275–282 (1998). http://pascalfrancis.inist.fr/vibad/index.php?action=getRecordDetail&idt=1658021
16. M.F. Zawrah, S. Meky, Ceram Int. **33**, 969 (2007)
17. M.F. Zawrah, N.M. Khalil, Improvement of physico-mechnical properties of self-forming MA-spinel castables. Brit. Ceram. Trans. **100**(3), 110–114 (2001). https://doi.org/10.1179/096797801681305
18. M.F. Zawrah, M.A. Serry, J. Schneider, K.-H. Zum Gahr, Microstructure and mechanical properties of non-stoichiometric $MgAl_2O_4$-spinel bodies, Cfi/Ber. DKG **77**(3), 22–26 (2000). http://pascalfrancis.inist.fr/vibad/index.php?action=getRecordDetail&idt=1330158
19. S.F. Krar, *Grinding Technology*, 2nd edn. (Delmar Publishers, Albany, 1995), p.334
20. G.S. Làzaro, S.C. Santos, C.X. Resende, E.A. dos Santos, Individual and combined effects of the elements Zn, Mg and Sr on the surface reactivity of a $SiO_2{\cdot}CaO{\cdot}Na_2O{\cdot}P_2O_5$ bioglass system. J. Non-Cryst. Solids **386**, 19–28 (2014). https://doi.org/10.1016/j.jnoncrysol.2013.11.038
21. W. Holland, G. Beall, *Glass-Ceramic Technology*, 1st edn. (The American Ceramic Society, Westerville, OH, 2002), p. 372. https://doi.org/10.2298/SOS0403216U
22. E.D. Zanotto, Am. Ceram. Soc. Bull. **89**, 19 (2010)
23. F.C. Serbena, I. Mathias, C.E. Foerster, E.D. Zanotto, Crystallization toughening of a model glass-ceramic. Acta Mater. **86**, 216–228 (2015). https://doi.org/10.1016/j.actamat.2014.12.007
24. E. Kleebusch, C. Patzig, T. Höche, C. Rüssel, The evidence of phase separation droplets in the crystallization process of a Li_2O-Al_2O_3-SiO_2 glass with TiO_2 as nucleating agent—an X-ray diffrraction and (S)TEM-study supported by EDX analysis. Ceram. Int. **44**, 2919–2926 (2018). https://doi.org/10.1016/j.ceramint.2017.11.040
25. I. Alekseeva, O. Dymshits, M. Tsenter, A. Zhilin, V. Golubkov, I. Denisov, N. Skoptsov, A. Malyarevich, K. Yumashev, Optical applications of glass-ceramics. J. Non Cryst. Solids **356**, 3042–3058 (2010). https://doi.org/10.1016/j.jnoncrysol.2010.05.103
26. L. Fu, H. Engqvist, W. Xia, Glass–ceramics in dentistry: a review. Materials **13**, 1–22 (2020). https://doi.org/10.3390/ma13051049
27. G.H. Beall, Milestones in glass-ceramics: a personal perspective. Int. J. Appl. Glas. Sci. **5**, 93–103 (2014). https://doi.org/10.1111/ijag.12063
28. M. Montazerian, E.D. Zanotto, Bioactive and inert dental glass-ceramics. J. Biomed. Mater. Res. Part A **1**, 619–639 (2016). https://doi.org/10.1002/jbm.a.35923
29. H.E.H Sadek, R.M. Khattab, Mahmoud F. Zawrah, Interceram **65**, 174 (2016)
30. R.M. Khattab, A. El-Rafei, M. F. Zawrah, In-situ formation of sintered cordierite-mullite nano-micro composites by utilizing of waste silica fume. Mater. Res. Bull. **47**(9), 2662 (2012)
31. Y. Zhong, L.L. Shaw, M. F. Zawrah, J. Am. Ceram. Soc. **93**(10), 3159(2010)
32. J. Suri, L.L. Shaw, M.F. Zawrah, Ceram. Int. **37**, 3477 (2011)
33. J. Suri, L.L. Shaw, M.F. Zawrah, Tailoring the relative Si_3N_4 and SiC contents in Si_3N_4/SiC nanopowders through carbothermic reduction and nitridation of silica fume. Int. J. Appl. Ceram. Technol. **9**, 291–303 (2012). https://doi.org/10.1111/j.1744-7402.2011.02710.x
34. S.E. Abo Sawan, M.F. Zawrah, R.M.Khattab, A.A. Abdel-Shafi, Mater. Chem. Phys. **239**, 121998 (2020)
35. P.N. Lemougna, J. Yliniemi, E. Adesanya, P. Tanskanen, P. Kinnunen, J. Roning, M. Illikainen, Reuse of copper slag in high-strength building ceramics containing spodumene tailings as fluxing agent. Minerals Engineering **155**, 106448 (2020). https://doi.org/10.1016/j.mineng.2020.106448

36. P.N. Lemougna, J. Yliniemi, A. Ismailov, E. Levanen, P. Tanskanen, P. Kinnunen, J. Roning, M. Illikainen, Recycling lithium mine tailings in the production of low temperature (700–900 °C) ceramics: Effect of ladle slag and sodium compounds on the processing and final properties. Constr. Build. Mater. **221**, 332–344 (2019). https://doi.org/10.1016/j.conbuildmat.2019.06.078
37. L.L. Hench, J.M. Polak, Third-generation biomedical materials. Science **295**, 1014–1017 (2002). https://doi.org/10.1126/science.1067404
38. O.P. Kulik, Powder Metall. Met. Ceram. **38**, 93 (1999)
39. P. Greil, Advanced engineering ceramics. Adv. Mater. **14**, 709–715 (2002). https://doi.org/10.1002/1521-4095(20020517)14:10%3C709::AIDADMA709%3E3.0.CO;2-9
40. R. Riedel, E. Ionescu, I.W. Chen, in *Modern Trends in Advanced Ceramics*, ed. by R. Riedel, I.W. Chen, Cerem. Sci. and Tech. (Wiley-VCH, Weinheim, Germany, 2008), pp. 3–38
41. R.B. Heimann, in *Introduction to advanced ceramics*, ed. by R.B. Heimann, Classic and advanced ceramics: from fundamentals to applications, (Wiley-VCH Verlag GmbH, 2010), pp. 157-174.
42. M.F. Zawrah, R.A. Essawy, H.A. Zayed, A.H. Abdel Fattah, M.A. Taha, Mechanical alloying, sintering and characterization of Al_2O_3–20 wt%–Cu nanocomposite. Ceram. Int. 40, 31-38 (2014). Doi: https://doi.org/10.1016/j.ceramint.2013.05.099.
43. M.A. Taha, A.H. Nassar, M.F. Zawrah, In-situ formation of composite having hard outer layer based on aluminum dross reinforced by SiC and TiO_2. Constr. Build Mater. **248**, 118638 (2020). https://doi.org/10.1016/j.conbuildmat.2020.118638
44. M.A. Taha, M.F. Zawrah, Mechanical alloying and sintering of a Ni/10 wt.%Al_2O_3 nanocomposite and its characterization. Silicon 10, 1351-1359 (2018). https://doi.org/10.1007/s12633-017-9611-4.
45. M.F. Zawrah, M.A. Taha, H. Abo Mostafa, In-situ formation of Al_2O_3/Al core-shell from waste material. Ceram. Int. **44**, 10693–10699 (2018). https://doi.org/10.1016/j.ceramint.2018.03.101.
46. M.F. Zawrah, H.A. Zayed, R.A. Essawy, A.H. Nassar, M.A. Taha, Preparation by mechanical alloying, characterization and sintering of Cu–20 wt.% Al_2O_3 nanocomposites. Mater. Des. **46**, 485-490 (2013). https://doi.org/10.1016/j.matdes.2012.10.032.
47. M.F. Zawrah, H. Ahmed, N.E. El-Baly, Fabrication of Al_2O_3-20 vol% Al nanocomposite powder using high energy milling. Mater. Res. Bull. **47**, 655–661 (2012). https://doi.org/10.1016/j.materresbull.2011.12.023
48. M.M.S. Wahsh, R.M. Khattab, M.F. Zawrah, Sintering and technological properties of alumina/zirconia/nano TiO_2 ceramic composites. Mater. Res. Bull. **48**, 1411–1414 (2013). https://doi.org/10.1016/j.materresbull.2012.12.024
49. M.F. Zawrah, N.M. Khalil, High alumina castable reinforced with SiC. Adv. Appl. Ceram. **104**, 312–317 (2005). https://doi.org/10.1179/174367605X73498
50. M.A. Taha, A.H. Nassar, M.F. Zawrah, Improvement of wetability, sinterability, mechanical and electrical properties of Al_2O_3-Ni nanocomposites prepared by mechanical alloying. Ceram. Int. **43**, 3576–3582 (2017). https://doi.org/10.1016/j.ceramint.2016.11.194
51. M.A. Taha, M.F. Zawrah, Effect of nano ZrO_2 on strengthening and electrical properties prepared by mechanical alloying. Ceram. Int. **43**, 12698–12704 (2017). https://doi.org/10.1016/j.ceramint.2017.06.153
52. J. Silvestre, N. Silvestre, J. de Brito, An overview on the improvement of mechanical properties of ceramics nanocomposites. J. Nanomater. **2015**, 1–13 (2015). https://doi.org/10.1155/2015/106494
53. M.F. Zawrah, H. Abo Mostafa, M.A. Taha, Effect of SiC content on microstructure, mechanical and electrical properties of sintered Al-20Si-xSiC nanocomposites fabricated by mechanical alloying. Mater. Res. Express **6**, 1–11 (2019). https://doi.org/10.1088/2053-1591/ab534e
54. F.S. Moghanlou, M. Vajdi, H. Jafarzadeh, Z. Ahmadi, A. Motallebzadeh, F. Sharifianjazi, M.S. Asl, M. Mohammadi, Spark plasma sinterability and thermal diffusivity of TiN ceramics with graphene additive. Ceram. Int. https://doi.org/10.1016/j.ceramint.2020.12.152

55. M. Sakkaki, F. Sadegh Moghanlou, M. Vajdi, M. Shahedi Asl, M. Mohammadi, M. Shokouhimehr, Ceram. Int. **46**, 4998 (2020)
56. S. Nekahi, F. Sadegh Moghanlou, M. Vajdi, Z. Ahmadi, A. Motallebzadeh, M. Shahedi Asl, Microstructural, thermal and mechanical characterization of TiB_2–SiC composites doped with short carbon fibers, Int. J. Refract. Met. Hard Mater. **82**, 129–135 (2019). https://doi.org/10.1016/j.ijrmhm.2019.04.005
57. M. Vajdi, F. Sadegh Moghanlou, S. Nekahi, Z. Ahmadi, A. Motallebzadeh, H. Jafarzadeh, M. Shahedi Asl, Role of graphene nano-platelets on thermal conductivity and microstructure of TiB_2–SiC ceramics. Ceram. Int. **46**, 21775–21783 (2020). https://doi.org/10.1016/j.ceramint.2020.05.289
58. Stuart Hampshire, in *Materials Science Forum*, vol. 606 (Scientific.Net, 2009), pp. 27–41. https://doi.org/10.4028/www.scientific.net/MSF.606
59. R.W. Harrison, W.E. Lee, Processing and properties of ZrC, ZrN and ZrCN ceramics: a review. Adv. Appl. Ceram. **115**(5), 294–307 (2016). https://doi.org/10.1179/1743676115Y.0000000061
60. P. Karvankova, M.G.J. Veprek-Heijman, M.F. Zawrah, S. Veprek, Thin Solid Films **467**, 133 (2004)
61. M. F. Zawrah, M A Zayed and Moustafa R. K. Ali. J. Hazard. Mater. **227–228**, 250 (2012)
62. M. F. Zawrah, Adel B. Shehata , E. A. Kishar, Randa N. Yamani, Synthesis, hydration and sintering of calcium aluminate nanopowder for biomedical applications. Comptes Rendus Chim. **14**(6), 611–618 (2011). https://doi.org/10.1016/j.crci.2010.11.004
63. J.P. Vacanti, R. Langer, Tissue engineering: The design and fabrication of living replacement devices for surgical reconstruction and transplantation. The Lancet **354**, 32–34 (1999). https://doi.org/10.1016/S0140-6736(99)90247-7
64. J. Elisseeff, W. McIntosh, K. Fu, R. Blunk, R. Langer, Controlled-release of IGF-I and TGF-β1 in a photopolymerizing hydrogel for cartilage tissue engineering. J. Orthop. Res. **19**, 1098–1104 (2001). https://doi.org/10.1016/S0736-0266(01)00054-7
65. J.E. Babensee, L.V. McIntire, A.G. Mikos, Growth factor delivery for tissue engineering. Pharmaceut. Res. **17**(5), 497–504 (2000). https://doi.org/10.1023/a:1007502828372
66. D.F. Williams, On the mechanisms of biocompatibility. Biomaterials **29**, 2941–2953 (2008). https://doi.org/10.1016/j.biomaterials.2008.04.023
67. H.E. Jazayeri, M. Rodriguez-Romero, M. Razavi, M. Tahriri, K. Ganjawalla, M. Rasoulianboroujeni, M.H. Malekoshoaraie, K. Khoshroo, L. Tayebi, Ceram. Int. **44**, 1 (2018)
68. A.F. Black, F. Berthod, N. L'Heureux, L. Germain, F.A. Auger, In vitro reconstruction of a human capillary-like network in a tissue-engineered skin equivalent. FASEB J. **12**, 1331–1340 (1998). https://doi.org/10.1096/fasebj.12.13.1331
69. M. Marcacci, E. Kon, V. Moukhachev, A. Lavroukov, S. Kutepov, R. Quarto, M. Mastrogiacomo, R. Cancedda, Stem cells associated with macroporous bioceramics for long bone repair: 6- to 7-year outcome of a pilot clinical study. Tissue Eng. **13**, 947–955 (2007). https://doi.org/10.1089/ten.2006.0271
70. P.K. Valonen, F.T. Moutos, A. Kusanagi, M.G. Moretti, B.O. Diekman, J.F. Welter, A.I. Caplan, F. Guilak, L.E. Freed, In vitro generation of mechanically functional cartilage grafts based on adult human stem cells and 3-D-woven poly(e-caprolactone) scaffolds. Biomaterials **31**, 2193–2200 (2010). https://doi.org/10.1016/j.biomaterials.2009.11.092
71. M.N. Rahaman, J.J. Mao, Stem cell-based composite tissue constructs for regenerative medicine. Biotechnol. Bioeng. **91**, 261–284 (2005). https://doi.org/10.1002/bit.20292
72. A. Tathe, M. Ghodke, A.P. Nikalje, Int. J. Pharm Sci. **2**, 19 (2010)
73. N.R. Patel, P.P. Gohil, Int. J. Emerg. Technol. Adv. Eng. **2**(4), 91 (2012)
74. L.L. Hench, Bioceramics: from concept to clinic. J. Am. Ceram. Soc. **74**, 1487–1510 (1991). https://doi.org/10.1111/j.1151-2916.1991.tb07132.x
75. P. Palmero, Structural ceramic nanocomposites: a review of properties and powders' synthesis methods. Nanomaterials **5**, 656–696 (2015). https://doi.org/10.3390/nano5020656
76. J. Venkatesan, S.K. Kim, Nano-hydroxyapatite composite biomaterials for bone tissue engineering—a review. J. Biomed. Nanotechnol. **10**, 3124–3140 (2014). https://doi.org/10.1166/jbn.2014.1893

77. R.A. Youness, M.A. Taha, A.A. El-Kheshen, N. El-Faramawy, M. Ibrahim, In vitro bioactivity evaluation, antimicrobial behavior and mechanical properties of cerium-containing phosphate glasses. Mater. Res. Express **6**, 1–13 (2019). https://doi.org/10.1088/2053-1591/ab15b5
78. R.A. Youness, M.A. Taha, M. Ibrahim, A. El-Kheshen, FTIR spectral characterization, mechanical properties and antimicrobial properties of La-doped phosphate-based bioactive glasses. Silicon **10**, 1151–1159 (2018). https://doi.org/10.1007/s12633-017-9587-0
79. S.M. Abo-Naf, E.M. Khalil, E.M. El-Sayed, H.A. Zayed, R.A. Youness, In vitro bioactivity evaluation, mechanical properties and microstructural characterization of Na_2O-CaO-B_2O_3-P_2O_5 glasses. Spectrochim. Acta A **144**, 88–98 (2015). https://doi.org/10.1016/j.saa.2015.02.076
80. E.M.A. Khalil, R.A. Youness, M.S. Amer, M.A. Taha, Mechanical properties, in vitro and in vivo bioactivity assessment of Na_2O-CaO-P_2O_5-B_2O_3-SiO_2 glass-ceramics. Ceram. Int. **44**, 7867–7876 (2018). https://doi.org/10.1016/j.ceramint.2018.01.222
81. R.A. Youness, M.A. Taha, H. Elhaes, M. Ibrahim, Molecular modeling, FTIR spectral characterization and mechanical properties of carbonated-hydroxyapatite prepared by mechanochemical synthesis. Mater. Chem. Phys. **190**, 209–218 (2017). https://doi.org/10.1016/j.matchemphys.2017.01.004
82. R.A. Youness, M.A. Taha, H. Elhaes, M. Ibrahim, Preparation, Fourier transform infrared characterization and mechanical properties of hydroxyapatite nanopowders. J. Comput. Theor. Nanosci. **14**, 2409–2415 (2017). https://doi.org/10.1166/jctn.2017.6841
83. R.A. Youness, M.A. Taha, M.A. Ibrahim, Effect of sintering temperatures on the in vitro bioactivity, molecular structure and mechanical properties of titanium/carbonated hydroxyapatite nanobiocomposites. J. Mol. Struct. **1150**, 188–195 (2017). https://doi.org/10.1016/j.molstruc.2017.08.070
84. R.A. Youness, M.A. Taha, A.A. El-Kheshen, M. Ibrahim, Influence of the addition of carbonated hydroxyapatite and selenium dioxide on mechanical properties and in vitro bioactivity of borosilicate inert glass. Ceram. Int. **44**, 20677–20685 (2017). https://doi.org/10.1016/j.ceramint.2018.08.061
85. R.A. Youness, M.A. Taha, M.A. Ibrahim, In vitro bioactivity, molecular structure and mechanical properties of zirconia-carbonated hydroxyapatite nanobiocomposites sintered at different temperatures. Mater. Chem. Phys. **239**, 122011 (2020). https://doi.org/10.1016/j.matchemphys.2019.122011
86. R.A. Youness, M.A. Taha, M. Ibrahim, In vitro bioactivity, physical and mechanical properties of carbonated-fluoroapatite during mechanochemical synthesis. Ceram. Int. **44**, 21323–21329 (2018). https://doi.org/10.1016/j.ceramint.2018.08.184
87. A. Doostmohammadi, Z.K. Esfahani, A. Ardeshirylajimi, Z.R. Dehkordi, Zirconium modified calcium-silicate-based nanoceramics: An in vivo evaluation in a rabbit tibial defect model. Int. J. Appl. Ceram. Technol. **16**, 431–437 (2019). https://doi.org/10.1111/ijac.13076
88. A.M. Fayad, A.M. Fathi, A.A. El-Beih, M.A. Taha, S.A.M. Abdel-Hameed, Correlation between antimicrobial activity and bioactivity of Na-mica and Na-mica/fluorapatite glass and glass-ceramics and their corrosion protection of titanium in simulated body fluid. J. Mater. Eng. Perform. **28**, 5661–5673 (2019). https://doi.org/10.1007/s11665-019-04296-5
89. T. Kokubo, H.-M. Kim, A.M. Kawashita, in *Ceramics for Biomedical Applications*, ed. by S. Somiya, F. Aldinger, R. Spriggs, K. Uchino, K. Koumoto, M. Kaneno, Handbook of advanced ceramics (Elsevier, 2013), pp. 385–412
90. L.C. Gerhardt, A.R. Boccaccini, Bioactive glass and glass-ceramic scaffolds for bone tissue engineering. Materials **3**, 3867–3910 (2010). https://doi.org/10.3390/ma3073867
91. A. Refaat, R.A. Youness, M.A. Taha, M. Ibrahim, Effect of zinc oxide on the electronic properties of carbonated hydroxyapatite. J. Mol. Struct. **1147**, 148–154 (2017). https://doi.org/10.1016/j.molstruc.2017.06.091
92. Z. Huan, J. Chang, Study on physicochemical properties and in vitro bioactivity of tricalcium silicate-calcium carbonate composite bone cement. J. Mater. Sci. Mater. Med. **19**, 2913–2918 (2008). https://doi.org/10.1007/s10856-008-3423-4

93. F. Martínez-Vázquez, M. Cabañas, J. Paris, D. Lozano, M. Vallet-Regí, Fabrication of novel Si-doped hydroxyapatite/gelatin scaffolds by rapid prototyping for drug delivery and bone regeneration. Acta Biomater. **15**, 200–209 (2015). https://doi.org/10.1016/j.actbio.2014.12.021
94. D. Li, X. Huang, Y. Wu, J. Li, W. Cheng, J. He, H. Tian, Y. Huang, Biomater. Sci. **4**, 272 (2016)
95. S.Y. Wang, H.Z. Hu, X.C. Qing, Z.C. Zhang, Z.W. Shao, Recent advances of drug delivery nanocarriers in osteosarcoma treatment. J. Cancer **11**, 69–82 (2020). https://doi.org/10.7150/jca.36588
96. A. Barbanente, R.A. Nadar, L.D. Esposti, B. Palazzo, M. Lafisco, J.J.J.P. van den Beucken, S.C.G. Leeuwenburgh, N. Margiotta, J. Mater. Chem. B **8**, 2792 (2020)
97. H. Aghaei, A.A. Nourbakhsh, S. Karbasi, R. Javad Kalbasi, M. Rafienia, N. Nourbakhsh, S. Bonakdar, K.J. Mackenzie, Investigation on bioactivity and cytotoxicity of mesoporous nano-composite mcm-48/hydroxyapatite for ibuprofen drug delivery. Ceram. Int. **40**, 7355–7362 (2014). https://doi.org/10.1016/j.ceramint.2013.12.079
98. S. Bhaskar, M. Kumar, A. Patnaik, A review on tribological and mechanical properties of Al alloy composites. Mater. Today **25**, 810–815 (2020). https://doi.org/10.1016/j.matpr.2019.09.032
99. A. William, Jr Gooch, An overview of ceramic armor applications, 35TH International Conference and Exposition on Advanced Ceramics and Composites Daytona Beach, Fl 32118, 23-28 January 2011, 1-15 (2011).
100. B. Alemour, O. Badran, M.R. Hassan, A Review of using conductive composite materials in solving lightening strike and ice accumulation problems in aviation. J. Aerosp. Technol. Manag. **11**, 1–23 (2019). https://doi.org/10.5028/jatm.v11.1022
101. V. kumar, H. Vasudev, N. Kumar, Int. J. Manage. Technol. Eng. **IX**, 2308 (2019)
102. E. Salernitano, C. Migliaresi, J. Appl. Biomater. Biom. 1, 3 (2003)
103. H.L.W. Chan, K. Li, C.L. Choy, Electroceramic fibres and composites for intelligent apparel applications. Wearable Electron. Photonics, 41–58 (2005). https://doi.org/10.1533/9781845690441.41
104. N. Setter, R. Waser, Electroceramic materials. Acta Mater. **48**, 151–178 (2000). https://doi.org/10.1016/S1359-6454(99)00293-1
105. Y. He, B. Ping, L. Lu, F. Wang, S. Hu, Electrically conductive ceramic composites prepared with printer toner as the conductive phase. Adv. Appl. Ceram **116**, 158–164 (2017). https://doi.org/10.1080/17436753.2016.1278339
106. D.M. Smyth, in *Electrical conductivity in ceramics: a review*, ed. by J.A. Pask, A.G. Evans, Ceramic Microstructures '86. Role of interfaces, vol. 21 (Plenum Press, New York and London, 1987), pp. 643–655
107. M.A. Ryan, J.P. Fleur, Where there is heat, there is a way. The Electrochem. Soc. Interface **30** (2002).
108. R.B. Heimann, *Plasma spray coating, principles and applications*, 2nd edn, (Wiley-VCH Verlag GmbH, Weinheim 2008), pp. 427
109. H. Henke, D. Adam, A. Köhler, R.B. Heimann, Development and testing of HVOF-sprayed tungsten carbide coatings applied to moulds for concrete roof tiles. Wear **256**, 81–87 (2004). https://doi.org/10.1016/s0043-1648(03)00348-x
110. S. Salman, R. Kose, L. Urtekin, F. Findik, An investigation of different ceramic coatings thermal properties. Mater. Des. **27**, 585–590 (2006). https://doi.org/10.1016/j.matdes.2004.12.010
111. Y. Adraider, Y.X. Pang, F. Nabhani, S.N. Hodgson, M.C. Sharp, A. Al-Waidh, Laser-induced deposition of alumina ceramic coating on stainless steel from dry thin films for surface modification. Ceram. Int. **40**, 6151–6156 (2014). https://doi.org/10.1016/j.ceramint.2013.11.067
112. Y. Adraider, S.N.B. Hodgson, M.C. Sharp, Z.Y. Zhang, F. Nabhani, A. Al–Waidh, Y.X. Pang, Structure characterization and mechanical properties of crystalline alumina coatings on stainless steel fabricated via sol-gel technology and fibre laser processing. J. Eur. Ceram. Soc. **32**, 4229–4240 (2012). https://doi.org/10.1016/j.jeurceramsoc.2012.07.012

113. K. Srinivasulu, V.S. Manisha, Int. J. Adv. Eng. Technol. **VII**, 126 (2016)
114. M.F. Zawrah, J. Schneider, K.-H. Zum Gahr **22**(2), 167 (2002)
115. M.F. Zawrah, J. Schneider and K-H Zum Gahr. J. Mater. Sci. Eng. A **332**(1–2), 167 (2002)
116. R. Pompe, Some recent promising developments in industrial application of advanced technical ceramics. Transact. Mater. Res. Soc. Japan B **19**, 709–718 (1994). https://doi.org/10.1016/B978-1-4832-8382-1.50156-7
117. N.L. Wang, X.Y. Zhang, H.T. Jiang, T.T. Dong, D. Yang, Fabrication of Er^{3+}/Yb^{3+} co-doped Y_2O_3 transparent ceramics by solid-state reaction method and its up-conversion luminescence. Mater. Chem. Phys. **135**(2012), 709–713 (2012). https://doi.org/10.1016/j.matchemphys.2012.05.048
118. S. Arabgari, R. Malekfar, K. Motamedi, Parameters effects on the surface morphology and structure of Nd:YAG nanopowders synthesized by co-precipitation method. J. Nanoparticle Res. **13**, 597–611 (2011). https://doi.org/10.1007/s11051-010-0055-7
119. X.B. Ji, J.G. Deng, B. Kang, H. Huang, X. Wang, W. Jing, T. Xu, Thermal decomposition of $Y_3Al_5O_{12}$ precursor synthesized by urea homogenous co-precipitation. J. Anal. Appl. Pyrolysis **104**, 361–365 (2013). https://doi.org/10.1016/j.jaap.2013.06.012
120. K. Guo, H.-H. Chen, X. Guo, X.X. Yang, F.F. Xu, J.T. Zhao, Morphology investigation of yttrium aluminum garnet nano-powders prepared by a sol-gel combustion method. J. Alloys. Compd. **500**, 34–38 (2010). https://doi.org/10.1016/j.jallcom.2010.03.037
121. J. Su, J.H. Miao, L.H. Xu, Y.Q. Ji, C.Q. Wang, Synthesis and characterization of nanocrystalline Nd^{3+}-doped gadolinium scandium aluminum garnet powders by a gel-combustion method. Mater. Res. Bull. **47**, 1709–1712 (2012). https://doi.org/10.1016/j.materresbull.2012.03.039
122. B.T. Huang, Y.Q. Ma, S.B. Qian, D. Zou, G.H. Zheng, Z.X. Dai, Luminescent properties of low-temperature-hydrothermally-synthesized and post-treated YAG:Ce (5%) phosphors. Opt. Mater. **36**, 1561–1565 (2014). https://doi.org/10.1016/j.optmat.2014.04.025
123. Y.H. Sang, H.M. Qin, H. Liu, L.L. Zhao, Y.N. Wang, H.D. Jiang, J.Y. Wang, Partial wet route for YAG powders synthesis leading to transparent ceramic: a core-shell solid-state reaction process. J. Eur. Ceram. Soc. **33**, 2617–2623 (2013). https://doi.org/10.1016/j.jeurceramsoc.2013.04.009
124. C.C. Barry, M.N. Grant, *Ceramic Materials, Science and Engineering*, 1st edn. (New York, NY, Springer, 2013)
125. D. Eliche-Quesada, L. Pérez-Villarejo, P.J. Sánchez-Soto, in *Introduction to ceramic materials: synthesis, characterization, applications and recycling*, ed. by J. Cuppoletti, Ceramic materials: synthesis, characterization, applications and recycling (Intech Open, 2019), pp. 1–5
126. C.H. Leong, A. Muchtar, C.Y. Tan, M. Razali, N.F. Amat, Sintering of hydroxyapatite/yttria stabilized zirconia nanocomposites under nitrogen gas for dental materials. Adv. Mater Sci. Eng. **2014**, 1–6 (2014). https://doi.org/10.1155/2014/367267
127. E. Champion, Sintering of calcium phosphate bioceramics. Acta Biomater. **9**, 5855–5875 (2013). https://doi.org/10.1016/j.actbio.2012.11.029
128. F.J.T. Lin, L.C. De Jonghe, Microstructure refinement of sintered alumina by a two-step sintering technique. J. Am. Ceram. Soc. **80**, 2269–2277 (1997). https://doi.org/10.1111/j.1151-2916.1997.tb03117.x
129. T.-S. Yeh, M.D. Sacks, Low-temperature sintering of aluminum oxide. J. Am. Ceram. Soc. **71**, 841–844 (1988). https://doi.org/10.1111/j.1151-2916.1988.tb07533.x
130. X. Kuang, G. Carotenuto, L. Nicolais, Adv. Perform. Mater. **4**, 257 (1997)
131. Z. He, J. Ma, Grain growth rate constant of hot-pressed alumina ceramics. Mater. Lett. **44**, 14–18 (2000). https://doi.org/10.1016/S0167-577X(99)00289-X
132. L. Gao, J.S. Hong, H. Miyamoto, D.D.L. Torre, Bending strength and micro- structure of Al_2O_3 ceramics densified by spark plasma sintering. J. Eur. Ceram. Soc. **20**, 2149–2152 (2000). https://doi.org/10.1016/S0955-2219(00)00086-8
133. N.J. Lóh, L. Simão, C.A. Faller, A.N. Jr, O.R.K. Montedo, A review of two-step sintering for ceramics. Ceram. Int. **42**, 12556-12572 (2016). https://doi.org/10.1016/j.ceramint.2016.05.065

134. M. Prakasam, J. Locs, K. Salma-Ancane, D. Loca, A. Largeteau, L. Berzina-Cimdina, Biodegradable materials and metallic implants-a review. J. Funct. Biomater. **8**, 1–15 (2017). https://doi.org/10.3390/jfb8040044
135. M.A. Taha, G.M. Elkomy, H. Abo Mostafa, E. Gouda, Effect of ZrO_2 contents and ageing times on mechanical and electrical properties of Al-4.5 wt.% Cu nanocomposites prepared by mechanical alloying. Mater. Chem. Phys. **206**, 116–123 (2018). https://doi.org/10.1016/j.matchemphys.2017.11.058
136. . M.A. Taha, A.H. Nassar, M.F. Zawrah, Effect of milling parameters on sinterability, mechanical properties of Cu-4 wt.% ZrO_2 nanocomposite. Mater. Chem. Phys. **181**, 26–32 (2016). Doi:https://doi.org/10.1016/j.matchemphys.2016.06.030
137. G.S. Upadhyaya, *Powder Metallurgy Technology*, 1st edn. (Camridge International Publishing 2002), p. 68
138. M.-Y. Chu, L.C. De Jonghe, M.K.F. Lin, F.J.T. Lin, Precoarsening to improve microstructure and sintering of powder compacts. J. Am. Ceram. Soc. **74**, 2902–2911 (1991). https://doi.org/10.1111/j.1151-2916.1991.tb06861.x
139. X.-H. Wang, P.-L. Chen, I.-W. Chen, Two step sintering of ceramics with constant grain-size, I. Y_2O_3, J. Am. Ceram. Soc. **89**, 431–437 (2006). https://doi.org/10.1111/j.1551-2916.2005.00763.x
140. S. Chakraborty, P.K. Das, D. Ghosh, Rev. Adv. Mater. Sci. **44**, 182 (2016)
141. M. Suárez, A. Fernández, J.L. Menéndez, R. Torrecillas, H.U. Kessel, J. Hennicke, R. Kirchner, T. Kessel, in Challenges and Opportunities for Spark Plasma Sintering: A Key Technology for a New Generation of Materials, Sintering Applications (In Tech, 2013), p. 319. https://doi.org/10.5772/53706
142. S. Bose, S. Dasgupta, S. Tarafder, A. Bandyopadhyay, Microwave-processed nanocrystalline hydroxyapatite: simultaneous enhancement of mechanical and biological properties. Acta Biomater. **6**, 3782–3790 (2010). https://doi.org/10.1016/j.actbio.2010.03.016
143. K.E. Haque, Microwave energy for mineral treatment processes-a brief review. Int. J. Miner. Process **57**, 1–24 (1999). https://doi.org/10.1016/S0301-7516(99)00009-5
144. N.K. Tolochko, Mechanisms of selective laser sintering and heat transfer in Ti powder. Rapid Prototyping J. **9**, 314–326 (2003). https://doi.org/10.1108/13552540310502211
145. T. Ikegami, J.G. Li, T. Mori, Y. Moriyoshi, Fabrication of transparent yttria ceramics by the low-temperature synthesis of yttium hydroxide. J. Am. Ceram. Soc. **85**, 1725–1729 (2002). https://doi.org/10.1111/j.1151-2916.2002.tb00342
146. Z. Xiao, S. Yu, Y. Li, S. Ruan, L.B. Kong, Q. Huang, Z. Huang, K. Zhou, H. Su, Z. Yao, W. Que, Y. Liu, T. Zhang, J. Wang, P. Liu, D. Shen, M. Allix, J. Zhang, D. Tang, Materials development and potential applications of transparent ceramics: a review. Mater. Sci. Eng. R **139**, 100518 (2020). https://doi.org/10.1016/j.mser.2019.100518
147. J. Zdeněk, A. Evelyn, C. Bolaňos, H. Monika, J. Petr, in *Physical and Metallurgical Characteristics of Fiber Reinforced Ceramic Matrix Composites*, ed. by C. Sikalidis, Advances in ceramics-characterization, raw materials, processing, properties, degradation and healing, (Intech Open, 2011), pp. 281–298
148. X.L. Zhao, Ceramics **102**, 3 (1993)
149. L.K. Zeng, G.Y. Zhang, Chin. J. Mater. Res. **8**(3), 245 (1994)
150. J.H Jeong, K.H. Auh, Modelling. Simul. Sci. Eng. **8**, 541 (2000)
151. B. Lin, F. Liu, X. Zhang, L. Liu, X. Zhu, in *Numerical simulations-applications, examples and theory*, ed. by L. Angermann. (Intech Open, 2010), pp. 401–414
152. H. Le Ferrand, Magnetic slip casting for dense and textured ceramics: a review of current achievements and issues. J. Eur. Ceram. Soc. 41, 24–37 (2021). https://doi.org/10.1016/j.jeurceramsoc.2020.08.030
153. A. De Zanet, V. Casalegno, M. Salvo, Laser surface texturing of ceramics and ceramic composite materials – a review. Ceram. Int. (2020). https://doi.org/10.1016/j.ceramint.2020.11.146
154. T.H. Maiman, Stimulated optical radiation in ruby. Nature **187**, 493–494 (1960). https://doi.org/10.1038/187493a0

155. M.F. Zawrah, J. Schneider, K.-H. Zum Gahr, Mater. Sci. Eng. A **332**, 167 (2002)
156. S. Grasso, M. Biesuz, L. Zoli, G. Taveri, A.I. Duff,D. Ke, A. Jiang, M.J. Reece, A review of cold sintering processes. Adv. Appl. Ceram. (2020). https://doi.org/10.1080/17436753.2019.1706825
157. European Comission, *Reference Document on Best Available Techniques in the Ceramic Manufacturing Industry* (2007), pp. 210–211
158. M. Cologna, B. Rashkova, R. Raj, Flash sintering of nanograin zirconia in <5 s at 850°C. J. Am. Ceram. Soc. **93**, 3556–3559 (2010). https://doi.org/10.1111/j.1551-2916.2010.04089.x
159. S. Ghosh, A.H. Chokshi, P. Lee, R. Raj, A huge effect of weak dc electrical fields on grain growth in zirconia. J. Am. Ceram. Soc. **92**, 1856–1859 (2009). https://doi.org/10.1111/j.1551-2916.2009.03102.x
160. Z. Chen, Z. Li, J. Li, C. Liu, C. Lao, Y. Fu, C. Liu, Y. Li, P. Wang, Y. He, 3D printing of ceramics: a review. J. Eur. Ceram. Soc. **39**, 661–687 (2019). https://doi.org/10.1016/j.jeurceramsoc.2018.11.013
161. L. Marcus, J.J. Beaman, J.W. Barlow, D.L. Bourell, Solid freeform fabrication-powder processing. Am. Ceram. Soc. Bull. **69**(6), 1030–1031 (1990)
162. E. Sachs, M. Cima, J. Cornie, Three-dimensional printing: rapid tooling and prototypes directly from a CAD model. CIRP Ann. Manuf. Technol. **39**(1), 201–204 (1990). https://doi.org/10.1016/S0007-8506(07)61035-X
163. T. Ohji, M. Fukushima. Mater. Rev. **57**(2), 115 (2012)
164. E.C. Hammel, O.L.-R. Ighodaro, O.I. Okoli, Processing and properties of advanced porous ceramics: an application based review. Ceram. Int. **40** (10), 15351–15370 (2014). https://doi.org/10.1016/j.ceramint.2014.06.095
165. M.A. Keane, Ceramics for catalysis. J. Mater. Sci. **38**, 4661–4675 (2003)
166. K. Usami, A. Okamoto, Hydroxyapatite: catalyst for a one-pot pentose formation. Org. Biomol. Chem. **15**, 8888–8893 (2017). https://doi.org/10.1039/c7ob02051a
167. S. ben Moussa, A. Mehri, B. Badraoui, Magnesium modified calcium hydroxyapatite: an efficient and recyclable catalyst for the one-pot Biginelli condensation. J. Mol. Struct. **1200**, 127111 (2020). https://doi.org/10.1016/j.molstruc.2019.127111
168. D. Milovac, I. Weigand, M. Kovacic, M. Ivankovic, H. Ivankovic, Highly porous hydroxyapatite derived from cuttlefish bone as TiO2 catalyst support. Process. Appl. Ceram. **12**(2), 136–142 (2018). https://doi.org/10.2298/PAC1802136M
169. S.C. Oh, J. Xu, D.T. Tran, B. Liu, D. Liu, Effects of controlled crystalline surface of hydroxyapatite on methane oxidation reactions. ACS Catal. **8**(5), 4493–4507 (2018). https://doi.org/10.1021/acscatal.7b04011
170. M. Shokouhimehr, S.M.G. Yek, M. Nasrollahzadeh, A. Kim, R.S. Varma, Palladium nanocatalysts on hydroxyapatite: green oxidation of alcohol and reduction of nitroarenes in water. Appl. Sci. **9**, 1–12 (2019). https://doi.org/10.3390/app9194183
171. J. Xu, T. White, P. Li, C. He, Y.F. Han, Hydroxyapatite foam as a catalyst for formaldehyde combustion at room temperature. J. Am. Chem. Soc. **132**(38), 13172–13173 (2010). https://doi.org/10.1021/ja1058923
172. https://www.ceramicindustry.com/ceramic-materials-properties-charts/. Accessed January 1, 2020

Recent Advances in Perovskite Ceramics: Synthesis, Processing, and Applications

José Luis Clabel Huamán, Nurul Akidah Baharuddin, Mohd Ambri Mohamed, Abdullah Abdul Samat, Hamimah Abd Rahman, and Euclydes Marega Junior

Abstract The perovskite-type structure has attracted interest in several applied and fundamental areas of solid-state science and advanced materials research. Perovskite ceramics exhibit, among others, a plethora of intriguing optical and electronic properties, where different techniques can characterize their properties. The fabrication process is essential for any device superimposed onto the physical structure. The device characterization deals directly with the finished product and is, thus, directly related to the final goal of all research efforts, namely, to produce a device efficiently. The challenge of the fabrication process lies not only in the synthesis and processing of the samples but also in the characterization methods applied for ceramics, which go hand in hand with the interpretation of data. This chapter reports on the synthesis, processing, characterization, and advanced applications of perovskite ceramics.

Keywords Ceramics · Synthesis · Processing · Perovskite · Physical properties

J. L. C. Huamán (✉) · E. M. Junior
São Carlos Institute of Physics, University of São Paulo, São Carlos, São Paulo 13560-970, Brazil
e-mail: jclabel@ifsc.usp.br

N. A. Baharuddin
Fuel Cell Institute, Universiti Kebangsaan Malaysia, UKM, 43600 Bangi, Selangor, Malaysia

M. A. Mohamed
Institute of Microengineering and Nanoelectronics (IMEN), Universiti Kebangsaan Malaysia, UKM, 43600 Bangi, Selangor, Malaysia

A. A. Samat
Faculty of Mechanical Engineering Technology, Universiti Malaysia Perlis, Pauh Putra Campus, 02600 Arau, Perlis, Malaysia

H. Abd Rahman
Faculty of Mechanical and Manufacturing Engineering, Universiti Tun Hussein Onn Malaysia, 86000 Batu Pahat, Johor, Malaysia

S. J. Ikhmayies (ed.), *Advanced Ceramics*, Advances in Material Research and Technology, https://doi.org/10.1007/978-3-031-43918-6_2

1 Introduction

Ceramics is an inorganic material known as one of the materials in human civilization. Pottery, brick, and toilet bowl are commonly known ceramics-based materials. After the industrial revolution, the development of technical ceramics propagated in parallel with the improved high-temperature furnace, which was initially built to melt the metals during that time. As technology improved, technical or advanced ceramics also expanded accordingly. There are several ways to categorize ceramics, whether by their composition, application, or main property. In general, this material can be categorized into traditional and advanced ceramics. Traditional ceramics include pottery, tiles, and bricks from natural-based ceramics, e.g., clay [1].

On the other hand, advanced ceramics are produced from raw chemicals via a control route that aims for specific characteristics of ceramics. A mixture of covalent and ionic bonds is generally found in ceramics. The atom's electronegativity determines the domination between these two primary bonds [2]. A high electronegativity will increase the tendency of the valence electrons to be transferred, forming the ionic bonding and producing the anions (−) and cations (+) such as NaCl, MgO, and Al_2O_3. In contrast, a low electronegativity value will cause the electrons to be shared, establishing the covalent bond formation as in diamond, BN, and SiC. These two bonds are considered primary bonds in ceramics that will determine the basic properties of the materials. Apart from that, the secondary bond exists in certain ceramic-based materials, as observed in graphite. Graphite is known for its multi-layers/multi-sheets configuration in which the primary bond is formed within the same sheet while inter-sheet bonding is called a secondary bond. However, the secondary bond is weak compared to the primary bond, allowing the sheet or layer to slide, which explains the lubricating mechanism of this material. It is also noticed that the mechanical behaviors of ceramics, such as hardness and strength, are mainly influenced by the primary bond instead of the second one [2].

The application of ceramics is usually closely related to the materials' behaviors. Ceramics can be divided into several applications, including abrasive-type applications, cement, clays, glasses-based products, and refractories. For abrasive applications in which ceramic products were used as cutting tools to cut softer targets, materials such as SiC, WC, and diamond are the prominent candidates. For this type of application, ceramic-based materials should possess excellent mechanical properties such as resistance toward wear, hardness, and toughness. The leading property must be formed into the desired shape for cement application and hardened accordingly. These two characteristics also allowed cement to be utilized as a bonding material to stick ceramic-based products such as bricks. Besides cement, clays are one of the famous ceramics categorized by their application. The mixture of clay with other materials produces various products, including pottery, tiles, and bricks. Glasses, a pretty famous category of ceramics, are made up of a mixture of non-crystalline silicates with oxides to form multicolor products with specific properties. These are glass-based products with different physical appearances and usage from glass containers, mirrors, and windows. The working temperature limits the application of glasses in our daily life. Ceramics from the refractories category can be utilized in high-temperature applications depending on the application. Refractory ceramics are heat-resistant engineered materials designed to withstand extreme temperatures, e.g., in solid oxide fuel cells, nuclear reactors, aerospace, and thermal insulators [1].

The desired characteristics are also related to the composition and crystal structure of the materials. For composition, ceramics are divided into several material compositions, such as carbides, fluorides, nitrides, oxides, and sulfides. The utilization of each composition is also related to the type of application, e.g., oxide-based ceramics is prominently known as electrolyte and electrode materials in solid oxide fuel cell application. In the crystal structure, ceramics are categorized into four main systems: magnesia (rock salt), fluorite, perovskite, and anisotropy. For the magnesia structure (MgO), the center ion, e.g., Mg^{2+}, is surrounded by six opposite charged ions, e.g., O^{2-}, leading to charge neutralization, which explains the higher packing formation of the structure. Due to this characteristic, ceramics with this structure possess a high melting point and suitable for a high-temperature application. The following structure, fluorite (CaF_2), is favorable in offering more oxygen vacancies via introducing a doping process that allows different elements or ions to be added into the structure. At high temperatures, it exhibits excellent O^{2-} anions. An example of ceramic materials made up of this structure is Urania (UO_2) and zirconia (ZrO_2), which allow yttrium doping to form yttria-stabilized zirconia composition. In addition, in the case of UO_2 utilized in a nuclear reactor, the fluorite structure helps in fission control due to its open space available in a unit cell. Next, the third and fourth structures are cubic perovskite and non-cubic perovskite (anisotropy) ceramics. In perovskites, the ideal structure is symmetrical cubic of ABO_x (or denoted as calcium titanium oxide, $CaTiO_3$ structure), but, in some cases, the central ion can be shifted to form non-symmetrical cubic, resulting in electrostatic dipole formation, which explains the ferroelectric property of perovskite ceramics. Barium titanate ($BaTiO_3$) is an example of ferroelectric-perovskite that is suitable to be used in ceramic-based capacitors. Figure 1 summarizes the classification of ceramics by application, composition, and structure.

Perovskite ceramics

Perovskite-structured ceramics are among the most famous ceramics available in various applications nowadays. This material was found by Gustav Rose (1839) in the mountain of Russia. It was named after a Russian nobleman and mineralogist, Count Lev Alekseyevich von Perovski [3, 4]. As mentioned in the previous section, the crystal structure of perovskite is ABO_x or can be denoted as ABX_3. By referring to the ABX_3, both A and B were made up of cations, where the A-site has a larger cation size than the B-site. On the other hand, the X-site is generally occupied by oxide ions (O^{2-}), but it is also possible to have the halide ions such as Cl^-, Br^-, and F^- at this site. For the ideal symmetrical cubic perovskite structure, the easiest way to understand this structure is by looking from the BO_6 octahedra corner sharing in x–y–z dimensions, as shown in Fig. 2a. From this visualization, the larger A-cation is placed at the center of every space created between the BO_6 networks. Due to the

Fig. 1 Classification of ceramics

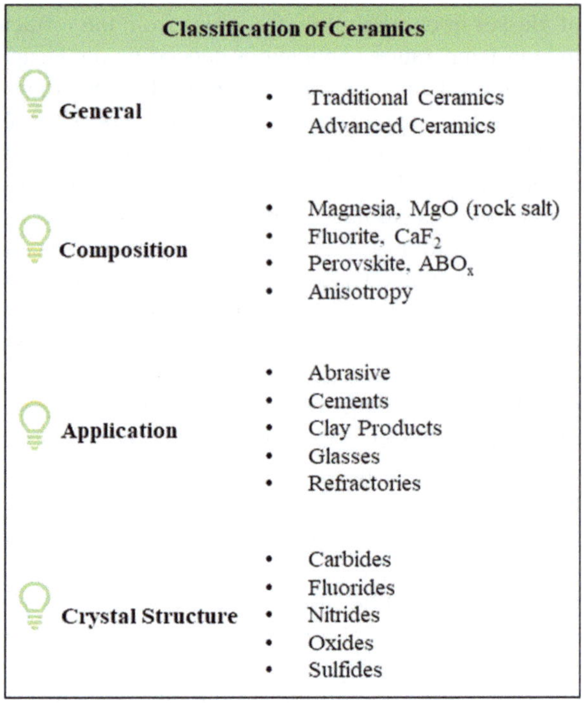

surrounding of the BO_6 octahedra network, the larger A-site cation is possessed 12-fold oxygen coordination, as shown in Fig. 2b. Distortion of this ideal symmetrical cubic forms a new cubic-based structure such as a tetragonal or orthorhombic structure. All cubic-based structures are commonly found in the oxide-type of perovskite or called perovskite oxide. Due to their electrical features of ferroelectricity or superconductivity, perovskite oxides are the most investigated perovskites compared to other perovskite-based structures [3].

Moreover, the structural transformation capability of perovskite lies in the A- and B-site cationic species across the periodic table. The size of the chosen cations at both sites affects the distortion from the ideal cubic structure, yet it is open to more interesting structure formation, which allows the alteration of perovskite's behaviors. Goldschmidt's tolerance factor, τ, is used to check the cation's compatibility in the perovskite crystal structure. It also indicates that the perovskite structure's stability and distortion depend on the existing ions. The τ-value is calculated using Eq. (1) below:

$$\tau = \frac{(r_A + r_O)}{\sqrt{2}(r_B + r_O)} \tag{1}$$

where r_A, r_B, and r_O are the radii of the A-site, B-site cations, and oxygen anion (usually), respectively.

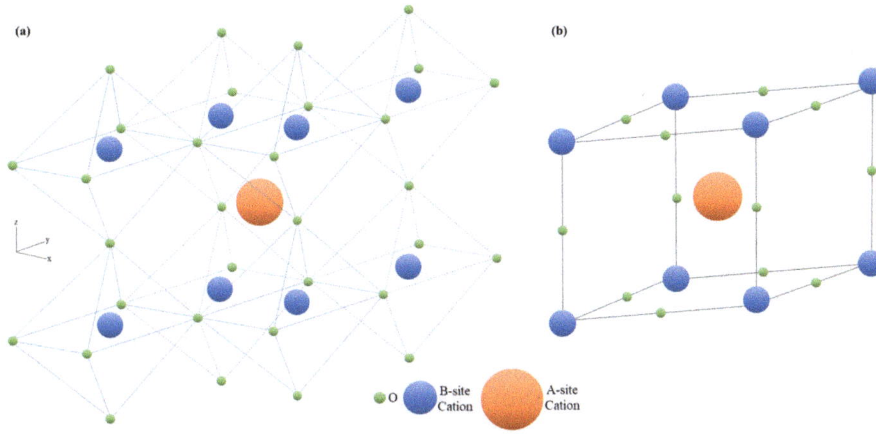

Fig. 2 Ideal perovskite structure visualized from **a** BO_6 octahedra, and **b** A-site cation with 12-fold oxygen surrounding view

The ideal size of A-site and B-site cations gives τ-value ~0.9–1.0, which results in ideal cubic perovskite structure formation. The τ-value below 0.9 means that the size of the A-site cation is too small to fit the hole or cavity of the BO_6 octahedra network, lowering the system symmetry and tilting the ideal cubic into rhombohedral or orthorhombic structures. The tilting in two directions usually forms the orthorhombic structure, while octahedral tilting in all directions forms the rhombohedral structure. On the other hand, the τ-value above 1.0 means that the size of the A-site cation is too large to fit into the cavity, suggesting tetragonal or hexagonal structures.

The versatility of perovskite in serving a wide range of applications depends on the variation of its properties. This situation is possible due to the structural displacement ability shown in tetragonal, rhombohedral, and octahedral-structured perovskite, which influences the bonding energy [5, 6]. The critical characteristics of perovskite include piezoelectric (ferroelectric material), thermoelectric, magnetic properties, and richness of oxygen vacancies. Perovskite is known for its high piezoelectricity (ferroelectric material). Progress in thin-film high-piezoelectric perovskites opens more potential for this type of materials in micro-electromechanical systems (MEMS), nanosensors, and nanoelectronic applications. Besides the piezoelectric properties, progress in improving the thermoelectric behavior of the perovskite aims to be applied in waste-heat conversion technology. With the utilization of the Seebeck effect mechanism, thermoelectric materials can convert heat-waste from industries into electricity. Using this approach, the voltage difference between two materials (whether it is an electrical conductor or semiconductor) is produced as a temperature difference product.

Many works have been done to improve the specific properties of perovskites by elemental doping in perovskite-based materials such as $SrTiO_3$ and $CaMnO_3$. The focus of these improvement works is on the thermoelectric characteristic or a high figure of merit ZT, a dimensionless unit to represent the heat to electricity conversion) of the *n*-type semiconductor oxides [7]. The calculation of ZT-value can be done as follows:

$$ZT = \frac{S^2 \sigma T}{k} \tag{2}$$

where S is the Seebeck coefficient, σ is the electrical conductivity, T is the absolute temperature, and k is the total thermal conductivity.

The oxygen vacancies in perovskite oxides ABO_3 have a crucial role in determining their properties. The ideal ABO_3 can be further altered to oxygen-deficient ($ABO_{3\text{-}\delta}$) than oxygen-rich ($ABO_{3+\delta}$) structures. In the case of oxygen deficiency ($O_{3\text{-}\delta}$), the lack of oxygen in the perovskite is also known as oxygen vacancy. Generally, the oxygen vacancies in the perovskite structure can be caused by two main reasons, (i) the cations substitution and (ii) the ion's non-stoichiometry. Each reason is narrowed down into two main factors, which are [8]:

(i) the cations substitution:

- the aliovalent substitution
- the isovalent substitution

(ii) the ions non-stoichiometry:

- the A-site cation non-stoichiometry
- the anion non-stoichiometry.

For the aliovalent substitution ($A_{1\text{-}x}A'_xBO_{3\text{-}\delta}$), the A-site cation commonly made up of rare-earth elements such as lanthanum (La) is substituted with the low valence alkaline-earth metal. For example, strontium (Sr) and calcium (Ca) are denoted as A'. These candidates are favorable compared to other group members, such as barium (Ba) and magnesium (Mg), because they require lower energy to form oxygen vacancies in the structure. This substitution will generate a negative charge, leading to neutrality attempts by the oxygen in the lattice (oxygen leaving its locations), hence forming vacancies in the structure [9, 10]. For the isovalent substitution, the partial substitution of A-site cation with the isovalent valence metal A' forms the oxygen vacancies. But compared to the aliovalent substitution, the isovalent substitution will not require a charge neutrality process as there is no charge difference between the A and A'. The oxygen vacancies form due to distorted structure due to significant differences in cation A and A' [9].

As for the ions non-stoichiometry factor, the A-site cation non-stoichiometry ($A_{1\text{-}x}BO_{3\text{-}\delta}$) produced vacancies without any introduction of A' cation, such as $(La_{0.85})(Fe)O_{3\text{-}\delta}$ and $(Ba_{0.95})(Co_{0.4}Fe_{0.4}Zr_{0.1}Y_{0.1})O_{3\text{-}\delta}$ [11, 12]. The introduction of A-site cation deficiency via the synthesis route increased the vacancies concentration, oxygen diffusion rate, and redox surface defects [12, 13]. As for the vacancy formation via the anion non-stoichiometry, the process is quite direct compared to the rest of the vacanciy formation mechanism or process. The vacancy usually forms by the heat treatment process in partial control pressure or working environment, for example, in a reduced atmosphere [8]. The oxygen vacancies behavior is essential in several fields, including solid oxide fuel cells, membrane reactors, and oxygen separation membrane applications.

2 Synthesis of Perovskite Ceramics

Synthesis routes play a crucial role in synthesizing and developing various perovskite ceramics originating from different precursor materials. Several aspects, including chemical and phase stability and compatibility, elemental homogeneity, shape and size of the final product, desirable properties, and types of applications, need to be thoroughly considered in selecting an appropriate synthesis route to yield the desired perovskite ceramics. Additionally, other parameters such as time, temperature, pressure, pH, chemical additives, and process complexity should also be considered in selecting a synthesis route. The selection of appropriate synthesis routes and parameters is essential for the performance of the produced perovskite ceramics. Some of the synthesis routes developed to synthesize various perovskite ceramics for different applications are solid-state reaction (SSR) method, ball milling, template-based synthesis, wet chemical methods (WCMs), and hydrothermal, solvothermal, sol–gel, solution combustion, co-precipitation, and microwave-assisted synthesis. Some of the general advantages and disadvantages of each method are summarized in Table 1.

2.1 Solid-State Reaction and Ball Milling

The most traditional and straightforward method of synthesizing perovskite ceramics is the solid-state reaction (SSR). Generally, the process is characterized by the mechanical mixing of precursor materials, including oxide powders, carbonates, or salts, in an agate mortar or a ball milling machine, followed by heat treatment processes. Ball milling is a simple top-down method where more extensive materials are reduced to smaller or finer materials via grinding and blending processes by colliding heavy metal or ceramic balls. It is also called the mechanochemical method and offers few remarkable advantages, such as producing fine sizes and well-dispersed powders, being cost-effective, and being applicable for grinding both dry and wet materials. Noise and extended milling and cleaning time are the significant disadvantages of this method. SSR and ball milling methods require a well-controlled processing temperature, and the final produced materials tend to be contaminated, resulting in low purity. For instance, Samat et al. [18] reported that $La_{0.6}Sr_{0.4}CoO_{3-\delta}$-$BaCe_{0.54}Zr_{0.36}Y_{0.1}O_{3-\delta}$ powders synthesized by the solid-state mixing method did not form a single-phase compound due to the presence of $BaCO_3$ and $SrCO_3$ secondary phases.

Table 1 Advantages and disadvantages of synthesis methods for synthesizing perovskite ceramics

Method	Advantages	Disadvantages	References
Solid-state reaction (SSR)	Inexpensive, simple, and large amount of output	Time-consuming (slow reaction), requires laborious mechanical milling and mixing of precursors, requires high temperature for sintering, and the material tends to contaminate (poor purity)	[14, 15]
Ball milling	Simple, cost-effective, reduces grinding time, produces fine and well-dispersed particles	Noise, long milling and cleaning times, difficult to control particle size and material possibly tends to contaminate	[14, 15]
Template-based synthesis	High surface area, large pore size and uniform pore size for hard template	Expensive for soft template, removal of template is costly, complicated, and harmful to the environment, damage the precursor materials and pore tends to collapse	[14, 15]
Hydrothermal or solvothermal	Requires low temperature and produce sample with high purity, better particle size, morphology, and homogeneity	Requires high pressures (up to 15 MPa) inside the pressure vessel and uncontrollable nucleation and aggregation during particle growth and development	[14–16]
Sol–gel	Produces high purity of ultrafine material, accurate control of the final material composition, able to eliminate impurities and agglomeration, and offers low processing temperature for material development	Precursor materials are quite costly, time-consuming (require long drying process), and request complete monitoring	[14, 15, 17]
Solution combustion	Fast reaction, time and energy savings, flexibility in the type of fuels used, high purity and homogeneity of final product and produces fine powders	Releases smog and ash, provides great loss of products and opportunity for impurity forming	[14, 17]

(continued)

Table 1 (continued)

Method	Advantages	Disadvantages	References
Co-precipitation	Able to easily control particle size and composition, simple and rapid preparation, cost-effective, offers low calcination temperature, does not require environmentally unfriendly chemicals such as alkoxides, and produces high purity of product with better chemical homogeneity, minimum agglomeration, and fine grain size	Time-consuming, poor crystallinity, not applicable to various types of metal oxides, does not work the best when the reactants have very different solubility and precipitation rate, and washing steps may cause deficiency of specific cations	[14, 15, 17]
Microwave-assisted synthesis	Cost-effective offers swift reaction (time-saving) at a shallow temperature, fast calcination of the material is carried out avoiding particle coarsening	Exaggerated growth existence of impurities	[15–17]

Heat treatment processes of SSR-based and ball milling-based powders are generally done at high calcination or sintering temperature, typically around 1200 °C and above. The heat treatment process at high temperatures is normally carried out for a long time, ranging between 8 and 24 h to allow the mobility of cations through the crystalline grains to form perovskite structure since the solid precursor materials do not interact with each other at room temperature [14, 15]. Kuzmin et al. [19] reported that the $La_{1-x}Sr_xScO_{3-\alpha}$ powder was obtained by grinding precursor powders of La_2O_3, Sc_2O_3, and $SrCO_3$ in zirconia mortar as annealed at 1200 °C. The final sintering of the sample to form a single-phase powder was done at 1700 °C for 10 h. Ding et al. [20] reported that the double-perovskite $Ba_2FeMoO_{6-\delta}$ powders prepared via the conventional SSR method need to be calcined at 1100 °C for 5 h and reduced at 1100 °C in 5% H_2 for 20 h to yield the phase-pure structure. Kong et al. [21] and Rosa Silva et al. [22] reported that the $Li_{0.375}Sr_{0.4375}M_{0.25}N_{0.75}O_3$ (M = Ti, Sn; N = Nb, Ta) powders and Y-doped $SrTiO_3$ prepared via ball milling solid-state mixing were calcined at 1100 °C for 15 h and 1300 °C for 10 h, respectively.

2.2 *Template-Based Synthesis*

Template-based synthesis is deemed a promising approach in synthesizing nanostructured perovskite ceramics with high surface area, uniform and large pore sizes such as nanowires, nanotubes, and nanorods. A uniform void space of porous materials is used as a template or host to keep the synthesized nanoparticles as guests.

The template also acts as a skeleton to organize the desired nanodevice, components, and interfaces. Habiballah et al. [23] have synthesized $Ba_{0.5}Sr_{0.5}Co_{0.8}Fe_{0.2}O_{3-\delta}$ (BSCF) perovskite nanowires via a template-based approach for the first time by varying spin coating rate (100, 500, and 1000 rpm), followed by drying at 150 °C for 15 h and calcination process at 400–900 °C for 4 h. Highly pure phase and dense BSCF nanowires with length ranging between 7 and 10 μm were obtained at a low spinning rate of 100 rpm at 900 °C. $Na_{0.5}Bi_{0.5}TiO_3$ (NBT) nanotube arrays were synthesized using a template-based approach using different sol-filling methods, namely immersion, deposition, negative pressure deposition, spin coating deposition, and combined negative pressure-spin coating deposition. Among all deposition techniques, the combined technique can produce the best features of NBT with wall thickness and outer diameter of 25 nm and 250 nm, respectively [24]. Template-based synthesis derived $LaNiO_3$ nanorods with an average diameter of (99 ± 10) nm were obtained at an annealing temperature of 900 °C. Other perovskite ceramics which have been synthesized via template-based synthesis are $Ba_{0.6}Sr_{0.4}TiO_3$ hollow spheres [25], $CeMnO_3$ nanofibers [26], $PbTiO_3$ nanostructured arrays [27], $LaBiFeO_3$ nanoparticles [28], and $BaCeTiO_3$ nanoshell tubes [29].

Generally, template-based synthesis can be classified as hard, soft, and colloidal templates. A defined porous rigid mold is filled with the precursor or targeted material for rigid templates. The primary mold is removed after the targeted material is formed. A soft template employs a nonrigid mold or template where the pore size and structure are controllable. The chance for the frameworks to collapse during crystallization and template removal is higher in the soft template than in the rigid template. The template contains inorganic nanoparticles that serve as the wooden template in the colloidal template. The colloidal template has controllable pore size and limited pore structures [14]. The growth of the targeted materials on rigid templates, such as polymer, silica, and carbon, can be removed via thermal decomposition or chemical etching or soft templates, such as emulsion micelles and even bubbles [30]. Although this method helps synthesize perovskite nanostructured ceramics for large-scale production, dealing with template removal using acid or alkaline etching is quite complicated [31]. Another drawback of template-based synthesis is residual template left in the final product affects its purity [28, 32–34].

2.3 *Hydrothermal or Solvothermal*

Hydrothermal or solvothermal refers to using an aqueous solution as a reaction medium in a closed reaction vessel to create high-temperature and high-pressure reaction environments by heating and pressuring the reaction medium. Water is used as a reaction medium in hydrothermal and organic solvents, including several types of alcohols used as a solvothermal reaction medium. The hydrothermal or solvothermal method mainly dissolves and recrystallizes a substance that is insoluble or poorly soluble under normal conditions. Generally, the method's advantages include an environment-friendly closed system, low processing temperature, high purity products, easily controllable morphology, crystal size, and low precipitation requirements [14, 32]. Normally, no calcination stage is required after the hydrothermal process, thus saving time to produce a specific perovskite ceramic. However, most of the compositions reported following a trial-and-error approach are considered extremely time-consuming [15]. There are some crucial parameters in determining the success of obtaining single-phase perovskite ceramics produced via the hydrothermal or solvothermal method. The essential parameters are molecular weight, boiling point, melting point, density, dipole moment and polarity of the solvent, medium alkalinity, system pressure, and synthesis duration [35–42].

François et al. [43] have synthesized $Ba_{1.01}Zr_{0.85}Y_{0.15}O_{3-\delta}$ material via a continuous hydrothermal process in supercritical conditions at 410 °C and 30 MPa using nitrate precursors and NaOH reactants. The as-synthesized powder formed an incomplete perovskite phase due to $BaCO_3$ and YO(OH) impurities. The as-synthesized powder was further calcined at 1000 °C for 1 h to yield pure phase powder. $La_{0.1}Sr_{0.9}Co_{0.9}Fe_{0.1}O_{3-\delta}$ powder synthesized via hydrothermal method has enhanced oxygen deficiency compared to the powder synthesized via the citrate method, as reported by Junior et al. [44]. $Ba_{0.85}Ca_{0.15}Ti_{0.9}Zr_{0.1}O_3$ powders synthesized via surfactant-assistant hydrothermal using different surfactants, namely, cetyltrimethylammonium bromide and sodium dodecyl sulfate, possess different relative densities and grain sizes after sintered at 1250 °C for 10 h. [45]. A non-aqueous solvothermal procedure was employed to obtain alkoxides of sodium, potassium, and niobium in 1-octadecene as the solvent and oleic acid as the capping agent $K_{0.5}Na_{0.5}NbO_3$ powder with an average particle size diameter of 50 nm at 600 °C.

2.4 Sol–Gel

A sol–gel method is one of the most popular syntheses of perovskite ceramics. This method has been traditionally used to produce inorganic ceramics and glass materials since the 1800s. Sol–gel is a process to form oxide linkages through an inorganic polymerization reaction, starting with a response between precursor material and solvent to form organic metal complexes. Then, the complexes undergo a polymerization reaction to produce colloid or sol, followed by hydrolysis to form a viscous gel. The gel will be subjected to drying, calcining, or sintering processes to yield ceramics powder.

Generally, there are two essential chemical reactions in the sol–gel method. The first reaction is the complexation between metal cations and a chelating agent such as citric acid and ethylenediaminetetraacetic acid (EDTA), which provides stable metal–chelate complexes in precursor solution using precursor solution preserving atomic-scale homogeneity. The second reaction is the polymerization of the complexes with polymerizing or esterification agents such as ethylene glycol and polyethylene glycol, forming three-dimensional network structures that hinder ion mobility and segregation. These two reactions help to produce high purity and homogeneity of perovskite ceramics and accurate control of the composition of the final produced material at low processing temperatures, which is regarded as the most remarkable advantage of the sol–gel method. Though the processing temperature of sol–gel gets reduced, the processing time remains the same. Longer time is undesirable both from the final product and economic point of view [46]. The formation of hard agglomerates of ultrafine crystallites or particles is also deemed one of the disadvantages of the sol–gel method [47]. Chemical agents play an essential role in the chemical reactions of the sol–gel process. The two essential chemical agents in a conventional sol–gel method are chelating agents and polymerizing or esterification. In a modified sol–gel method, surface-active agents (surfactant) and dispersing agents (dispersant) are introduced as alternative chemical agents, mainly to improve the microstructure properties of the final product. The roles of each chemical agent in the sol–gel method are summarized in Table 2.

Osman et al. [49] synthesized cerate-zirconate ceramics, $BaCe_{0.54}Zr_{0.36}Y_{0.1}O_{2.95}$ via the sol–gel method using various chelating agents from different functional groups, which are carboxylic (citric acid, glycolic acid, tartaric acid), polyamine carboxylic (EDTA, nitriloacetic acid), and polyamine (triethylenetetramine). They found that different groups of chelating agents did not affect the morphology, but did affect the phase formation and elemental composition of the final calcined powders. They also found that the chelating agent from the polyamine group aided in suppressing the formation of $BaCO_3$, which is the main impurity phase in synthesizing cerate-zirconate ceramics synthesized via the sol–gel method to yield high purity product at low calcination temperature (1100 °C). Mazlan et al. [50] also synthesized the same material using the same method. Still, they use different types of surfactants, which are cationic (benzalkonium chloride), anionic (sodium dodecyl sulfate), and non-ionic (polyoxyethylene (10) alkyl ether). They found that different surfactants influenced the phase formation and morphology of the calcined powders. The powders prepared using anionic and cationic surfactant have cubical, and spherical particles, respectively. A single-phase powder was obtained using non-ionic surfactant at a relatively lower calcination temperature (950 °C) than the previous work done by Osman et al. [49].

Sol–gel derived $La_{0.6}Sr_{0.4}CoO_{3-\delta}$ powders have been synthesized using different chemical agents, namely dispersing agent (activated carbon), polymerizing agent (ethylene glycol) and non-ionic surfactant (Tween80, polyethylene glycol, Brij97 and Triton-X-100). Among all chemical agents, only dispersing agent and polymerizing agent aided to form a single-phase $La_{0.6}Sr_{0.4}CoO_{3-\delta}$ powders at calcination temperature below 1000 °C [51, 52]. Furthermore, polymerizing agent-based $La_{0.6}Sr_{0.4}CoO_{3-\delta}$ powder has better electrochemical performance than dispersing agent-based $La_{0.6}Sr_{0.4}CoO_{3-\delta}$ powder [53]. Other perovskite ceramics which have been synthesized via sol–gel method are $RBaFeTiO_6$ (R = Pr, Nd) [54], $BaTiO_3/La_{0.7}Ba_{0.3}MnO_3$ [55, 56], $GdMO_3$ (M = Fe, Co) [57], $LaMg_xCr_{1-x}O_{3-\delta}$ ($x = 0.0, 0.2, 0.4, 0.6, 0.8$, and 1) [58], Gd-doped $SrZrO_3$ [59], $BaFeO_{3-\delta}$, and $BaM_{0.5}Fe_{0.95}O_{3-\delta}$ (M = Ti, Zr, Ce) [60].

Table 2 Chemical agents in a sol–gel method [48]

Chemical agent	Role	Remark	Example
Chelating agent	Binds all metal cations of the precursor materials and controls the rate of the hydrolysis reaction, phase transition, particle size, and powder morphology	A combined chelating agent is better than a single chelating agent as it is able to bind almost all metal cations, to form more stable complexes, it is less sensitive to pH, and it is able to reduce processing temperature	Citric acid, EDTA, triethylenetetramine, glycolic acid, nitriloacetic acid, tartaric acid
Polymerizing or esterification agent	Aids to form a stable polymer resin of metal–chelate complexes by creating a rigid network that controls the movement of metal cations in the complex solution during the heat treatment process	Formation of polymer resin avoids formation of particle agglomeration	Ethylene glycol, polyethylene glycol (PEG)
Dispersant	Controls the nucleation process, phase development, and particle growth	Has high degree of microporosity, high surface area and well adsorption ability	Activated carbon, carbon black, carbon nanotubes
Surfactant	Increases miscibility, colloidal stabilization, and particle dispersion in a material consisting of various components and controls the shape and particle size of the produced ceramics	Surfactants are classified into three groups: ionic (anionic and cationic), non-ionic and zwitterionic. The micelle formation which helps to control the shape and size of ceramics is influenced by processing temperature, pH, surfactant concentration, and surfactant composition	Pluronic F127 (tri-block copolymer), Triton-X-100, polyethylene glycol (PEG), polyvinyl alcohol, polyvinyl pyrrolidone

2.5 Solution Combustion

Solution combustion synthesis involves three steps, which are (1) formation of combustion mixture, (2) formation of gel, and (3) combustion of the gel [61]. It is a simple, fast, and versatile method to produce perovskite ceramics with high phase purity and superior powder properties, such as high surface area, narrow particle size distribution, and better sintering properties. For instance, spherical shape and uniform nanocrystalline (23–28 nm) $Ba_{1-x}Sr_xTiO_3$ powders were successfully synthesized via the solution combustion method using lemon juice as a fuel [62]. Fine-grained nanoparticles in sizes ranging from 50 to 200 nm and high surface area

(12.8 m^2/g) of $La_{0.6}Sr_{0.4}Co_{0.2}Fe_{0.8}O_{3-\delta}$ powder were also successfully synthesized via acetylacetone-assisted solution combustion method [63].

The solution combustion method is characterized by an exothermic reaction between oxidants (i.e., metal nitrates) and organic fuel (i.e., urea, citric acid, and glycine). Both soluble and insoluble precursor materials can be used in this method. The mixture of oxidants and fuel in an aqueous solution is subjected to relatively mild heating at a temperature ranging between 300 and 500 °C. The main limitations of this method are powder agglomeration and uncontrollable powder morphologies. Ravichandran et al. [64] reported that the grain distribution was not homogeneous, and agglomeration was observed in pure phase $Bi_{1-x}Al_xFeO_3$ powder synthesized via solution combustion method using urea as fuel after sintered at 550 °C for 2 h. Besides, the formation of micron-size agglomerates with a specific surface area of 4.8 m^2/g was also found in the $BaCeO_3$ powder synthesis via hydrazine-nitrate combustion [65]. Another limitation of the solution combustion method is the presence of leftover organic impurities due to incomplete combustion affects the phase purity of the product. Thus, after the combustion process, the calcination step is generally required to remove excess organic impurities from the as-synthesized powders and improve the final powders' crystallinity [61, 66].

The success of obtaining single-phase perovskite ceramics through this method depends on few factors, including the type of fuel [67–71], fuel-to-oxidizer ratio or precursor ratio [72–76], mixing time [77, 78], salts [79, 80], and other special considerations [81–84].

2.6 Co-precipitation

The co-precipitation method is expected to produce material with smaller particle sizes and better homogeneity. Generally, this method requires supersaturation conditions where solutions with soluble metal cations are mixed with a precipitation agent solution. It involves simultaneous nucleation and agglomeration processes to form precipitates. Then, the precipitates will be filtered and dried before being subjected to the calcination process at an appropriate low temperature to yield the desired product. Lower heat treatment temperature helps to reduce energy consumption [85, 86]. $NdFeO_3$ [87], $CaTiO_3$ [88], $Ba_{0.6}Sr_{0.4}TiO_3$ [89] materials were obtained after heat-treated at 750 °C for 1 h, 750 °C for 3 h, and 800 °C for 2 h, respectively. Other advantages of the co-precipitation method are simple and rapid preparation, cost-effectiveness, high phase purity, good chemical homogeneity product, and fine particle or grain size [14, 15, 90]. Multicomponent phase-pure perovskite lead strontium calcium titanate synthesized via reverse co-precipitation has a near-equiatomic and homogeneous distribution of elements in the calcined powders [91]. $SrBi_{1.8}Y_{0.2}Nb_2O_9$ powders prepared via co-precipitation have smaller (90 nm) crystallite size than the one prepared via SSR method (675 nm) [92]. $LaCoO_3$ powders prepared via co-precipitation have a larger surface area (8.121 m^2/g) than the one prepared via the sol–gel method (3.757 m^2/g) [93].

It is crucial to control some of the important parameters in the co-precipitation method, such as temperature, pH, and concentration to yield pure phase perovskite ceramics at low temperatures. Chen et al. [90] have studied the effect of NaOH concentration on powder formation of $Ba_{0.85}Ca_{0.15}Zr_{0.1}Ti_{0.9}O_3$. NaOH has two functions: to neutralize H^+ arising from hydrolysis and ensure the sufficient precipitation of various ions in the co-precipitation process. The NaOH concentration was varied from 4 to 10 mol/L. At low NaOH concentrations (4 and 6 mol/L), an incomplete perovskite phase was formed, while no pure perovskite phase was formed at high NaOH concentrations (10 mol/L). The optimal NaOH concentration to form a high-purity perovskite phase was found at 8 mol/L. Sharma et al. [91] have studied the effect of different concentrations of deionized water and ethylene glycol as a solvent on the phase formation of lead strontium calcium titanate. They reported that the sample prepared using deionized water formed a single perovskite phase at a lower temperature (850 °C) than the sample prepared using ethylene glycol (950 °C). The effect of co-precipitation pH value on the powder properties of $LaAlO_3$ was studied by Sim et al. [94]. They found that as the pH value increased from 6 to 10, the calcined powder at 950 °C changed its main phase from $LaAlO_3$ (pH = 6) to La_2O_3 (pH = 9) and La $(OH)_3$ (pH = 10) and surface area reduced from 8.3 m^2/g (pH = 6) to 1.2 m^2/g (pH = 10).

2.7 *Microwave-Assisted Synthesis*

Nowadays, microwave-assisted synthesis is considered a powerful approach in materials science, including synthesizing various perovskite ceramics for different applications. It uses microwave radiation as a heating source instead of conventional heating in a furnace to achieve uniform heating. The ability of microwave radiation to penetrate sample bulk makes microwave-assisted synthesis a rapid and cost-effective method to obtain a crystalline structure material at a lower processing temperature. Microwave-assisted synthesis is also known as the hybrid synthesis method, as it is normally combined with other synthesis methods. The hybrid method significantly reduced the time for synthesizing perovskite ceramics.

Kostyukhin et al. [95] synthesized $LaFeO_3$, which is frequently used as a catalyst in oxidation processes, electrodes, and sensors through one-step hydrothermal microwave-assisted at 240 °C and 60 bar. They reported that microwave irradiation during heating helps produce high crystallinity of $LaFeO_3$ at a shorter time (3 h) than conventional heating (48 h). Ji et al. [96] synthesized $Ba_{0.85}Ca_{0.15}Zr_{0.1}Ti_{0.9}O_3$ powders to be used in piezoelectric or ferroelectric fields through a novel microwave-assisted sol–gel-hydrothermal method. They managed to obtain the powders at a temperature of 180 °C and 300 W after 60 min reaction time. The powders were well-crystallized with a homogenous composition and fine grains (85 nm). Asefi et al. [97] synthesized $BiFeO_3$ powders, which has potential applications in optical devices and information storage, sensors, spintronics, and photocatalysts through microwave-assisted solution combustion. Precursor solution with different fuel content (φ =

0.5, 0.75, 1, and 2) was poured into a beaker and placed in a microwave oven (2.45 GHz, 750 W) for 2 min. Single-phase $BiFeO_3$ powders were obtained at low fuel content ($\varphi = 0.5$ and 0.75) without further calcination. The crystallite size of the microwave-combusted $BiFeO_3$ powders is smaller than as-calcined powders, 32 nm, and 54 nm, respectively. The results from Asefi et al. work are in line with a job done by Maziviero et al. [98]. They reported that the formation of microwave-assisted combustion-derived γ-Al_2O_3 material was favored by the combination of low microwave power and low fuel content. Therefore, fuel content plays a crucial role in the microwave-assisted combustion synthesis method.

In some cases, a different heat treatment process (calcination or sintering) is required after microwave-assisted synthesis. For instance, $CaZrO_3$ powders synthesized by microwave-assisted hydrothermal at 180 °C with a heating rate of 140 °C/min, a pressure of 4 bar, and synthesis times of 20 and 160 min did not transform into crystalline phase. $CaCO_3$ impurity was observed in the powder prepared at a lower synthesis time. After calcined at 800 °C, the powders formed a crystalline phase [99]. In another example, $ZnTiO_3$ powders were also synthesized by microwave-assisted hydrothermal. The sample without conventional heat treatment resulted in an amorphous material, while the sample with conventional heat treatment began a crystallization process at 500 °C [100]. In both cases, it can be concluded that microwave-assisted hydrothermal only provides energy for the crystallite nucleation process and not for phase growth. Even though further heat treatment is sometimes required, sintering temperature or time is significantly reduced. Sahu and Kumar [101] found that the sintering temperature and time needed for the formation of single-phase $Sr_{0.8}Bi_{2.15}Ta_2O_9$ ceramic synthesized via microwave-assisted SSR is lower (1050 °C, 30 min) than the conventional sintering method (1200 °C, 4 h). Deepa et al. [102] also reported similar results. SSR-derived $Na_{0.5}Bi_{0.5}TiO_3$ material formed a single-phase after calcined at 800 °C for 10 min and at 1000 °C for 3 h by microwave sintering and conventional sintering, respectively.

3 Processing of Perovskite Ceramics

Perovskite ceramics with a porous structure, for instance, are often processed or fabricated as a bulk or a thin film to be used in various applications such as supercapacitors [103], catalyst for methane combustion [104], photocatalysts [105], gas sensors [106], filters [107], and solar cells [108]. In addition, porous perovskite ceramics using a mixed ionic and electronic conductor (MIEC) are frequently used as electrodes or substrates for applications in batteries and solid oxide fuel cell (SOFC) [109–114]. Thus, it is essential to design and fabricate proper porosity, microstructure, and mechanical properties of the porous perovskite ceramics according to their respective applications. In MIEC porous structure, the properties are crucial to improving network structure and surface area, thus enhancing the electrochemical performance of batteries and SOFC [112, 115, 116]. Typically, the fabrication of perovskite ceramics involves few steps such as powder processing, ink or slurry preparation, deposition, and sintering. Some of the methods are simple, cost-effective, and time-saving, and some of them are vice versa. The standard fabrication techniques used in fabricating bulk or thin film of perovskite ceramics are dry pressing, screen printing, and spin coating. General advantages and disadvantages of the standard fabrication techniques and other fabrication techniques which have been developed to process the produced perovskite ceramics powders are summarized in Table 3.

Dry pressing is a simple, reproducible, and fast technique to produce perovskite ceramics' bulk or thick film. The process of the dry pressing technique is also simple. The perovskite ceramics powder is grounded in an agate mortar or milled using a ball milling machine. The grounded or milled powder is placed between two stainless steel die in a mold and pressed under the desired pressure to obtain a pressed powder or green pellet. The obtained green pellet is then subjected to sintering or other heat treatment process at a high temperature for few hours to produce a dense pellet. Usually, the density of the final pellet is affected by the dry pressing pressure and sintering or heat treatment temperature. Apart from density, these two parameters also need to be properly designed to avoid sample cracking or breaking during pressing and sintering steps [117]. Danilov et al. produced a high density (95%) $BaCe_{0.5}Zr_{0.3}Dy_{0.3}O_{3-\delta}$ pellet via dry pressing technique after sintered at 1450 and 1600 °C for 5 h. The dense pellet with 30 μm thickness was used as an electrolyte for a protonic ceramic fuel cell application, and the cell recorded a power density of 290 mW/cm^2 at 700 °C [118]. Bazin et al. produced porous ceramic membranes from Sayong ball clay powders for nanofilter application. The powders were mixed with starch and pressed at various compaction pressures (50–200 MPa). The sample compacted at 200 MPa and sintered at 1000 °C has the best properties as a nanofilter membrane [119]. Later, they fixed the compaction pressure at 200 MPa and varied the sintering temperatures (900–1200 °C). They reported that bulk density increased (1.5–1.9 g/cm^3), porosity decreased (40.2–0.34%), and flexural strength increased (1–21 MPa) as the sintering temperatures increased. The optimum properties (bulk density of 1.6 g/cm^3, porosity 18%, and flexural strength of 6 MPa) of the nanofilter membrane were obtained for the sample sintered at 1050 °C [120]. Barinova et al. studied the impact of compaction pressure (20–200 MPa) on the density of green pellets and combusted pellets of $Y_2Ti_2O_7$ [121]. They found that the compaction had no impact on the green pellet since no obvious crack was observed in all compacted green pellets. However, after combusted, only the compacted green pellet at 20 MPa showed crack formation.

Screen printing is a simple and cheap technique that has been extensively used to fabricate homogeneous-porous perovskite ceramics, thin or thick film. This technique can produce an exact final product based on the desired design and properties by preparing and adjusting a screen printer's pattern, size, and distance between the plate and the print head. The screen printing technique consists of a screen or mesh made from synthetic fiber or metal wire connected to a rigid aluminum/steel frame having a rectangular plane in most applications. An amount of ink or paste or slurry is put on the frame and then is swiped across the surface of the screen or mesh at a steady pace by a squeegee. Through a pattern drawn on the mesh, the squeegee

Table 3 A summary of fabrication techniques to process perovskite ceramics powders (adapted from [14])

Method	Advantages	Disadvantages	Deposition rate/film thickness
Dry pressing	Simple, reproducible, fast	Unsuitable for mass production, crack issue before or after sintering	(1–100) μm
Screen-printing	Simple, cost-effective, precise final output	Crack issue or delamination after sintering	(10–100) μm
Spin-coating	Cost-effective, great uniformity, reproducible	Requires flat and smooth substrate	(1–1000) μm
Tape casting	Simple, cost-effective, suitable for mass production	Slow drying process, easily crack	(10–1000) μm
Impregnation/infiltration	Escalate in catalytic activity, improved triple-phase boundary length, lower operating temperature	Requires highly porous support, tedious steps, nonhomogeneous distribution of particles	–
3D-printing	Fast, reproducible, cost-effective, flexible	Size limitation, high-cost equipment, imperfection output image	–
Pulsed laser deposition (PLD)	Good for multicomponent fabrication, easy, reproducible, better morphology	Expensive, slow, low accuracy, limited deposition scale	(60–150) μm
Electrophoretic deposition (EPD)	Simple, fast method, suitable for any shape of substrate	Solvent media must be carefully chosen	(1–200) μm
Magnetron sputtering	Low temperature, composition can be controlled	Expensive equipment, slow deposition rate	(0.25–2.5) $\mu m\ h^{-1}$

(continued)

Table 3 (continued)

Method	Advantages	Disadvantages	Deposition rate/film thickness
Plasma spray	Suitable for large-scale production, requires no sintering step	Expensive equipment	(100–500) μm h^{-1}
Atomic layer deposition (ALD)	Variety choices of precursors, uniform deposition, flexible in determining deposition factors	Expensive, very slow, and use of corrosive products	(1–100) μm h^{-1} (rate), <0.001 μm(thickness)
Dip-coating	Simple, cost-effective, able to coat on irregular shape of substrate	Time consuming	(25–200) μm
Brush-painting	Simple, cost-effective	Unsuitable for mass production, hard to control uniformity of final product	–
Tape calendaring	Vary in thickness, good interfacial structure between substance and substrate	Involves multiple layers for lamination	(5–200) μm
3D-printing stereolithography	Good surface finishing, flexible material options, excellent mechanical performance, high-resolution printer	Requires postprocessing	(0.5–5) μm (grain size)
3D-printing fused deposition modeling	Easy remove of printed object/ paste, rapid reproduction	Poor mechanical properties, poor surface quality, not good in small features, low-resolution printer	~100 μm(thickness), (1–5) μm (grain size)

pressure is used to move the ink material onto a substrate. Finally, a definite picture of the pattern is transferred to the substrate, and a layer of ink is deposited [122]. The disadvantages of screen printing technique include the requirements for rheological properties of slurries, particle size of solid perovskite ceramics, material and life of screen, difficulty in cleaning screen plate, and low density of the produced film [123]. To enhance the quality and properties of the fabricated thin film through screen printing techniques, few critical parameters, which are ink rheology, printing number, and sintering process, need to be carefully designed [124]. Ink rheology is affected by a few factors, including viscosity, yield stress, thixotropy, and viscoelasticity [125]. Usually, the ink is a mixture of perovskite ceramics powder, binder, and solvent. Somalu et al. [48] conducted rheological studies of nickel/scandia-stabilized zirconia screen printing inks prepared via triple-roll milling technique for solid oxide fuel cell anode fabrication. In preparing the inks, they varied the amount of binder (1–3%) and solid content loading (25–30 vol.%). They found that the ink with 3% binder and 28 vol.% solid content loading has the best balance between thixotropic and viscoelastic properties.

Besides, hexanol-based ink appeared better than terpineol-based ink due to a minimum number of film defects. Furthermore, the fabricated film showed no crack or delamination even after 15 cycles of electrical and electrochemical performance testing [126, 127]. Abdul Samat et al. optimized the number of the screen-printed film layers in evaluating the electrochemical performance of $La_{0.6}Sr_{0.4}CoO_{3-\delta}$. They reported that the $La_{0.6}Sr_{0.4}CoO_{3-\delta}$ film with a thickness of (3.90 ± 0.18) μm and porosity of (23.09 ± 0.42)% recorded the lowest area-specific resistance value (0.11 Ω cm^2) as compared to the other films with thickness ranging between (4.36 ± 0.30) μm and (5.88 ± 0.40) μm [53]. Baharuddin et al. studied the effects of sintering temperatures ranging from 900 to 1300 °C on the structural and electrochemical properties of the screen printed $SrFe_{0.5}Ti_{0.5}O_{3-\delta}$ perovskite ceramic film. They reported that the thickness and porosity of the printed films decreased while the hardness and electrical conductivity of the printed films increased. Sufficient porosity is essential to enhance the fabricated film's active electrochemical area, thus increasing its electrochemical properties [128].

Spin coating is a simple, reproducible, and fast technique for depositing thin coating onto a flat substrate. The substrate to be coated is held by a rotatable fixture, often using a vacuum to clamp it in place. The coating solution or slurry is deposited onto the surface of the substrate by the action of rotating or spinning, which causes the solution to spread out, forming a very uniform coating of the chosen material on the surface of the substrate. It is well known that rotation speed, time, and solid content determine the final thickness of the deposited material on the substrate. Huo et al. showed that solid content affects the microstructure and electrical properties of $Co_{2.77}Mn_{1.71}Fe_{1.10}Zn_{0.42}O_8$ ceramic prepared by slurry spin coating technique at a spin speed of 3000 rpm for 30 s for the thermistor device. They found that an appropriate increase in solid content from 50 wt.% to 80 wt.% enhanced the uniform dispersion of the ceramic particles in slurries and grain densification. The thick ceramic film with 70 wt.% solid content shows higher density and relatively high stability with a resistance drift rate of 19.4%. The resistance drift rate for all samples is less than 1% per 100 h after pre-aging at 125 °C for 400 h [123]. The impact of spin speed on the film thickness of C–N–TiO_2 co-doped nanocatalysts for photocatalysis or as an electron transporting layer for perovskite solar cells was done by Mouele et al. [129]. They reported that the film thickness decreased from 560 to 160 nm as the spin speed increased from 2000 to 2900 rpm at a constant spin time of 90 s after annealing at 350 °C.

4 Characterization Techniques of Perovskite Ceramics

4.1 X-ray Powder Diffraction (XRD)

For this part, structural characterization via the X-ray diffraction method is discussed. At the same time, the basic perovskite ABO_3 (large A-site cation, small B-site cation), also known as ternary ABO_3 type perovskite oxide, is taken into the spotlight. The easiest way to confirm and study the structural parameters (lattice parameter, density, crystallite size) is via the X-ray diffraction (XRD) analysis, which is relatively easy to identify for the ideal cubic perovskite as the XRD-sharp peak spectrum without any splitting was observed (Fig. 3).

Suppose the perovskite-structured sample is a known material for the researcher. In that case, the diffraction peaks of XRD may be used to determine the phase structure properties, e.g., plane distances and miller indices. In the case of unknown

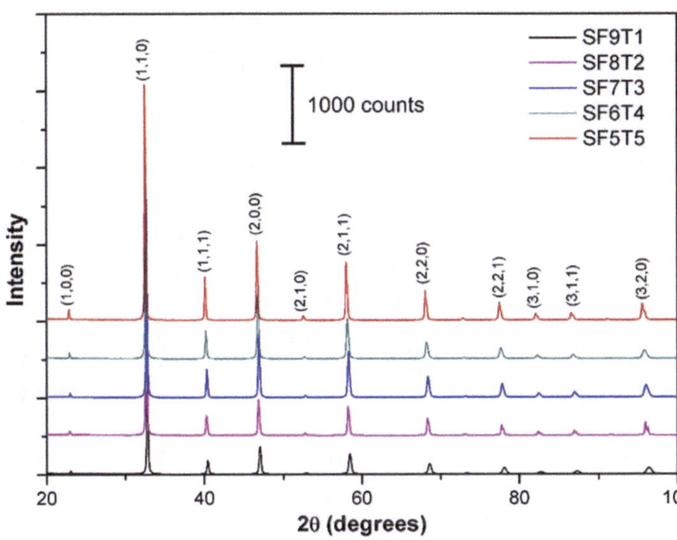

Fig. 3 XRD patterns for cubic-perovskite $SrFe_{1-x}Ti_xO_{3-\delta}$. Reprinted from Baharuddin et al. [130], with permission from Elsevier. Copyright (2019)

perovskite-structured materials, when the sample has an unknown structure, the XRD can acquire the values of the plane distances, which effectively stimulates the structural investigation. In addition, XRD is also important to help identify any distortions in the perovskite structure. Distortion in the phase structure of perovskite is vital to be evaluated, which later will influence the materials' properties such as catalytic, ferromagnetic, and piezoelectric of perovskite [4, 131, 132].

The perovskite structure can be changed due to structural alteration and processing/working environment (pressure, temperature) factors. Before reporting the phase transformation detection via XRD, one must know how to differentiate the phase change from the secondary phase formation (phase purity). A straightforward trick to differentiate these two is by looking at the plane position or simply 2θ (°-value). For the phase structure transformation, such as cubic to tetragonal or vise versa, the peak will be split into two but still from the same base, or simply the area below the peak or 2θ range remains the same, as shown in Fig. 4 [133].

Generally, the transformation from cubic into the tetragonal structure can be observed via peak splitting; for example, the single peak will be split into two peaks as a is longer equivalent to c ($a \neq c$) in the unit cell. As for the transformation to orthorhombic, a single sharp cubic peak will be split into three peaks as $a \neq b \neq c$. While for the secondary phase, new peaks are formed at different degrees of 2θ. An example of this finding can be seen in Song et al. [134] report. In this work, the parent B-site doped $La_{0.4}Sr_{0.6}Fe_{0.7}Ti_{0.3\text{-}x}Co_xO_{3\text{-}\delta}$ (LSFTC) was managed to maintain its cubic structure even after 0.2 mol ratio ($x = 0.2$). However, once the samples

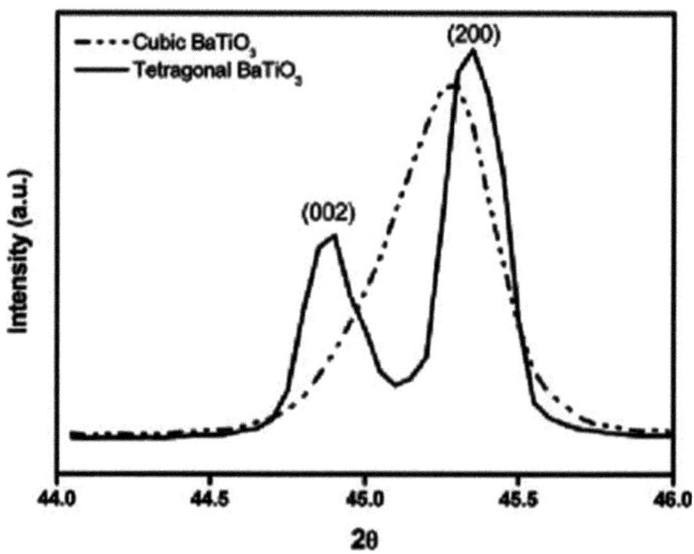

Fig. 4 Example of the XRD pattern emphasized on peak splitting due to crystal system transformation. Reprinted from Kwon and Yoon [133], with permission from Elsevier. Copyright (2007)

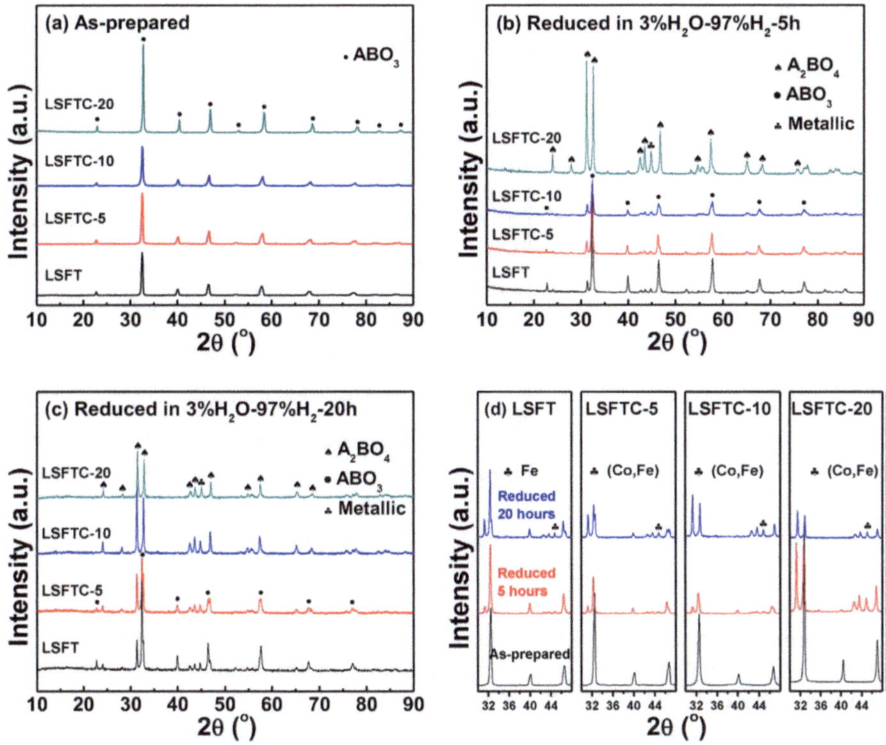

Fig. 5 Example of the XRD patterns for materials with secondary phases. Reprinted from Song et al. [134], with permission from Elsevier. Copyright (2020)

were tested under different durations for a reduced wet-H_2 treatment environment (to be used as an anode for symmetrical solid oxide fuel cell), the formation of the secondary phases of K_2NiF_4 layered perovskite and metallic phases can be seen as in Fig. 5 [134].

In the case of structure distortions, a basic understanding of the cause of diffraction and data confirmation from the XRD will be beneficial. The environment-induced distortion from other structures to ideal cubic is also possible. The perovskite $(1-x)(Na_{1/2}Bi_{1/2})TiO_3-xBaNbO_3$, or called NBT-BT05, in which $x = 0.05$, was proven to undergo a direct transformation from rhombohedral (space group: R3c) phase to the ideal cubic (space group: Pm3m) with the change of temperature [135]. In this work, besides the prominent peaks for NBT-BT05, the superlattice peak, namely, S1 located at 2θ around 38.5° (see Fig. 6), is also evaluated to confirm the transformation. Interestingly, the S1 peak has entirely disappeared from a temperature of 473 K and was consistent until the highest temperature of 900 K. This confirmed that neither rhombohedral nor tetragonal phase existed in the perovskite due to A- and B-site modification that causing the difference in the lattice strain, hence distorting or tilting the structure into cubic Pm3m.

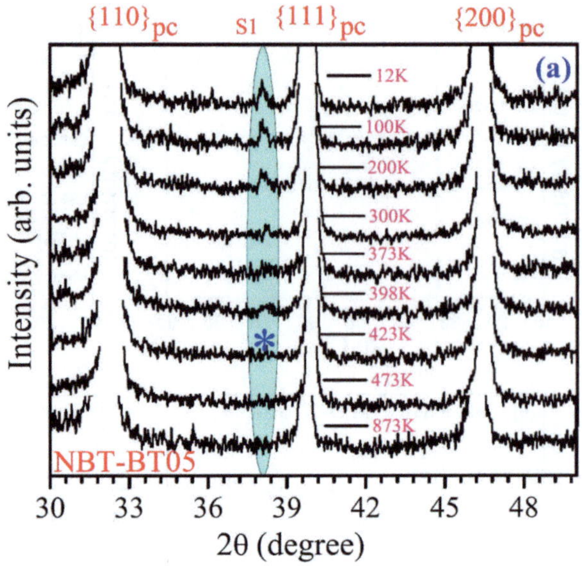

Fig. 6 Changes in XRD pattern of NBT-BT05 focusing on S1 area at different temperatures. Reprinted from Jayakrishnan et al. [135], with permission from Elsevier. Copyright (2021)

The temperature-related factor for a phase transition is also extensively reported by Oku [136]. In this work, perovskite-structured methylammonium trihalogenoplumbates (II), $CH_3NH_3PbX_3$ (X = Cl, Br, and I) for solar cell materials application are reviewed. With the variation in dopant species studied, the crystal systems and transition temperature (temperature when the phase structure changes accordingly) do show different values and groups. But, in general, all materials of $CH_3NH_3PbX_3$ (X = Cl, Br, and I) possessed a cubic perovskite at high temperature. Table 4 summarizes the crystal system and transition temperature for reported materials. Figure 7 shows the XRD pattern for one of the compositions ($CH_3NH_3PbI_3$).

Besides both sites doping, some researchers distort to allow phase transformation by manipulating the B-site. The selection of B-site doping is not because of the insignificancy at A-site doped cations; however, the A-site usually tends to have a stable valence causing not much structural manipulation that later influenced the perovskites' structural, physical and chemical properties. Yet, in particular cases,

Table 4 Crystal systems and transition temperatures of $CH_3NH_3PbX_3$ (X = Cl, Br and I). Repoduced from Oku [136], under the terms and conditions of the Creative Commons Attribution 3.0 Unported (CC BY 3.0) license. https://creativecommons.org/licenses/by/3.0/

Material	$CH_3NH_3PbCl_3$	$CH_3NH_3PbBr_3$	$CH_3NH_3PbI_3$
Crystal system	Cubic	Cubic	Cubic
Transition Temperature (K)	177	236	330
Crystal system	Tetragonal	Tetragonal	Tetragonal
Transition Temperature (K)	172	149–154	161
Crystal system	Orthorhombic	Orthorhombic	Orthorhombic

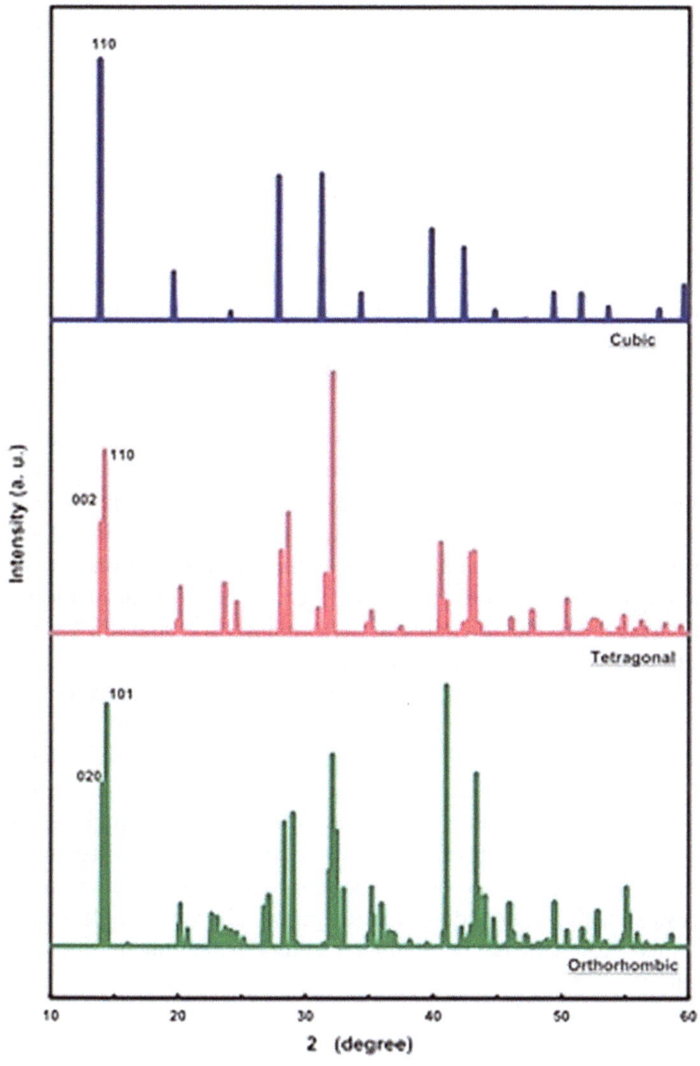

Fig. 7 Calculated X-ray diffraction patterns of $CH_3NH_3PbI_3$ with cubic, tetragonal, and orthorhombic structures. Reproduced from Oku [136] under the terms and conditions of the Creative Commons Attribution 3.0 Unported License. https://creativecommons.org/licenses/by/3.0/

the A-site deficiency also plays an essential role in enhancing material properties, as mentioned in the previous part (e.g., oxygen vacancies). Going back to the ideal cubic with BO_6 network, the species of B-site cation that fill up the space will determine the perovskite-structural changing flexibility. Considering cations B-site only, the distortion from cubic to tetragonal, orthorhombic, or trigonal may happen if the cation species are too small. There are three possible distortions at the B-site or at the BO_6 network, namely B-site cation displacement, BO_6 tilting, and BO_6 distortion [4], which will not be covered in this current discussion.

4.2 Transmission Electron Microscopy (TEM)

As the XRD will help confirm the phase structural parameters, crystal system, and purity of the perovskite materials, the crystallite size to the extent of nanoparticles can also be estimated via XRD. In this aspect, the full width at half maximum (FWHM) in XRD analysis becomes the primary function in determining the crystallite size. Usually, this value can be obtained directly from the built-in XRD fundamental analysis using the Scherer equation to calculate the size.

One can further analyze the microstructure of the perovskite via electron diffraction tools such as Transmission electron microscopes (TEM). Besides the structural parameter (crystal system, phase purity, etc.), microstructure analysis is crucial to determine the defects at the surface, interlayer, and intragrain, which influence the performance of the perovskite in their respective application. For example, in solid oxide fuel cell application, microstructure failure like delamination at the inter-component layer led to electrochemical performance degradation. The perovskites' microstructure can vary from atomic to micro size depending on the applications.

On the other hand, TEM is best known for its capability to capture particles as small as nano with a resolution below 0.1 nm at 30–1500 kV [137], explaining the high usage of this analysis to confirm the microstructure of any samples. Due to the higher voltage range compared to the scanning electron microscope (SEM), electrons obtained enough energy to allow transmission holistically throughout the samples. Thus, the results obtained from TEM will cover the surface microstructure and deeper until reconfirming the phase structure or atomic scale, with minimal sample volume utilizing as small as 0.1 nm incident electron probe. The incident electron will scatter at several trials before leaving the sample, resulting in imaging or diffraction patterns up to the atomic structure level [138]. Multi-directional electron diffraction on the crystal was applied to understand the structural parameters (e.g., crystal system, lattice parameters). Moreover, an added value analysis such as high-resolution TEM and energy-dispersive X-ray spectroscopy will be handy in constructing the atomic structure model. In the case of a known sample, it is easier to recognize the perovskite-structured material by comparing TEM images and diffraction patterns [136, 139]. Figure 8 shows the overview of the signals generated by TEM.

Electron diffraction, known as the selected area/area electron diffraction (SAD/SAED), is one of the main functions in the TEM that helps in a crystallographic study. SAD or SAED allows part of the electrons to be scattered (without any deflection) to specific angles that correspond to the crystal structure of the perovskite sample, yielding the series of spots called the selected area diffraction pattern (SADP) that depends on the users' area selection. Any tilting or distortion in the crystal structure will be reflected in the resulting SADP, proving the versatility of TEM analysis beyond microstructure imaging purposes. Generally, the SAED is quite similar to

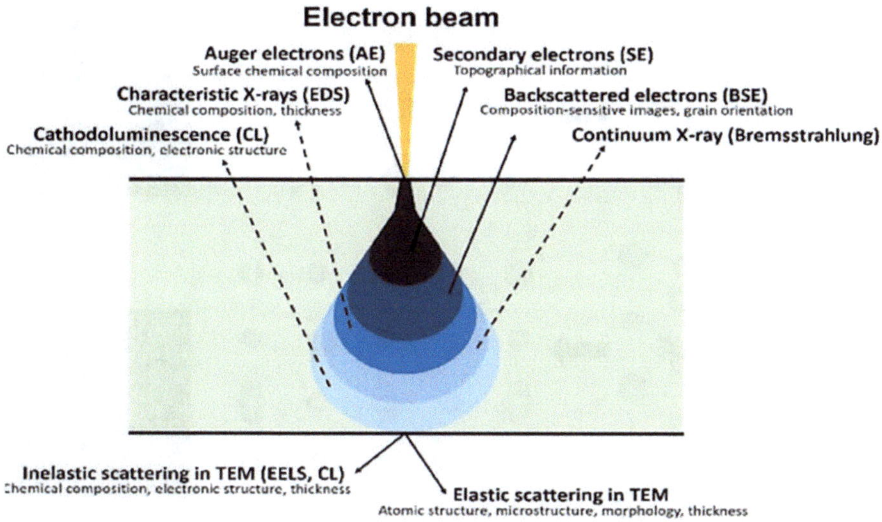

Fig. 8 Overview of the signals generated by the TEM. Reprinted from Rothmann et al. [138], with permission from John Wiley and Sons. Copyright (2017)

XRD as both help to confirm the crystal system. However, the samples' area for this electron diffraction experiment can be extremely small (in the nanometer range). Talking about the tilting/distortion observation from the SADP, the example of phase transformation detection via TEM has been discussed extensively again by Oku [136]. Figure 9 shows the rearranged figure from the source on the cubic and tetragonal perovskite $CH_3NH_3PbI_3$'s atomic structure model with the SADP results along [110] direction.

As a remark, microstructural analysis via TEM is a powerful tool that is underestimated, which is usually believed to give images and particle size, especially for the perovskite-structured materials. Interestingly, TEM and XRD can complement each other in confirming the crystal system and lattice parameters and monitoring the structural transformations. Extensive work that specifies the TEM capabilities on microstructure, morphology, crystal system, and physical characteristics of the perovskite materials should be done to achieve more accurate data on the as-produced materials.

5 Physical Properties

Perovskite structure is of interest due to its significant physical properties connected with technological devices. The physical properties of several perovskite materials are different from their intrinsic and extrinsic properties and the interaction between them. In this Review, we summarize and critically assess the most recent advances

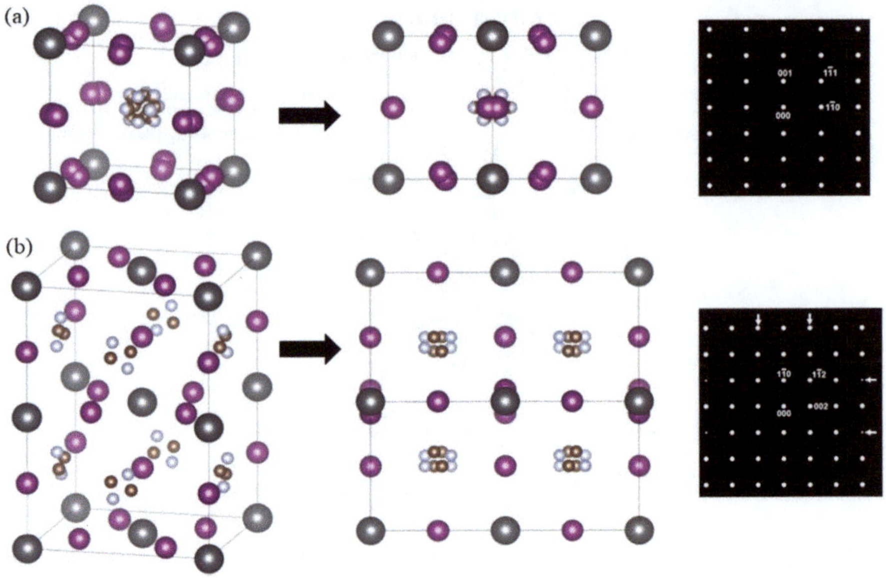

Fig. 9 SADP at [110] direction for **a** cubic $CH_3NH_3PbI_3$, and **b** tetragonal $CH_3NH_3PbI_3$. Reproduced from Oku [136] under the terms and conditions of the Creative Commons Attribution 3.0 Unported License. https://creativecommons.org/licenses/by/3.0/

in understanding the physical properties of perovskite structures, such as optical, dielectric, ferroelectric, and magnetic properties. For example, it has been reported that the optical properties of many materials, such as the refraction, absorption, reflection, and scattering of light, among the most important, are different at the nanoscale compared with their values at the macroscale. Besides, the rich variations in physical properties arising from carrier doping are not yet thoroughly understood. The scale effect (obtained from the different synthesis processes) and the doping effect are discussed, including the physical reasons. Dielectric materials exhibit linear polarization behavior as a function of the applied field. Ferroelectric ceramics are dielectrics with a permanent electric dipole, which can be oriented upon using an electric field.

In contrast, ferromagnetic ceramics with magnetic moment can be oriented upon applying a magnetic field. Although various properties arise from the crystal symmetry adopted by these materials, they are sensitive to variables such as chemical composition, purity, densities of surface and bulk defects, grain size, and sintering conditions. Hence, critical assessment and discussion on these parameters are vital for effective quality control of the devices produced from these materials.

5.1 *Optical Properties*

Optical spectroscopy has been widely used for the characterization of the perovskite materials. The experimental procedure for analyzing perovskites through optical spectroscopy is fundamental to the development of optical devices.

5.1.1 UV–Visible Spectroscopy

Optical spectra in the ultraviolet and visible (UV–Vis) regions are of fundamental importance for understanding the structure of electronic bands of a solid. Besides, absorption is an important characteristic for its application in an optoelectronic device. It determines the optical bandgap, which is a key to calculate the theoretical limit for various parameters of a photonics device. The measurement spectrometer consists of a light source, a monochromator (or polychromator), and a detector, as shown in Fig. 10. Optical components such as mirrors and lenses are used to optimize the light beams. A monochromator is a device with mechanism that produces monochromatic light from a light source, while a polychromator produces different wavelength components simultaneously.

Absorption occurs when an atom can react with incoming light (which is a stream of photons). It can either scatter the light or absorb it. During scattering, light may be redirected or alter the direction. In the case of absorption, the atom makes a quantum jump to one of its higher energy levels after absorbing light, i.e., specific wavelengths cause excitation in the electronic energy levels. The wavelength (λ) of light required for electronic transitions is typically in the ultraviolet (λ = 200–390 nm) and visible region (λ = 390–780 nm) of the electromagnetic radiation spectrum, i.e., it can cause excitation in the electronic energy levels. Electronic absorption spectroscopy is frequently called UV–Vis spectroscopy. Within each electronic transition, there are also vibrational and rotational transitions. This absorption of energy can be detected by UV–Vis spectrometry using a charged-coupled device (CCD) camera. A (CCD) camera detects incoming light photons from the sample and converts them to electrical charge to create a spectroscopic spectrum [140].

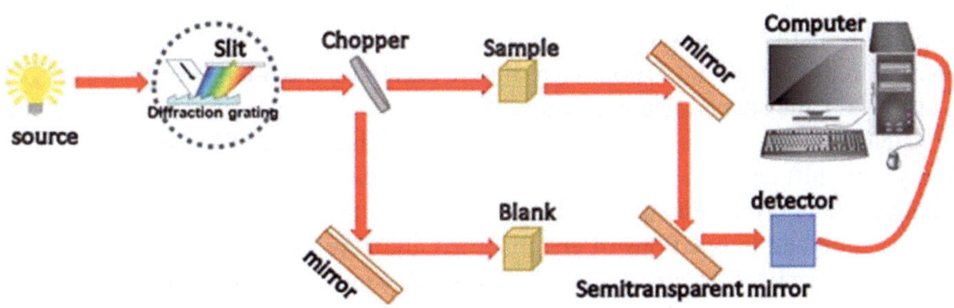

Fig. 10 Schematic diagrams showing UV–Visible spectrophotometer

The absorption coefficient (α) is the property of a material that gives the amount of light absorbed. The inverse of the absorption coefficient is the average distance traveled by a photon before it gets absorbed. The absorbance (A) is given by $A = \varepsilon cl = \alpha c$, where ε is the molar absorptivity, c is the concentration of a solution, l is the path length of the sample, and α is the absorption coefficient. The ratio of transmitted light intensity and incident light (I/I_0) is called the transmittance (T). The relation between A and T is $A = -\log T = \log\left(\frac{1}{T}\right) = \varepsilon cl = \alpha c$. Generally, in the literature, one can find the absorption measurement shown as the absorption coefficient or absorbance. Both I and I_0 can be measured, so A can be determined experimentally. The absorption coefficient is wavelength dependent, and the spectroscopy provides a spectrum between α and λ for a given sample; solution or solid. The solid sample should be thin to avoid absorption saturation due to the blurring of the optical path. The band gap energy E_g of a material can be calculated by the method proposed by Wood and Tauc [141]. According to these researchers, E_g is related to the absorption coefficient and energy of photons by the following equation $\alpha h\nu = A(h\nu - E_g)^n$, where h is Planck's constant, ν is the frequency of light, ($h\nu$) is the energy of the photon, A is a constant, and n is a variable that depends on the nature of electronic transition (n = 1/2, 2, 3/2 or 3 for direct allowable, indirect allowable, direct prohibited, or indirect prohibited, respectively).

The UV–Vis spectrum of nanomaterials is generally different from their bulk counterparts. Figure 11a shows the UV–Vis spectra of organometal halide perovskite Pe's) materials having different particle sizes [142]. The average nanocrystallite sizes calculated from the XRD results are ~76, ~52, ~24, ~15, and ~10 nm for samples 4,4-bis(N-carbazolyl)-1,1-biphenyl (CBP) and organometal halide perovskites (Pe's) with weight ratios of 1:1, 3:1, 4:1, 10:1, and 15:1, respectively. It can be seen that the nanoparticles show broad absorption spectrum due to quantum confinement effects and the absorption peaks shift toward higher energies as the average nanocrystallite sizes of the nanoparticles decrease, indicating a positive change in the material's band gap. According to Brus [145], the band gap consists of the contributions of the band gap of the bulk material, and the band gap due to quantum confinement i.e., $E_g = E_{g,\text{bulk}} + E_{g,\text{conf.}}$, the $E_{g,\text{conf.}}$ for a cubic box takes the form of $3\hbar^2\pi/2md^2$, where $\hbar$ is the reduced Planck's constant, m is the effective mass of carriers (reduced mass for electrons and holes), and d is the size of the nanocrystallites. Peak widths of the nanoparticles also depend on the size distribution of the nanoparticles [143, 146], doping [144, 147], and temperature [143, 148]. In $BaTiO_3$ (BT) pure samples with thermal treatment, BT1-1 (1000 °C) to BT1-2 (1100 °C) and BT3-1 (1000 °C) to BT3-2 (1100 °C), the absorbance decreases, and increases, respectively, with the increase of calcination temperature (see Fig. 11b). The enhancement of the light absorption and slight shift can be associated with the strain, particle size, and the presence of chemical defects on their surface [149–152]. In addition, the effect of free carriers in the structure and interactions of chemisorbed alkoxide ions on the surface with BT can create localized electronic states within the band gap, thus influencing optical absorption. On the other hand, small variation in the doping content of the doped samples led to modification of the absorbance due to the presence of doped ions. For example, $(Ba_{0.95}Ca_{0.05})(Ti_{0.92}Sn_{0.08\text{-}x}Zr_x)O_3$ (abbreviated as $BCTSZ_x$) ($0 \leq x \leq$

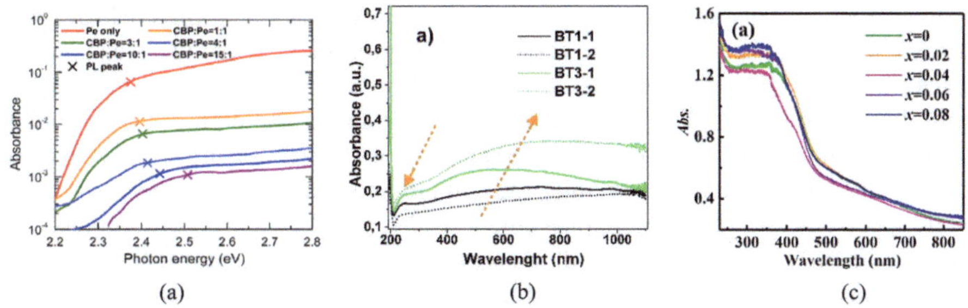

Fig. 11 UV–Vis absorption spectra of **a** CBP/Pe. Reproduced from Di et al. [142] under the terms and conditions of the Creative Commons Attribution license (CC-BY). https://creativecommons.org/about/cclicenses/ **b** $BaTiO_3$. [143], and **c** $(Ba_{0.95}Ca_{0.05})(Ti_{0.92}Sn_{0.08-x}Zr_x)O_3$. Reprinted from Pan et al. [144], with permission from Elsevier. Copyright (2021)

0.08) perovskite ceramics [144]. All samples exhibit effective light absorption when the wavelength was below 750 nm, along with enhanced absorption at 450 nm, which is associated with the electronegativity of the doping elements and electronegativity difference between the A (or B) cation and the oxygen anion (see Fig. 11c).

The quantitative analysis from UV–Vis spectroscopy is used to determine the analyte concentrations in a solution, which is related to the absorbance using Beer's Law. Also, it makes it possible to determine the component of interest based on the area of its absorption spectra and aggregation state of the constituents and functional groups in molecules (or confirm the identity of a compound). Besides, the UV–Vis spectrum is size and composition dependence. However, environmental effects can affect the spectroscopic properties and should be considered, such as solvent effects, pressure, and temperature. Such considerations allow us to elucidate, a change in the energy/wavelength indicates a change in the degree of geometric interaction, and a state affected (ground or excited), which will be indicated by the curve involved (absorption or emission).

5.1.2 Fourier Transform Infrared (FTIR) Spectroscopy

Infrared spectroscopy grew out of the need for a simple, reliable analytical technique used to identify organic, polymeric, and, in some cases, inorganic materials in pure compounds and mixtures. The method for performing IR spectroscopy is Fourier transform infrared spectroscopy (FTIR), which uses an interferometer. In FTIR, an interference pattern of the source is measured, and the spectrum is transformed from time domain to frequency domain by Fourier transformation. A typical FTIR spectrometer consists of a source, a Michelson interferometer with a beam splitter (BS), and a detector. Interferometer-based operation offers three main advantages; multiplex, throughput, and precision [153]. The first FTIR spectrometer detects the entire spectrum at once compared to the grating monochromator that only sees a narrow spectral region and requires a long time to scan the wavelength making it a slow process. This advantage enables many scans can be completed faster and hence a much higher signal–noise response. The multiplex advantage means many scans can be completed and averaged on an FTIR in a shorter time than one scan on most dispersive instruments. The second, the FTIR spectrometer, has a high optical throughput due to the lower use of necessary mirrors, which means fewer reflective losses. This means more energy hits the sample and, therefore, the detector in an FTIR spectrometer. Thus, the improvement of the signal-to-noise ratio, means that the instrument's sensitivity to small absorptions will be more significant, which is reflected in a more transparent and distinguishable spectrum. Lastly, the use of laser in the FTIR spectrometer enables the wavelength in a measured spectrum to be measured very accurately, since using a laser to control the velocity of the moving mirror and for timing the collection of data points throughout the mirror stroke.

It is important to point out that different FTIR instruments require to be either vacuum pumped or nitrogen (or dry air) purged, to avoid the strong absorption of IR radiation by molecular vibrations of H_2O and CO_2 in air, before the measurement process [153]. Likewise, the spectral range covered by an FTIR instrument depends on the specific combination of optical components (light sources, beam splitters, optical windows, mirrors, detectors, and polarizers) used. Light in the infrared (IR) region (14,000–10 cm^{-1}) is of lower energy than that in the ultraviolet and visible regions; therefore, it excites the vibrational and rotational motions of a molecule rather than the electronic. Three main zones have been used to distinguish different portions in the IR range, far-infrared (FIR, 400–10 cm^{-1}), "mid-infrared" (MIR, 4000–400 cm^{-1}), and "near-infrared" (NIR, 14,000–4000 cm^{-1}). When light waves (electromagnetic energy) interact with a molecule, if the frequency of the light is the same as the frequency of the molecular vibration a peak will appear in the IR spectrum. Since each vibrational mode will correspond to a bond within the molecule, the structure of the molecule can be determined. In other words, each molecule or chemical structure will produce a unique spectral fingerprint, making FTIR analysis a great tool for chemical identification. The key with IR spectroscopy is that the molecule must have a change in the dipole moment. If this is not present, the molecule is not "IR active," and an IR spectrum cannot be collected, however, it can be obtained by Raman analysis.

FTIR analysis in transmission mode in solid sample powder needs a prior treatment and is examined by mixing it with an alkali halide (i.e., potassium bromide KBr). FTIR analysis in reflectance mode enables application on a more significant assortment of sample forms without additional treatment. Reflections may be totally internal, specular, or diffuse. Powdered samples contain strong interferences from scattered light, which are affected by mean particle size, particle size distribution, and packing density. However, it is an effective tool in detecting the shape of nanometer-sized materials. As particle size increases, the width of the peak decreases, and intensity increases [146]. The decrease in the width of the peaks of the spectra suggests a change in a lattice structure and an increase in crystallinity. The MIR range is the most commonly used in FTIR to analyze perovskite materials since it contains the fingerprint region. Different types of compounds such as double perovskite Ba_2BiVO_6 [154], perovskite transparent ceramics $(1-x)K_{0.52}Na_{0.48}NbO_3$–$x$$BaBiO_3$ [155], and

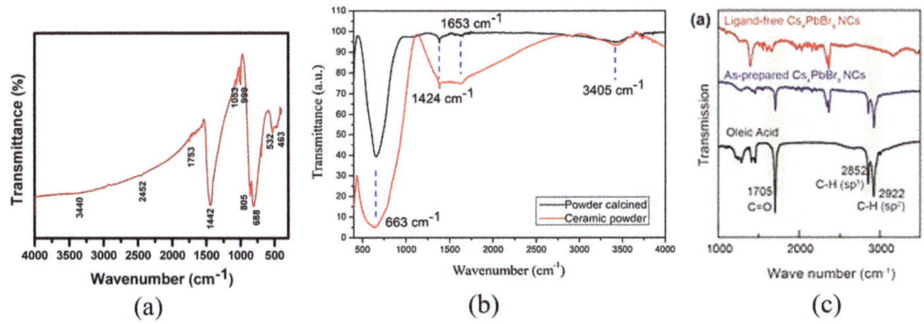

Fig. 12 FTIR spectrum of the **a** polycrystalline ceramic Ba_2BiVO_6. Reprinted from Parida et al. [154], with permission from Elsevier. Copyright (2019), **b** perovskite transparent ceramics $(1-x)$ $K_{0.52}Na_{0.48}NbO_3$–$x$$BaBiO_3$ for $x = 0.064$. Reprinted from Jin et al. [155], with permission from Springer Nature. Copyright (2018). **c** oleic acid and as-prepared and ligand-free Cs_4PbBr_6 NCs, clearly showing the removal of oleic acid after the washing procedure. Reprinted from Zhang et al. [156], with permission from the American Chemical Society. Copyright (2018)

Cs_4PbBr_6 perovskite nanocrystals [156], were analyzed using FTIR spectroscopy on small cleaved surfaces and confirmed the fingerprint of atoms present in the compound (see Fig. 12). Among the double perovskites, Ba_2BiVO_6 was analyzed, where the anti-symmetric stretching mode of the cations for VO_6 and BiO_6 were identified, respectively, and the asymmetric bending modes of vibrations for O–V–O and O–Bi–O were determined. Also, the symmetric and asymmetric stretching vibrations of O–Ba–O were identified. In Cs_4PbBr_6, perovskite nanocrystals were found to be associated with oleic acid molecules, which act as ligands on the abundant surface of the nanocrystals. By washing the nanocrystals, the presence of oleic acid was completely eliminated, resulting in the disappearance of all three peaks.

5.1.3 Ellipsometry Characterization Technique

The molecular orientation can be analyzed by variable angle spectroscopic ellipsometry [157]. This method detects the anisotropy of this orientation, even in a sample without long-range structural order. Ellipsometry (polarization analysis) is an analytical method to model the sample's optical properties using the interference of obliquely incident light, i.e., changes in the polarization state of light, as shown in Fig. 13. The incident light (linearly polarized), with the *s*- and *p*-polarizations depending on the oscillatory direction of the electric field (E_s and E_p in Fig. 13), have the same amplitudes and phases. The oscillatory direction of the p-polarization is parallel to the incident plane, while that of the s-polarization is perpendicular. After light incidence on the sample surface, the phase and amplitude of the incident light change by the sample because of the multiple interferences, and all components superimpose to form the reflected light from the sample surface with altered state of polarization, i.e., of different polarization state from that of the incident light. The

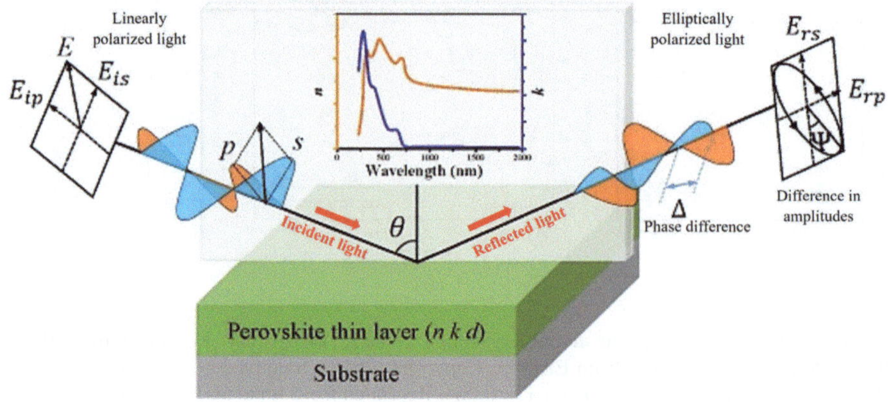

Fig. 13 Measurement principle of ellipsometry. Reprinted from Li et al. [158], with permission from Elsevier. Copyright (2020)

change in the amplitudes and phases of the *s*- and *p*-polarizations in the light makes the plane of the polarization rotate as the light propagates, as shown in Fig. 13, leading the light to be elliptically polarized light. The difference in amplitudes between the *s*- and *p*-polarization in the reflected light is monitored as the ellipsometric parameter Ψ. The difference in the phases is monitored as the parameter Δ. Such parameters depend on the refractive index $\boldsymbol{n}$, extinction coefficient $\boldsymbol{k}$, thickness of sample $\boldsymbol{d}$, and also on the wavelength, and incident angle $\boldsymbol{\theta}$. Usually, multiple wavelengths are applied in the sample, getting different values of Ψ and $\boldsymbol{\theta}$, consequently, thus $\boldsymbol{n}$ and $\boldsymbol{k}$ are dependent on the wavelength. From the experimental values of Ψ and $\boldsymbol{\theta}$, it is possible to determine the samples' $\boldsymbol{n}$, $\boldsymbol{k}$, and $\boldsymbol{d}$.

The Ψ and $\boldsymbol{\theta}$ measured from ellipsometry are defined from the ratio of the amplitude reflection coefficients for *s*- and *p*-polarizations (r_p and r_s),[1] as follows [157]:

$$\rho \equiv \tan\Psi e^{i\Delta} \equiv \frac{r_p}{r_s} = \left(\frac{E_{rp}}{E_{ip}}\right) / \left(\frac{E_{rs}}{E_{is}}\right) \tag{3}$$

where ρ is called the complex reflectance ratio and $\Delta = \delta_p - \delta_s$ is the *p-s* phase difference of the field components.

Applying the relationships r_p, r_s, and ρ to obtain

$$\tan\Psi = |r_p|/|r_s| \tag{4}$$

[1] The amplitude reflection coefficient for p-polarized light is defined as $r_p \equiv (E_{rp}/E_{ip}) = (n_t\cos\theta_i - n_i\cos\theta_t)/((n_t\cos\theta_i + n_i\cos\theta_t)) = |r_p|e^{i\delta_p}$, similarly, the amplitude reflection coefficient for s-polarized light is expressed by $r_s \equiv (E_{rs}/E_{is}) = (n_i\cos\theta_i - n_t\cos\theta_t)/((n_i\cos\theta_i + n_t\cos\theta_t)) = |r_s|e^{i\delta_s}$, where $|r|$ and δ show the amplitude ratio and phase difference between the incident and reflected waves, respectively.

Since the reflectance for the *p*- and *s*-polarizations are given by $R_p = |r_p|^2$ and $R_s = |r_s|^2$, respectively, $\tan\Psi$ can be interpreted as the reflectance ratio, i.e., $\tan\Psi = (R_p/R_s)^{1/2}$. Thus, through the spectroscopic ellipsometry method under variable angles, it is possible to analyze samples' optical constants and thickness.

Variable angle spectroscopic ellipsometry has been used to study electronic properties of $BiFeO_3$, which were performed at incidence angle $\theta_i = 65°$, in the UV–Vis spectral range [159]. The pseudo-dielectric function $\langle\varepsilon(E)\rangle$ is obtained directly from the measurements of ellipsometry angles Ψ and Δ. The $\langle\varepsilon(E)\rangle$ was calculated directly from the complex reflectance ratio $\rho \equiv \tan\Psi e^{i\Delta}$ as follow: $\langle\varepsilon(\mathrm{E})\rangle = \sin^2\theta_i\left[1 + \tan^2\theta_i\left(\frac{1-\rho}{1+\rho}\right)^2\right]$. The extinction coefficient k of the $BiFeO_3$ powder was determined from $\langle\varepsilon(E)\rangle$ which is obtained from the measurement of ellipsometry angles Ψ and Δ, and plotted against photon's energy $E = h\nu$ (see the inset in Fig. 14a). The optical properties of pure and Fe^{3+}-doped CeO_{2-y} samples were investigated by spectroscopic ellipsometry [160]. The direct band gap energy was determined from the Tauc plot shown in Fig. 14a, where $(\alpha h\nu)^2$ was plotted against photon's energy ($E = h\nu$), then a linear fit was performed in the linear region, and the value of band gap energy was represented by the intercept with the energy axis.

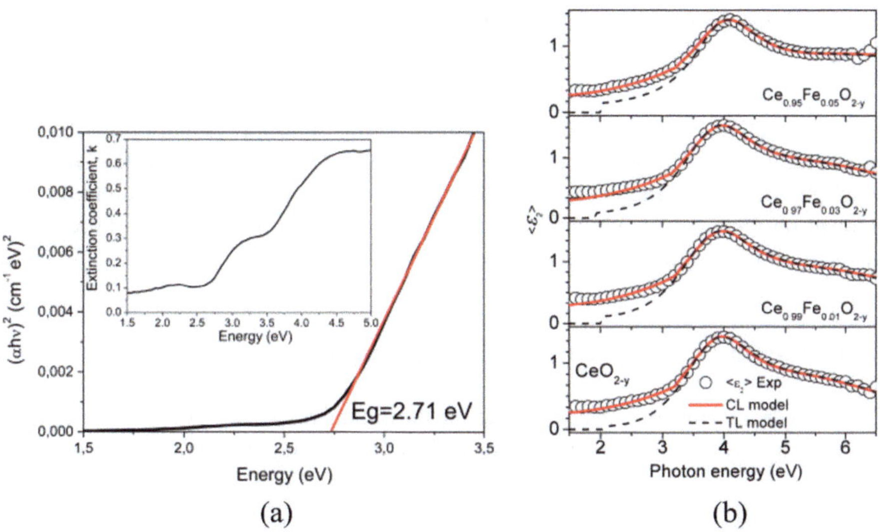

Fig. 14 **a** Tauc plot of $(\alpha h\nu)^2$ versus photon's energy $E = h\nu$ to obtain the direct band gap energy (E_g) from the intercept with energy axis obtained from a linear fit for the linear part of the curve. Inset shows the extinction coefficient (k) of the $BiFeO_3$ powder obtained from spectroscopic ellipsometry. Reprinted from Čebela et al. [159], with permission from Elsevier. Copyright (2017). **b** Imaginary part of pseudo-dielectric function (open symbols) together with numerical fits using Tauc–Lorentz (dashed line) and Cody–Lorentz (full line) model. Reprinted from Radović et al. [160], with permission from Elsevier. Copyright (2013)

Among all applied models, Cody–Lorentz (CL) and Tauc–Lorentz (TL) models,[2] the best agreement between theoretical and experimental data was obtained using the CL model (see Fig. 14b). The imaginary part of the pseudo-dielectric function, $\langle \varepsilon_2(E) \rangle$,is deduced from the ellipsometric measurements. All the nanosize-related effects (variation of band gap and defects within the band gap) are reflected in the spectral dependence of the complex dielectric function, which is related to the imaginary part $\langle \varepsilon_2(E) \rangle$, related to the energy-band structure. Figure 14b presents the spectra of $\langle \varepsilon_2(E) \rangle$ obtained by ellipsometry for pure and Fe doped samples (open circles) against photon's energy in the range (1.5–6.5) eV. TL model failed to describe adequately the behavior of $\langle \varepsilon_2(E) \rangle$. This is due to the fact that the TL model is based on the assumption that $\langle \varepsilon_2(E) \rangle$ vanishes below the band gap. While the fits based on the CL model showed very good agreement with experimental spectra, due to considering the defects and/or disordered states.

Spectral dependence of optical parameters of Fe and W co-substituted $BaTiO_3$ perovskite ceramics, such as real (ε_1) and imaginary (ε_2) parts of the dielectric function, ***n***, ***k***, and absorption coefficient was carried out for photon energy in the range (1.4–4.96) eV. All were analyzed using the ellipsometry experiments [161]. The plots of evaluated of ε_1, ε_2, n, and k are shown in Fig. 15. The ε_2 presents a monotonic variation (strictly increasing), although this behavior is also increasing for ε_1 a decrease was observed around 3.15 eV. Such behavior is because there is a relationship between dielectric function (ε_1, ε_2) and optical constant (***n***, ***k***): $n = \left(1/\sqrt{2}\right)\left(\sqrt{\sqrt{\varepsilon_1^2 + \varepsilon_2^2} + \varepsilon_1}\right)$, and $k = \left(1/\sqrt{2}\right)\left(\sqrt{\sqrt{\varepsilon_1^2 + \varepsilon_2^2} - \varepsilon_1}\right)$. The value of the refractive index n in the UV–visible region is in the range 1.38–1.61, and the refractive index shows a peak around 3.75 eV (330 nm). The k-values increase when photon energy increases, besides, values of the extinction coefficient are low at lower photon energy, which indicates that the absorption loss is low in the higher wavelength range studied here.

5.1.4 Luminescence Spectroscopy Measurements

The extrinsic and intrinsic properties define the optical response in perovskites. Luminescence spectroscopy is a very sensitive technique that explores the interaction of

[2] The Tauc–Lorentz oscillator model is expressed as $\varepsilon_2(E) = \begin{cases} 0, E \le E_g \\ \frac{(E-E_g)^2}{E^2} \cdot \frac{AE_0\Gamma E}{(E^2-E_0^2)^2+\Gamma^2E^2}, E > E_g \end{cases}$, where A, E_0, and Γ are the peak amplitude, position, and width, respectively. The Cody–Lorentz oscillator model is expressed as $\varepsilon_2(E) = \begin{cases} \frac{E_1}{E} e^{\left\{\frac{(E-E_t)}{E_u}\right\}}, 0 < E \le E_t \\ \frac{(E-E_g)^2}{(E-E_g)^2+E_p^2} \cdot \frac{AE_0\Gamma E}{(E^2-E_0^2)^2+\Gamma^2E^2}, E > E_t \end{cases}$, where parameter E_1 ensures continuality of dielectric function at $E = E_t$, E_t is the demarcation energy between Urbach tail region and interband transitions. Energy E_p separates the absorption onset behavior from the Lorentz oscillator behavior.

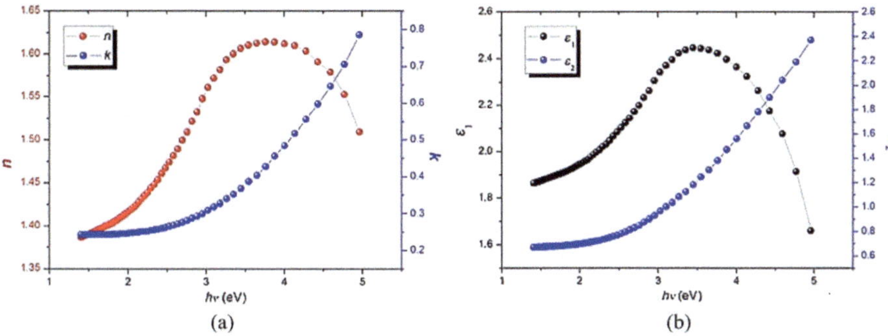

Fig. 15 Optical parameters of Fe and W co-substituted $BaTiO_3$ perovskite ceramics as functions of $h\nu$ **a** n and k. **b** ε_1 and ε_2. Reprinted from Bourguiba et al. [161], with permission from Springer Nature. Copyright (2016)

matter and photons, and investigates the electronic states in the band gap. The monitoring of material optical response, after interaction, as a function of the energy is essential, i.e., obtaining the number of photons emitted by the sample as a function of the photon energy. Luminescence is a non-destructive technique together with other techniques, such as absorption, holds valuable information on the interaction of matter and photons.

The active materials in samples (pure or doped) interact with their neighborhood ions/atoms. Luminescence in these materials arises from optical stimulation sources, that allows changing the state of the material from ground state to the excited state. The luminescence emitted is measured by a light detection system which comprises a photomultiplier tube and detection filters. Luminescence response presents a functional dependence on three parameters that can be changed during experiment; sample temperature, excitation wavelength, and excitation power.

A schematic of the luminescence system is presented in Fig. 16a. The source can be any laser whose photon energy exceeds the band gap of the sample to be examined and whose power is sufficient to excite an adequate signal. For example, a Nd:YV04 solid-state laser operating on line of wavelength 532 nm, with power-up to 250 mW (produces photons at 2.54 eV), is used as the excitation source. This laser can be generated at continuous electric powers of watts, whereas tens of milliwatts are often adequate to give good signals. This depends on the sample to be examined (as well as the system throughput and detector sensitivity), taking into account that too high laser intensity at the focused spot can damage the sample. The laser beam from the excitation source is filtered using a narrow bandwidth filter before entering a 532 nm free-space isolator, which stops any light from straying from the path. The optical path is then redirected using silver-coated mirrors (>98% reflectivity) to alter the direction of the light towards the sample. The laser is then reflected off a 45° mirror directed to the beam splitter. The optical beam is divided in half to measure the optical power in real-time. The remaining beam is focused using a High-Power MicroSpot Focusing Objective (LMH-10X-532) with a center

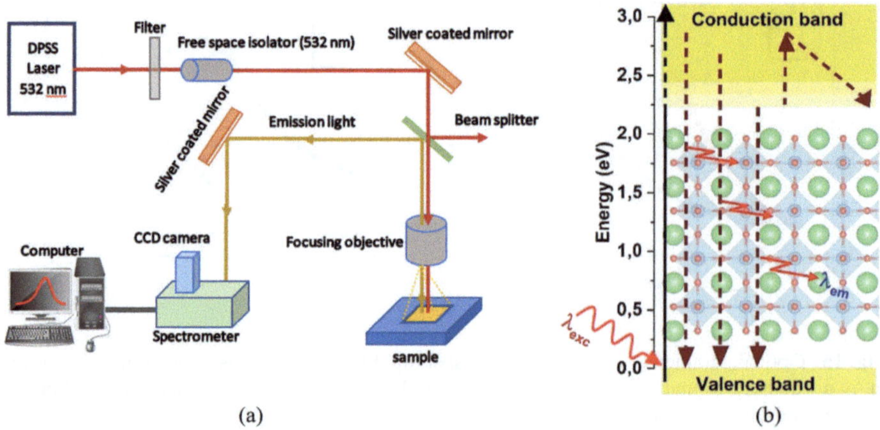

Fig. 16 **a** Schematic of luminescence measurement system. **b** Electronic transitions involving electron–hole recombination in the perovskite

wavelength of 532 nm. The light hits the sample after being focused on the objective, and the focused beam path is typical. The subsequent luminescence signal is collected and transmitted through the 0.55 m long monochromator and into a cooled-Synapse 1024 × 256 Open-Electrode CCD Detector in the spectrometer measures the luminescence intensities. The luminescence spectrum can significantly improve samples from samples held below room temperature. Lower temperatures reduce the thermal broadening of the excited carrier energies, which at temperature Γ is roughly k_BT, where k_B is Boltzmann's constant [162]. In addition, low temperatures tend to reduce the role of competing for nonradiative paths for recombination, giving a higher efficiency for the luminescence process, which results in an improved signal-to-noise ratio. Virk [163], provide a brief but trenchant summary of luminescence characterization and its instrumentation.

In ceramics materials, the valence and conduction band are separated by a band gap, as shown in Fig. 16b. The optical stimulation from the excitation source (λ_{exc}) induces valence electrons to be excited from the valence band to unoccupied states in the conduction band. The electron–electron and electron–phonon scattering processes result in rapid thermalization and relaxation to energy levels in the conduction band. Recombination of these excess conduction band electrons with excess valence band holes results in transitions to valence band and radiative luminescence emission (λ_{em}).

Recent studies have shown that incorporating earth-rare ions into the perovskite structure can tune optical band gap, defect, ion migration, surface morphology, and optical response [164–166]. Enhanced up-conversion emissions in Er^{3+}:$BaTiO_3$ (BTO) ferroelectric glass–ceramics via electric-stimulated polarization technique are discussed by Xie et al., [167]. Under the modulation of the electric field and upon 980 nm light excitation excites the electrons of $^4I_{15/2} \rightarrow {}^4I_{11/2}$ of Er^{3+}, Ti^{4+} moves toward the electric field, and O^{2-} is in the opposite direction, promoting the structural asymmetry of the BTO host (see Fig. 17a). Such asymmetry increases the dopant ions' 4f–4f electric dipole transition probabilities. Thus, temperature and electric field factors can alter crystal-field structure properly in glass–ceramics, consequently the optical response. Up-conversion optical properties and the corresponding energy transfer mechanism in polarized glass–ceramics are shown in Fig. 17b, c. The increased luminescence intensity of Er^{3+}:$BaTiO_3$ samples poled by increased voltage was observed, indicating that the ferroelectric domain is uniformly orientated as the applied electric field, which is modification of the ligand field around rare-earth ions. The energy transfer mechanism of Er^{3+} results in a dynamic process as ground state absorption (GSA), the excited state absorption (ESA), and energy transfer up-conversion processes (ETU). Similar observations were observed in others types of perovskite samples Er^{3+}:$BaTiO_3$ [168], Pr^{3+}:$BaTi_{0.9}(Yb_{0.5}Nb_{0.5})_{0.1}O_3$ [169], Er^{3+}:$(Ba_{0.95}Sr_{0.05})(Zr_{0.1}Ti_{0.9})O_3$ [170], and $K_{0.5}Na_{0.5}NbO_3$–$SrTiO_3$ [171], Er^{3+}:$BaTiO_3$ [165, 172].

Luminescence modulation properties in $(1-x)Na_{0.5}Bi_{0.5}TiO_3$–$xNa_{0.5}K_{0.5}NbO_3$:0.002Er ceramics are used in information storage by the photoluminescence contrast before and after light irradiation [173]. The mechanisms of up-conversion and luminescence modulation are represented in Fig. 17d. Upon 980 nm light excitation, two emission peaks from Er^{3+} luminescent centers are observed due to the transition of $^2H_{11/2}$ and $^4S_{3/2} \rightarrow {}^4I_{15/2}$. In high-temperature sintering, the defect energy levels

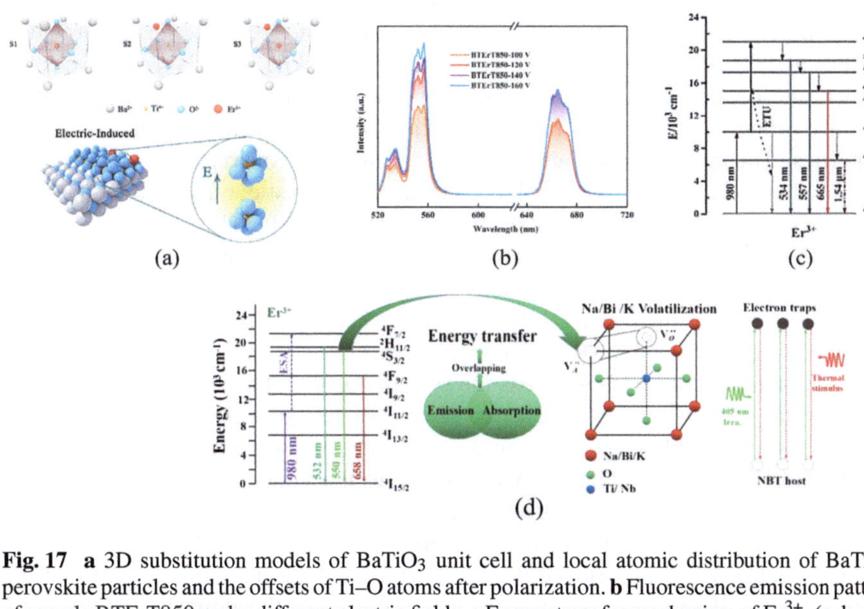

Fig. 17 **a** 3D substitution models of $BaTiO_3$ unit cell and local atomic distribution of $BaTiO_3$ perovskite particles and the offsets of Ti–O atoms after polarization. **b** Fluorescence emission pattern of sample BTErT850 under different electric fields. **c** Energy transfer mechanism of Er^{3+}. (a, b, and c) reprinted from Xie et al. [167], with permission from Elsevier. Copyright 2019. **d** Mechanism of UC emission and its modification for Er^{3+} in $(1-x)Na_{0.5}Bi_{0.5}TiO_3$–$xNa_{0.5}K_{0.5}NbO_3$:0.002Er ceramics. Reprinted from Li et al. [173], with permission from the American Chemical Society. Copyright (2018)

formed from the volatilize (K, Na, and Bi). Under a wavelength of 405 nm, electrons are excited to the defect energy levels and trapped by the oxygen vacancies. As a result, color centers form, which disappear when thermally stimulated. With the increase of $Na_{0.5}K_{0.5}NbO_3$ content $x \geq 0.08$, the degree of ergodicity is enhanced, while concentrations to $x = 0.02$, 0.04, and 0.06% a strong nonergodicity. In this last, the luminescence contrast is greatly enhanced through an electric field poling because of forming a long-ranged order ferroelectric state in them.

Among other advantages, the luminescence spectrum contains information about the band edge and presence of excitonic peaks in perovskite ceramics [174, 175], trap states, and the degree of electron–phonon interactions [176–178]. Exciton binding energy can be estimated from the temperature-dependent PL spectrum [179, 180]. These parameters are crucial for the characterization and optimization of photonic devices.

5.2 *Dielectric Properties*

Common perovskites such as $SrTiO_3$ experienced several conduction mechanisms under doping or reduction, from a transparent insulator to semiconducting or even metallic behaviors depending on the operational temperature. In the case of behavior under strain and electric field, this type of perovskite shifts from the paraelectric (zero net polarization after external field removal) to ferroelectric (polarization remains after the field removal) characteristics. Perovskite materials properties are temperature-dependent due to the phonon scattering effect, which explains the carrier mobility property. Apart from that, perovskites are also known for their dielectric properties. The dielectricity is considered a traditional property of any perovskite-structured ceramics alongside its piezoelectric characteristic, making it suitable for energy conversion (mechanical to electrical) applications [181].

In general, dielectric materials are also known as insulators, but with a particular characteristic of polarizability they are called polarized insulators. Figure 18 helps to understand the polarizability of dielectric materials. When a conductor is placed between two oppositely charged plates, the charge moves freely to the positive terminal, creating current flow across the system. On the other hand, when a dielectric material was inserted between two plates connected to a voltage source, this material was incapable of conducting the free electrons when an external load was supplied, as the valence electrons were bound to the nuclei of their parent atoms. Nevertheless, the bounded valence electrons tend to position themselves near the positively charged plate. This process is called polarization. At the same time, the valence electrons still experience the force from the external load, resulting in the electrons in a pull-and-push situation; electrons push onto the atoms by the attraction force (electrostatic force) and are pulled away from the nucleus by the external force. But it is still possible for the electrons to be released from the nucleus attraction force when the external force is too large compared to the nucleus attraction force. This phenomenon is known as a dielectric breakdown.

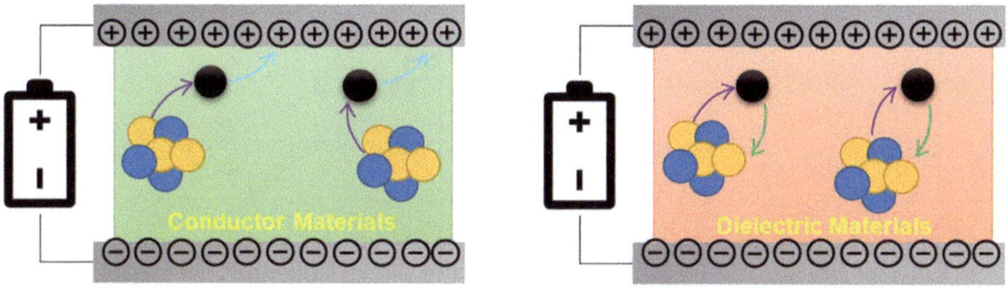

Fig. 18 Comparison between conductor and dielectric materials

As mentioned in the previous section, perovskite-based materials are suitable for photonic device applications due to their optical behavior. Apart from this, perovskite-based materials are also ideal for energy applications due to their dielectric characteristics [181]. The conductivity and permittivity of a material determine how it interacts with an applied electric field. The fundamental parameter describing a dielectric is its relative permittivity. As shown in the following equation, the material's dielectric permittivity ε_r is the combination of the real and imaginary parts of complex dielectric constant under alternating current conditions. The real part defines the capability of dielectric materials to store energy, while the imaginary part represents the ability to lose energy [181].

$$\varepsilon_r = \varepsilon_r' - j\varepsilon_r'' \tag{5}$$

where ε_r' is the real part of the complex dielectric constant, ε_r'' is an imaginary part of the complex dielectric constant, and $j = \sqrt{-1}$.

The presence of polarizable constituents in the crystal results in the dielectric properties of bulk perovskites. These include cation distributions, octahedral tilting, and distortions, as well as any defects such as grain boundaries and various point defects [182]. As polarizable constituents become more mobile, relative permittivity increases with temperature and decreases as the intensity of the applied electric field increases. In perovskites, many factors possibly influence the resulting dielectric response. Besides the microstructure-related factors, doping at both A- and B-sites of these materials will significantly alter the dielectric behavior. In addition, a material's dielectric constant can be refined by changing its composition. Alkali halide perovskite, for example, is already used to control the dielectric properties of metal halide perovskite materials owing to its large polarizability and bonding strength [183]. Figure 19 shows the relation between composition alterations with the dielectric constant of perovskites. From these findings, previous work concluded that the addition of sodium iodide (NaI) and rubidium iodide (RbI) had no effect on the frequency-dependent dielectric constant, whereas potassium iodide (KI) had a significant effect on the real part of the dielectric constant as shown in Fig. 19a, and it is also confirmed that the response is independent of the halide ions (Fig. 19b) [183].

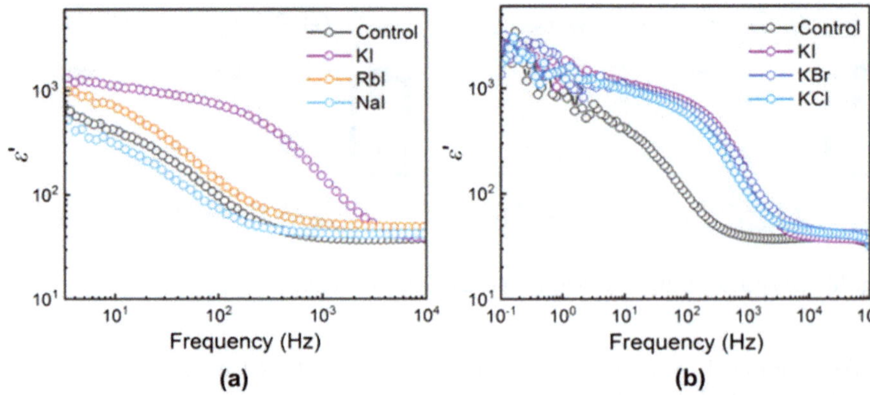

Fig. 19 Frequency-dependent real part of the complex dielectric constant for perovskite films incorporating different **a** alkali iodide, and **b** potassium halide species. Reprinted from Su et al. [183], with permission from Springer Nature. Copyright (2021)

The complex dielectric constant is frequency dependence, which is obvious in Fig. 19a and b, and explained in detail in a supplementary note by R. Su et al. [183]. In addition, Fig. 20, shows the relationship between the real part of the dielectric constant and frequency, where three main regions are emphasized and labeled as A (below 1 Hz, low frequency), B ((1–10^4) Hz, moderate frequency), and C (above 10^4 Hz, high frequency). For the real-part analysis, in region B, the target sample has a significantly higher dielectric response than the control, while changes in regions A and C are negligible. The trend recorded for the real part follows the Cole–Cole equation, the most commonly used model in dielectric materials to describe dielectric relaxation [183–185]. The Cole–Cole equation can describe the dielectric behavior of ideal insulators. However, due to the effect of DC conductance of perovskite films on the quasi-device used in the impedance spectroscopy measurement, the sandwiched structure indium tin oxide (ITO)/Perovskite/Au could be viewed as an ideal dielectric capacitor connected in parallel with resistance in perovskite photovoltaics [183].

$BaTiO_3$ is one of the most commonly used ceramic materials in the industry, particularly as a capacitor, due to its excellent dielectric properties [186]. Most electronic parts contain a ceramic-based capacitor known as a multi-layer ceramic chip capacitor (MLCC). To measure the capability of MLCCs, the equation used is as follows:

$$C = \varepsilon_0 \varepsilon_r \frac{NA}{d} \tag{6}$$

where ε_0 is the absolute permittivity of a vacuum (8.85 pFm^{-1}), ε_r is the relative dielectric constant of the materials, N is the number of MLCC layers, A is the area of the electrode, and d is the thickness of dielectric.

According to Eq. (6), in order to have a higher capacitance C, in addition to reducing the MLCC's thickness, the perovskite material used should have a high

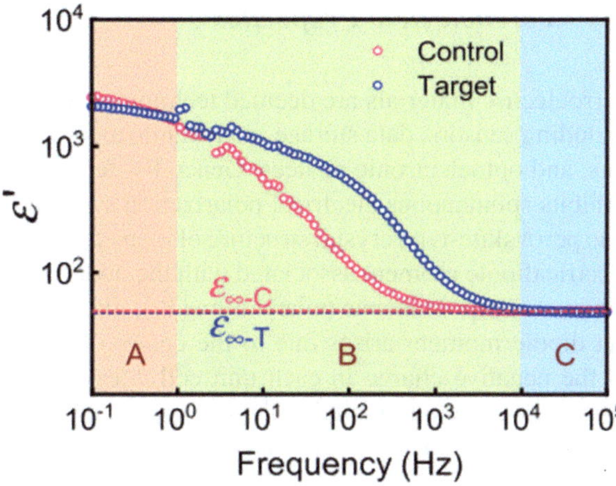

Fig. 20 The real part of dielectric constant for perovskite films as a function of frequency: range A (low frequency), range B (moderate frequency), and range C (high frequency)]. Reprinted from Su et al. [183], under a Creative Commons Attribution 4.0 International License, http://creativecommons.org/licenses/by/4.0/

dielectric constant. A high capacitance value of $BaTiO_3$ led to the domination of these materials in the field of capacitors. One of the most important parameters in producing a high capacitance of $BaTiO_3$-based capacitors is the controllable particle size [186]. As the particle size decreases, the tetragonality that is needed as the source of the dielectric constant will also decrease. In other words, the industry requires very fine $BaTiO_3$ powders with high dielectric constants to manufacture high capacitance MLCCs [186]. Several studies, both experimental and theoretical, have been published to determine the dielectric constant of perovskites. For example, the capacitance spectroscopy techniques were used to probe the alternating current response of semiconductor materials and devices with respect to bias-voltage and frequency at different temperatures [187]. Capacitance-based techniques provide vital information, including defect activation energy and density, carrier concentration, and dielectric constant, for evaluating and understanding the electrical properties of semiconductor materials [187].

The measurement of dielectric properties of bulk perovskite is not particularly reliable experimentally. In general, the dielectric properties are measured experimentally by comparing the polarization differences between two polarization states generated by hysteresis loops created by switching the direction of polarization currents. The results of this method are typically validated using theoretical models, such as the modern electron wavefunction response, which employs a geometric phase. This method is available in computational density functional theory today (DFT) [188]. As for the experimental studies, many approaches were used, depending on the potential application of the perovskite materials, because the complete analysis of dielectric properties requires a variety of measuring methods. For example, scientists usually report the relative permittivity of a perovskite solar cell because it is crucial as it determines physical constants, such as the absorption coefficient, that affects the efficiency of the device [188].

5.3 Ferroelectric Properties

Ferroelectric materials are deemed technologically important in many applications, including sensors, data storage, nonvolatile memories, solar cells, nanoscale transistors, and optoelectronic devices. Generally, ferroelectric is a dielectric material that exhibits spontaneous electrical polarization switchable by an external electric field. The perovskite-type crystal structure of many ferroelectric materials has a permanent electric dipole moment associated with the underlying ionic unit cell. Consequently, it possesses spontaneous polarization (P_s), the dipole moment per unit volume. The net dipole moment arises due to the center of the positive charge, and the center of the negative charge in each unit cell does not overlap. The crystal structure is non-polar cubic and paraelectric above the Curie temperature (T_C). Commonly, T_C is up to 450 °C in some ferroelectric perovskite materials but significantly lower in others.

The ferroelectric material tends to form a ferroelectric domain in which the polarization or the unit cells within a domain are aligned. A domain wall is a boundary separating two adjacent domains. The orientation of the polarization and the domain walls are determined by crystal symmetry. Upon cooling from above the T_C, the domains are randomly oriented in the ferroelectric. This results in zero net polarization. Under an external electric field in excess of the coercive field (E_C), the dipoles within the ferroelectric material are gradually aligned with the applied field until saturation, in which they are completely aligned. When the external electric field is subsequently reduced to zero, the dipoles become less aligned. However, they do not return to their original orientation. A very high degree of alignment remains, and the material remains polarized at a level lower than the saturation polarization. This is called remanent polarization (P_r). When a bipolar external electric field in excess of the E_C is cycled, an electric polarization–electric field (*P-E*) hysteresis loop (*P-E* hysteresis loop) is generated. This is analogous to ferromagnetic materials under an external magnetic field. The name "ferroelectric" was given to the materials because their electric behavior is equivalent to the magnetic behavior of ferromagnetic materials [189].

Studies of the response of ferroelectric domains to an external electric field and mechanical stress offer valuable information for applications of ferroelectric materials in energy storage and power generation devices. The hysteresis loop can be used as one of the effective tools to draw this information. Many types of ferroelectric hysteresis loops have been observed, as shown in Fig. 21. Figure 21a represents *P-E* hysteresis loops and energy storage properties for a linear dielectric, while Figs. 21b–d represent *P-E* hysteresis loops and energy storage properties for nonlinear dielectrics, which include a classical ferroelectric, antiferroelectric and relax or ferroelectric, respectively. The *P-E* hysteresis loop shown in Fig. 21b is the most common. This type of hysteresis loop is achievable in a classical ferroelectric material and characterizes a wide variety of ferroelectric materials.

Perovskite ferroelectrics may exist in different phases as a function of temperature, stress, electric field, and composition. Barium titanate, $BaTiO_3$ (BTO), is considered

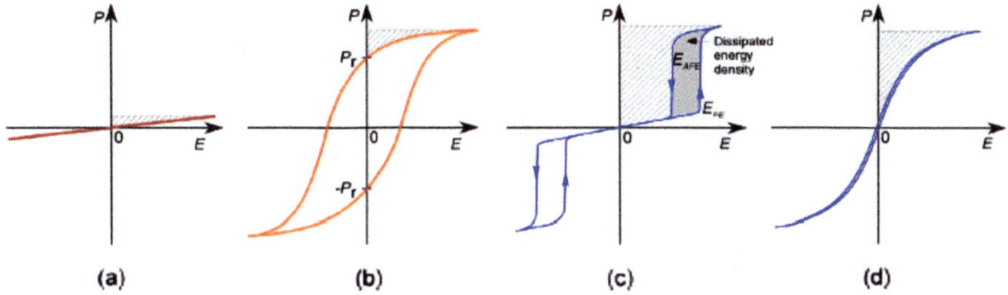

Fig. 21 *P-E* hysteresis loop for **a** linear dielectric; **b** classical ferroelectric; **c** antiferroelectric, and **d** relaxor ferroelectric. The gray area represents the dissipated energy density and the hatched areas represent the recoverable energy density of dielectric, and ferroelectric capacitors during the positive charge–discharge cycle. Reprinted from Shkuratov and Lynch [189], with permission from Elsevier. Copyright (2011)

the first polycrystalline perovskite material, which has both properties of ferroelectric and dielectric at room temperature. It has been found to be a suitable material for various ferroelectric-based applications, such as optical waveguides, dielectric memories, dielectric resonators, and electromechanical elements. BTO belongs to the ABO_3 perovskite family, with Ba at A-site and Ti at B-site. Ba atoms are located at corner positions, O atoms at face-centered positions, and Ti atoms are located at a central position, as shown in Fig. 22. Variation in crystallographic orientation affects the ferroelectricity of BTO. It was found that the phase-transition temperature in BTO is grain size-dependent, and the ferroelectric phase is unstable at room temperature when particle diameter decreases below a threshold value [190]. The properties of doped BTO are determined by the grain size, doping, preparation root, and texturing [191]. For instance, a change in structure and electrical and magnetic properties was found when BTO was doped with iron (Fe) at B-site. Fe^{3+} ions were replaced by Ti^{4+} ions and magnetic properties were found to be dependent on temperature. A decrease in magnetization was noticed with an increase in the concentration of Fe [192].

Figure 23 shows the ferroelectric hysteresis loops (*P-E* loop), and Table 5 shows the values of saturation polarization (E_s), coercivity, and remanence polarization for pure and Fe-doped $Sr_{0.5}Ba_{0.5}Ti_{1-x}Fe_xO_3$ ($0.00 \leq x \leq 0.025$ in the step of 0.05) nanoceramics prepared via sol–gel combustion method. The *P-E* hysteresis loops signify that all samples showed ferroelectric behavior. From Table 5, it can also be observed that the presence of ferroelectricity in the prepared materials may be due to the long-ranged polar order of the dipoles. The positive oxygen vacancies created by the acceptor dopants, the iron (Fe), may be neutralized by introducing the negative charges. The dissimilarity in ionic radii of Ti^{4+} and Fe^{3+} ions varies the elastic behavior in the materials by doping defect. By application of the external electric field, stretching and reorientation of the dipoles in the direction of polarization happens, which reduces the potential energy of dipoles. The occurrence of the lattice defects and the grain walls fixes the direction of polarization on every side of a domain

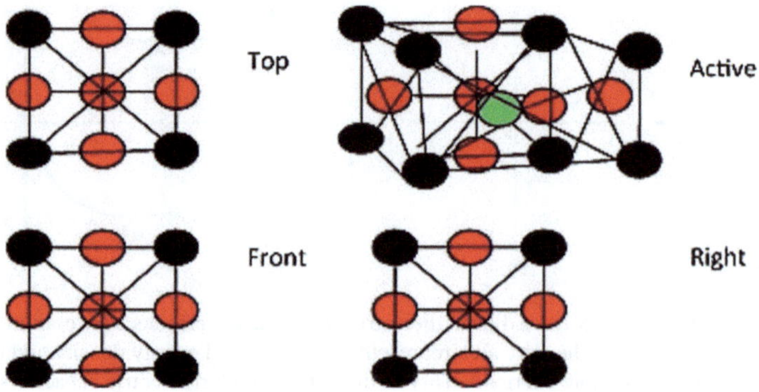

Fig. 22 ABO_3 perovskite structure of barium titanate, $BaTiO_3$ (BTO). The black circle represents A-site (Ba), the green circle represents B-site (Ti), and the red circle represents oxygen (O). Reprinted from Tewatia et al. [191], with permission from Elsevier. Copyright (2021)

wall, which confines the movement of dipoles. Thus, such types of *P-E* hysteresis loops were observed with an increase in Fe^{3+} ion concentration.

In general, ferroelectric perovskite materials have great charge carrier separation properties and possess voltages that can exceed the band gap. However, most ferroelectrics have a wide band gap above 3.0 eV, which leads to the absorption part of the solar spectrum only in the ultraviolet regain [194, 195]. This is a huge obstacle to improving solar energy conversion efficiency. Thus, some measurements, like metal doping, have been taken in order to reduce the band gap. For instance, Co-doped lead-free $(Bi_{0.5}Na_{0.5})Ti_{1-x}Co_xO_3$ perovskite ceramics have been synthesized by the solid-state reaction method. XRD, Raman spectroscopy, and SEM patterns confirmed the lattice structure of the ceramics influenced by Co ions. The optical band gaps of as-prepared ceramics decreased from 3.0 to 1.43 eV due to the changing grain size and the electronic transitions of the doped metal (Co). Additionally, the P_r increases from 5.75 $\mu C/cm^2$ to 11.47 $\mu C/cm^2$ according to the electric hysteresis loops, indicating that precise metal doping enhances the ferroelectric properties of the ceramic samples. The ceramics can be used for great potential applications in solar cells and memory devices [196].

5.4 *Magnetic Properties*

Over the years, perovskite structured materials have gained increasing interest from scientific researchers due to their unique structural and compositional flexibility with the general formula, ABO_3, in which A and B are cations, represent an important class of functional materials oxides that provide promising magnetic and electrical behaviors [197, 198]. The atomic arrangement of perovskite is very thermodynamically stable. They have the ability to accommodate a wide range of elements in the periodic table due to the flexibility of their structure, which makes them a perfect hosting

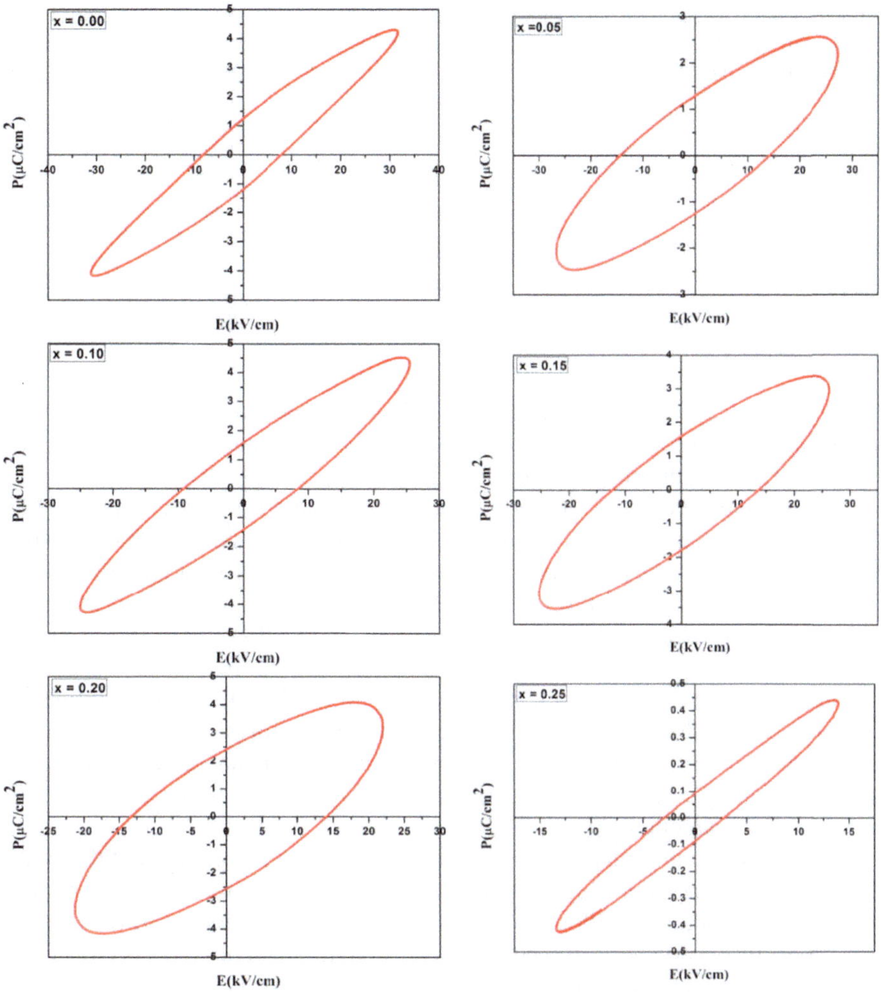

Fig. 23 *P-E* loops of $Sr_{0.5}Ba_{0.5}Ti_{1-x}Fe_xO_3$ ($0.00 \le x \le 0.025$ in the step of 0.05) nanoceramics measured at a frequency of 50 Hz. Reprinted from Bhoyar et al. [193], with permission from Elsevier. Copyright (2020)

structure for adapting different macroscopic characteristics according to the desired applications [199, 200]. In the ideal cubic perovskite structure, for example, $LaMnO_3$, rare-earth atoms (and alkaline-earth atoms) occupy the A-site, and manganese (transition metal) occupies the central B-site inside the oxygen octahedron. The source of electric crystal field splitting comes from the Mn^{3+} ion inside the O^{2-} octahedron and the MnO_6 complex is formed with Mn ion at the central position surrounded by six O ions. The magnetic properties of manganites are mainly determined by the transfer of electrons between the manganese and the oxygen orbitals [201].

Table 5 Values of saturation polarization (E_s), coercivity (E_c) and remanence polarization (E_r) of $Sr_{0.5}Ba_{0.5}Ti_{1-x}Fe_xO_3$ ($0.00 \leq x \leq 0.025$) nanoceramics prepared via sol–gel combustion method. Reprinted from Bhoyar et al. [193], with permission from Elsevier. Copyright (2020)

x	E_s (esu/g)	E_c (Oe)	E_r ($\times 10^{-2}$)
0.00	4.294	8.047	1.222
0.05	2.557	14.154	1.277
0.10	4.524	8.863	1.562
0.15	3.383	13.221	1.580
0.20	4.079	14.132	2.404
0.25	0.440	2.902	0.090

The ceramics perovskite, for example, $BaTiO_3$, has structural characteristics which make it appropriate for doping with 3d transition metals (Mn, Fe, Co, Ni) that have been predicted to enhance magnetic properties [202]. The $BaTiO_3$ has been used as piezoelectric multi-layer devices [203], energy storage capacitors [204], and piezoelectric and ultrasonic actuators. Meanwhile, perovskite manganites have a wide range of magnetic properties, such as colossal multiferroic properties, magnetocaloric effect, and magnetoresistance. Magnetic refrigeration, bioprocessing, and miniaturized magnetic sensor are among various applications that can be obtained from the magnetism characteristics. Besides that, the application of an external magnetic field on perovskite manganites such as $Ln_{1-x}M_xMnO_3$ can strongly enhance the electrical conductivity and raise up colossal magnetoresistance (CMR) effect [201]. CMR is a property of some materials, mostly manganese-based perovskite oxides, that enables them to dramatically change their electrical resistance in the presence of a magnetic field.

The enhancement of magnetic properties has brought the formation of the double perovskite material. Double perovskites are formed by a "doubling" of the perovskite unit cell with an additional (double) A-site ($A'A''B_2O_6$) or double B-site ($A_2B'B''O_6$) [205, 206]. In the double B-site perovskite, different magnetic fields between rare-earth cations and transition metal cations have their own basis in oxygen-mediated cationic exchange interactions. Furthermore, transition metal ions often produce mixed-valence states, which are believed to significantly influence magnetic properties [207]. Magnetic double perovskite materials have been identified to offer multifunctional characteristics such as multiferroicity, magnetoelectric coupling, magnetocapacitance, magnetoresistance, etc.

The modification of perovskite materials' magnetic properties also focuses on the development of perovskite composite materials. The definition of a composite is a material that is made up of two or more constituents that will produce different properties which are distinct from the parent source materials. In the last decade, perovskites (ABO_3) and spinels (AB_2O_4) have gained attention as structured composite materials with multifunctional applications such as electro-ceramics perovskite that have potential use in fuel cells, catalysis, superconductors, electrochemical sensors, etc. [208]. Composite multiferroic materials that have a combination of ferroelectricity, ferromagnetism, and ferroelasticity are formed by physically mixing ferroelectric and ferromagnetic materials and are expected to exhibit strong magnetoelectric coupling at ambient temperature.

A study was performed by Moustafa et al. [208] on the magnetic properties of multiferroic composite based on perovskite–spinel revealed that the composites exhibit typical ferromagnetic hysteresis loops, indicating the presence of ordered magnetic structure. In this study, the magnetic properties were measured using a vibrating sample magnetometer. Figure 24 displays the magnetic field dependence magnetization (M-H curve) of $La_{0.8}Gd_{0.2}FeO_3$ (LGFO), $Mn_{0.5}Cu_{0.5}Fe_2O_4$ (MCFO), and LGFO-MCFO composites at room temperature where LGFO shows no saturation occurred within the applied magnetic field. The M-H curve exhibit an increment of the MCFO contents resulting in an increase in the saturation magnetization (M_s) and the remnant magnetization. Magnetization can be defined as a vector quantity that gives a measure of the density of permanent or induced dipole moment in a given magnetic material. The enhancement of the magnetization may be contributed by the interfacial strain in the composites that can cause changes in the orientation of the spin.

Besides the composite composition, other processing parameters of perovskite materials, such as doping amount, calcination temperature, and calcination time, have been found to affect its microstructure and magnetic properties. This has been proven by the study done by Lin et al. [209]. The increment of Mg^{2+} doping amount has increased the saturation magnetization of $LaFeO_3$. However, the saturation magnetization and the residual magnetization (M_r) of the $LaFeO_3$ were reduced when the calcination temperature increased from 400 to 600 °C. These phenomena occurred due to the increment of particle size at the increment of calcination temperature, which brought the reduction of the specific surface area. The decrease of the specific surface area will then reduce the magnetic interaction that affects the magnetization characteristics of the sample. Other than that, higher sintering temperature has the

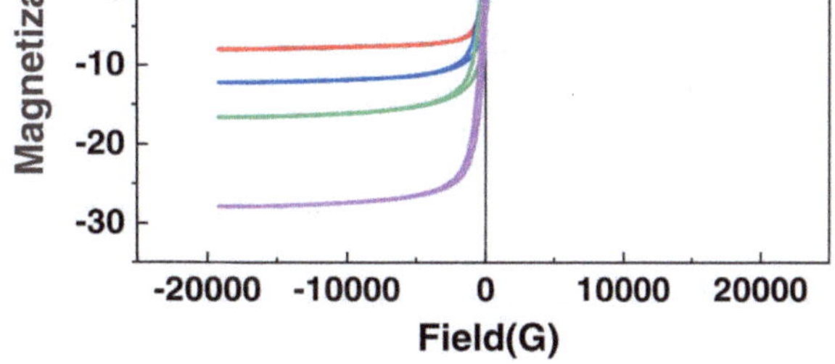

Fig. 24 Hysteresis loops of composites. Reprinted from Moustafa et al. [208], with permission from IOP. Copyright (2022)

possibility to form impurity phases which can affect the magnetic properties of the perovskite composite materials. For example, the influence of calcination temperature and calcination time on the M_r, M_s, and coercivity (H_c) of the perovskite ($La_{0.85}Mg_{0.15}FeO_3$) material is displayed in Fig. 25, where H_c is the ability of a ferromagnetic material to withstand an external magnetic field without becoming demagnetized. Based on the graph [209], 600 °C seems to be the best calcination temperature for $La_{0.85}Mg_{0.15}FeO_3$ to obtain good magnetic properties.

It can be clearly seen that perovskite materials have unusual and excellent magnetic properties. The magnetic properties of double perovskite such as $A_2B^{'}B^{''}O_6$ are controlled by modifying their cations, such as magnetic order, leading to a wide range of possibilities and interesting, useful new materials. The fundamental insight obtained in this work will be of great value in the future use of perovskite or double perovskite oxides in emergent technological applications. Focus on high

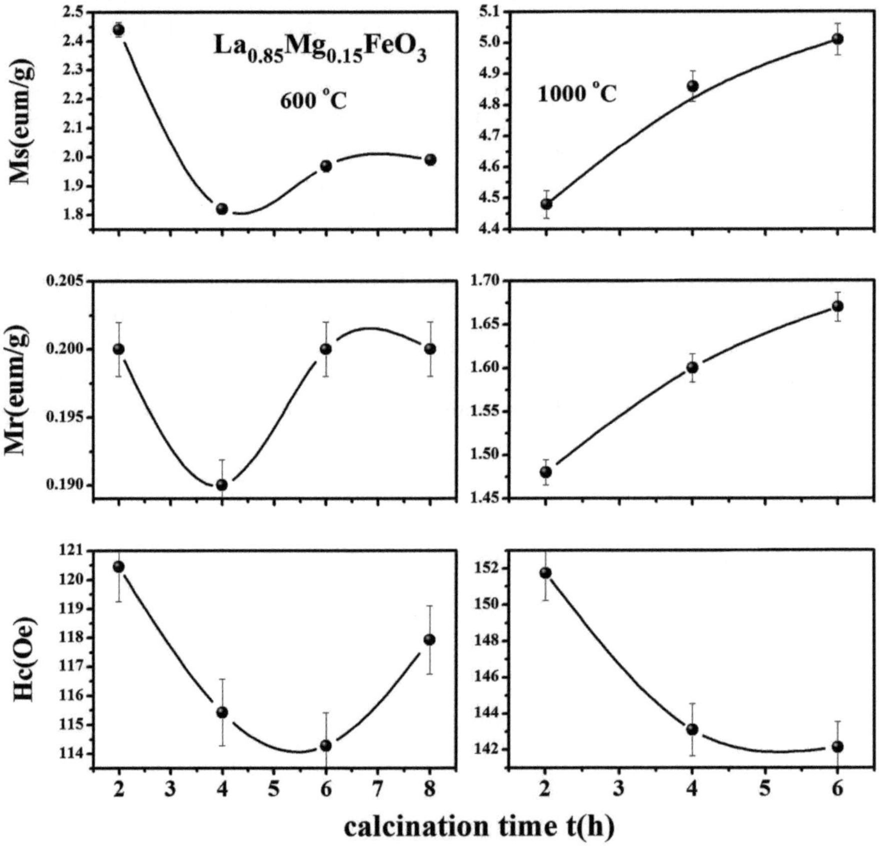

Fig. 25 The variation trend diagram of magnetic parameters of $La_{0.85}Mg_{0.15}FeO_3$ sample calcined at 600 and 1000 °C for different times. Reprinted from Lin et al. [209], with permission from Elsevier. Copyright (2018)

stability and long-term lifetime is primary and essential to ensure the successful commercialization of perovskite materials.

6 Application of Perovskite Structures

6.1 Perovskite Ceramics in Energy Applications

Perovskite oxides, complex oxides with the general formula ABO_3, where A is a large-radius rare-earth ion, and B is a transition metal ion, have elicited increasing interest and shown great potential as an efficient oxygen evolution reaction (OER) electrocatalysts to replace noble-metal-based catalysts, especially in alkaline media [210]. Perovskite materials are essential, effective, low-cost energy materials used in electronic systems, photonic devices, storage, and energy utilization. In this section, the recent advances in perovskite oxide applications in oxygen reduction reaction (ORR), OER, electrochemical water splitting, metal-air batteries, solid-state batteries (SSBs), oxygen separation membranes, and solid oxide fuel cells (SOFCs) are highlighted (see Fig. 26). The discussion focuses on the recent progress of perovskite oxide's structure, properties, and applications in energy conversion and storage. Perovskite oxides are promising oxygen electrode catalysts because of their low cost, flexibility, and tailorable properties. A wide investigation of perovskite oxide materials over the last few decades mainly aims to overcome the issue of high overpotentials and slow kinetics of oxygen electrocatalysis (i.e., ORR and OER) for further development of fuel cells and electrolyzers.

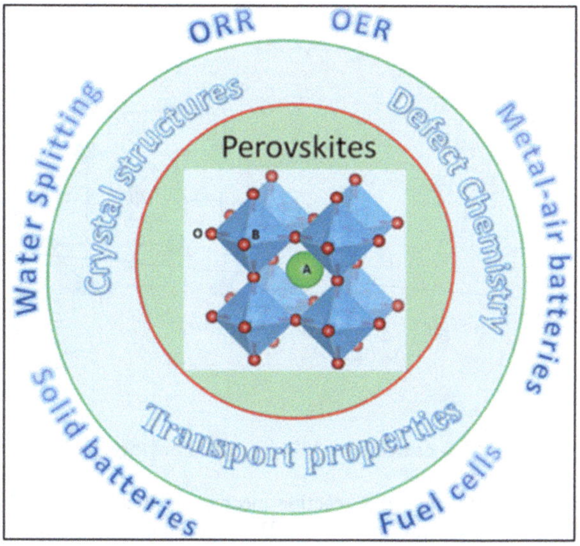

Fig. 26 Recent advances in perovskite-type oxides for energy conversion and storage applications. Reprinted from Sun et al. [211], with permission from John Wiley and Sons, Copyright (2020)

Low electrical conductivity is a drawback of many perovskite oxides. To enhance electrode performance by increasing electrical conductivity pathways, ORR perovskite-based electrodes are commonly catalyzed by carbon-based catalysts. However, carbon materials can cause serious problems such as carbon corrosion and influence perovskite properties. Nanostructured transition metal perovskite oxides can substantially lower activation energy during ORR and are regarded as efficient noble-metal-free catalysts to reduce oxygen. They also offer high chemical activity, stability, versatility, and low cost [212]. Their ability to allow partial substitution at the A and B sites ($A_{1-x}A'_xB_{1-y}B'_yO_3$), introducing changes in the valence states of metal cations and oxygen vacancies, offers a space for continuous enhancement to discover the highest performance and the most economic electrocatalysts. The improvement of perovskite oxide properties has focused on finding the most active materials and establishing activity descriptors such as anion doping, replacing A site with rare-earth metals, orbital filling, metal–oxygen covalency, numbers of d electrons, the p-band center of bulk, and B-site cation redox [211, 213]. Those stated improvement approaches aim to develop and achieve effective bifunctional ORR/OER electrocatalysis perovskite electrodes. Figure 27 displays the reaction mechanism of ORR and OER.

The need for renewable energy and storage technologies has demanded clean hydrogen from renewable sources that directly split water using solar thermal energy [214]. Thermochemical cycles involve two or more steps in which a solid oxide is thermally reduced and oxidized, and their net result is the dissociation of water. However, it occurs through two or more endothermic and exothermic reactions with lower temperature requirements. Perovskite oxides are widely studied as electrodes in electrochemical water splitting due to their excellent electrochemical

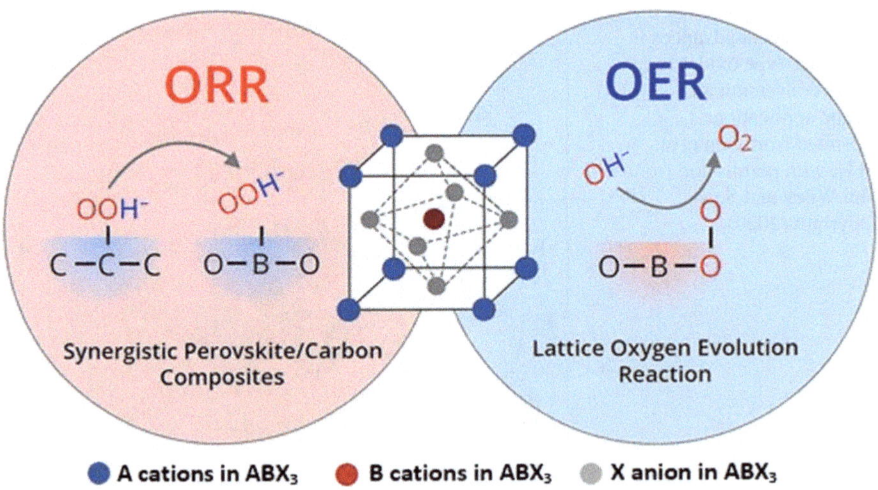

Fig. 27 ORR and OER reaction mechanism Beall et al. [212], with permission from American Chemical Society. Copyright (2021). https://pubs.acs.org/doi/10.1021/acscatal.0c04473

stability, composition: structural flexibility, and low cost. Water splitting reactions on perovskite oxides involve the redox reaction [215, 216].

Photoelectrochemical (PEC) water splitting from perovskite materials is a promising device as an environmentally friendly hydrogen production method. It enables very active and stable water splitting using light-sensitive semiconductors and molecular catalysts. Additionally, a perovskite has a high absorption coefficient, long electron and hole diffusion length, and excellent electrical and optical property operation [217]. Figure 28 shows the configuration of PEC that consists of a perovskite light absorption layer (fluorine-doped tin oxide), an electron-transporting layer, and a hole-transporting layer to extract the generated electron and hole, and a passivation layer to prevent perovskite degradation in an aqueous solution.

The performance of the perovskite-based device is influenced by diffusion length, lifetime, and mobility of the generated carriers. Discovering a new perovskite photocatalyst with a higher hydrogen production rate remains a challenge in the field of PEC or photocatalytic water splitting. Therefore, Tao et al. developed a learning machine models (Fig. 29) that can be used to predict hydrogen production rate and band gap for ABO_3-type perovskite, where atomic parameters and experimental conditions are considered variables in the development [218].

Recent advances in porous perovskite nanostructures for electrochemical energy conversion and storage systems, such as metal-air batteries, have attracted considerable attention during the past decade [109, 219]. Perovskite oxides with suitable composition and nanostructure exhibit low overpotentials, extended surface

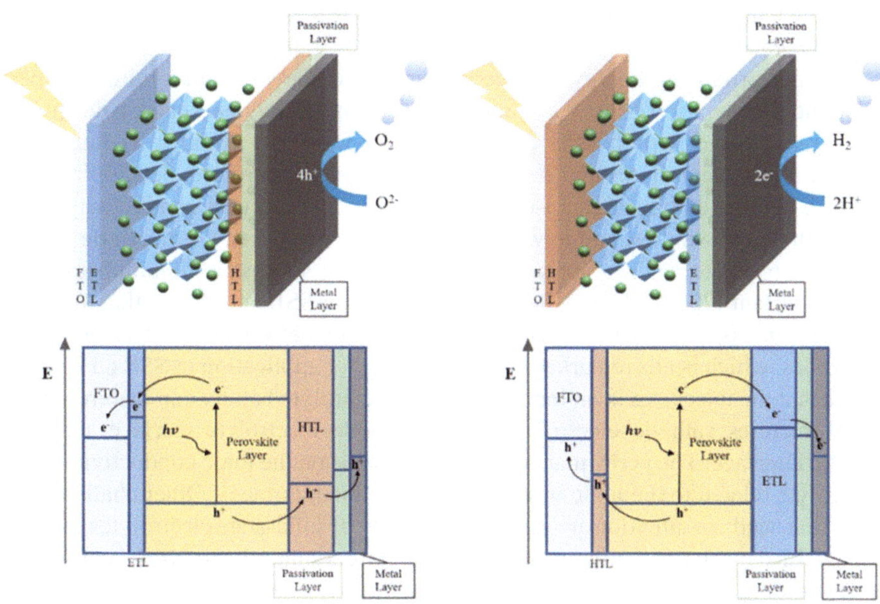

Fig. 28 Operating mechanism of perovskite-based photoelectrochemical device. Reproduced from Kim et al. [217] under the terms and conditions of the Creative Commons Attribution License

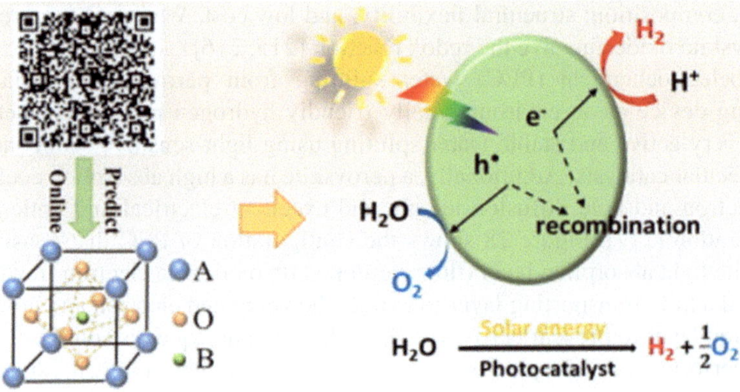

Fig. 29 Machine learning models to predict hydrogen production rate and band gap of ABO_3-type perovskite for photocatalytic water splitting. Reprinted from Tao et al. [218], with permission from Elsevier. Copyright (2021)

area, and excellent bifunctionality (ORR and OER processes) are adequate for this application. Porous $La_{0.6}Sr_{0.4}Co_{0.8}Mn_{0.2}O_3$ oxide nanofibers integrated with RuO_2 nanosheets, $La_{0.7}Sr_{0.3}MnO_3$ with $Ce_{0.75}Zr_{0.25}O_2$, $La_{0.56}Li_{0.36}Ti0_{.97}Al_{0.03}O$, $La_{0.8}Sr_{0.2}Co_{1-x}Mn_xO_3$ (LSCM), and $PrBa_{0.5}Sr_{0.5}Co_{2-x}Fe_xO_{5+\delta}$ are amongst the perovskite materials that have been developed and showed promising potential as catalysts for the metal-air batteries [210, 211]. The porous structure for perovskite metal-air batteries is essential because ORR or OER performance depends on its surface area.

Another promising energy storage technology is the solid-state lithium battery as the next-generation battery system due to its high safety and energy density. A solid-state system is preferable due to safety concerns by eliminating flammable organic liquid electrolytes and limited energy density. Therefore, developing solid electrolytes that exhibit high Li-ion conductivity, low electron conductivity, wide electrochemical window, and low interface resistance between the electrode and the solid electrolyte to realize the practical application of solid-state lithium batteries is very important [220, 221]. Challenges in developing SSBs are fewer than in liquid batteries. One issue is the low interfacial compatibility between solid electrolytes and electrodes, which needs remarkable improvement for application in SSBs. Figure 30 shows that the interfaces in an inorganic solid–electrolyte battery can feature several basic structures: cathode–electrolyte interface, anode–electrolyte interface, and inter-particle interface. The performance of SSBs depends on the ionic conductivity of the electrolyte related to the ionic transport of Li^+ at the interfaces. Other challenges in SSBs that need our attention are ionic conductivity of solid-state electrolytes, prevention of Li dendrite penetration in high-strength ceramic electrolytes, and a clear understanding of the correlation of microstructure, dynamic behavior, and chemical reaction of solid–electrolyte interphase.

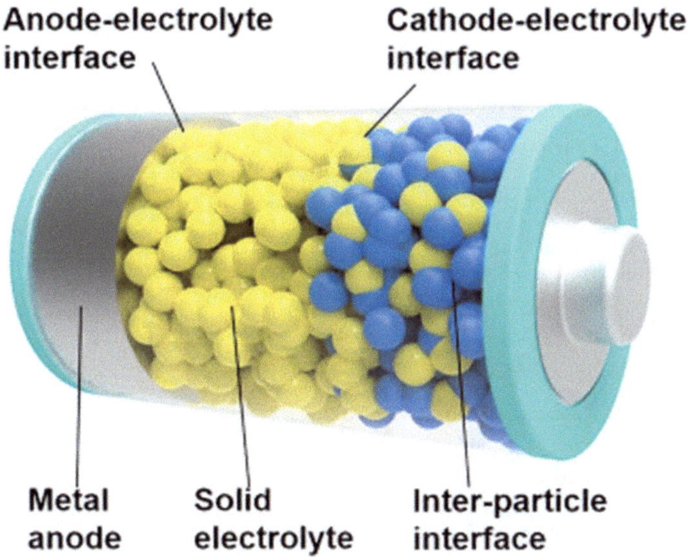

Fig. 30 Schematic of different interfaces in solid-state batteries (The schematic only presents the general structures of inorganic solid–electrolyte batteries, and several specific components, notably conductive additives and polymer electrolytes in other batteries, are not shown). Reprinted from Xu et al. [221], with permission from Elsevier. Copyright (2018)

Perovskite ceramics also have been widely used as a cathode, electrolyte, or anode component for SOFCs. SOFCs are essential devices that convert chemical energy in hydrocarbons or other fuels into electricity at high efficiency. Each principal component has its function and characteristics. Figure 31 illustrates the working principles and the process for each element of SOFC. A cathode material of SOFC should have sufficient high ionic and electronic conductivities and stability in the cathode's oxidizing atmosphere fed with air. Lanthanum manganite-based cathodes, lanthanum cobaltite, ferrite cathodes, nickelate cathodes, and K_2NiF_4-type structure cathode materials have been reported [223, 224]. The composition and microstructure of cathode materials have a large effect on the performance of SOFCs. The mixed ionic–electronic conductor perovskite oxides containing Ba, La, and Sr at the A site and transition elements Fe, Cr, and Co at the B site have attracted much attention for cathode SOFC application, including $Ba_xSr_{1-x}Co_yFe_{1-y}O_{3-\delta}$, $La_{1-x}Sr_xCo_yFe_{1-y}$ and $SrTi_{1-x}Fe_xO_{3-\delta}$.

The electrolyte in SOFC is placed between a porous anode and a cathode and is responsible for performing a vital function of separating the anode and the cathode gasses. Its functions are to conduct ions, separate reacting gases, and block internal electronic conduction. In addition, it must be stable in oxidizing and reducing environments [225]. Oxygen ion and proton conductors are commonly used as electrolytes in SOFCs. Perovskite lanthanum gallate oxides ($LaGaO_3$), especially $La_{1-x}Sr_xGa_{1-y}Mg_yO_{3-\delta}$, show high oxygen ion conductivity [226]. The four main

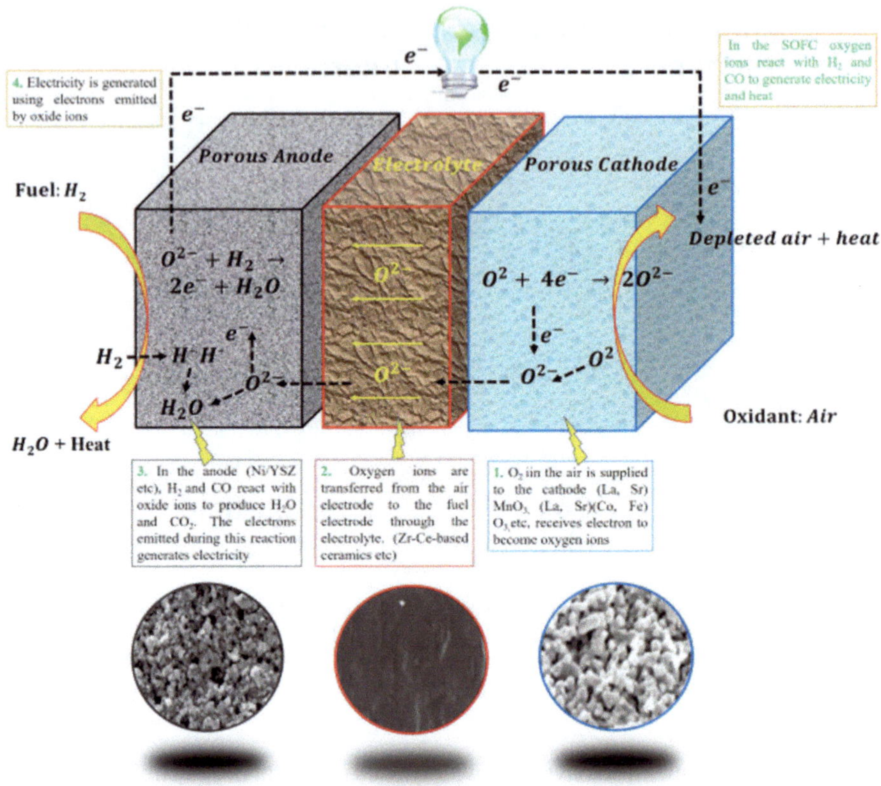

Fig. 31 Working principles of SOFC. Reprinted from Hanif et al. [222], with permission from Elsevier. Copyright (2022)

groups of electrolyte materials are zirconia-based, ceria-based, $BaZrO_3$ and $BaCeO_3$ based, $LaGaO_3$ based, and bismuth-based electrolytes [222]. The stated perovskite electrolytes can work at high to intermediate.SOFC operating temperature.

Another SOFC component is the porous anode that directly exposes the catalyst to fuels. Hence, it is one of the most important components. Perovskite oxide for anode SOFCs must have sufficient high ionic and electrical conductivity and excellent thermal and mechanical stability in the reducing conditions of the anode, typically in contact with H_2 or other fuels. It should also be chemically and thermally compatible with the dense electrolyte layer [227]. Ni-based anodes in SOFCs have excellent electrochemical properties and low cost but a few drawbacks, such as nickel sintering, carbon surface deposition, and sulfur poisoning when dirty hydrocarbon fuels are used, leading to component degradation. Thus, perovskite anode materials such as LSCM and $La_4Sr_8Ti_{11}Mn_{0.5}Ga_{0.5}O_{37.5}$ have been introduced as anode materials for hydrocarbon-fuelled SOFCs [228, 229]. Perovskite $La_{0.8}Sr_{0.2}Cr_{0.97}V_{0.03}O_3$, LSCM and $La_{1-x}Sr_xCr_{1-y}Ru_yO_3$, $La_{1-x}Sr_xCr_{1-y}Ni_yO_{3-\delta}$ and

$La_{0.6}Sr_{0.4}Fe_{0.8}Co_{0.2}O_3$–$Ce_{0.8}Gd_{0.2}O_{1.9}$, and doped $SrTiO_3$ have also been investigated as anodes for hydrocarbon-fuelled SOFCs [227]. In summary, perovskite oxide ceramic materials represent the exploited application in energy storage and conversion devices.

6.2 *Perovskite Ceramics in Optical Device Applications*

Owing to unique magnetic, mechanical, optical, and electrical properties, perovskite material has been exploited mainly for potential optical device applications. Such as waveguide devices, second harmonic generation, optical modulators, UV blocking and detection for next-generation wearable devices, powerless sensors for easy detection of UV radiation, optical coatings, and filters for Microwave/IR/UV, among others.

Ferroelectric materials such as $BaTiO_3$ and $SrTiO_3$ are widely investigated due to their physical properties and potential applications. They have been found to possess enhanced physical properties, exhibiting significant non-linear optical coefficients, large dielectric constants, and low charge loss. For the application in optical devices, the stability of such materials is a big challenge due to the enhanced properties of strong electron–electron interactions involving highly localized sites. However, such effects are sensitive to variables such as chemical composition, purity, number of surface and bulk defects, grain size, shape, sintering conditions, crystalline structure, and transition temperature [230]. Thus, considerable effort has been spent on rational synthesis and process agents in pure and doped structures [231, 232]. Erbium is an essential metal for doping $BaTiO_3$ since Er^{3+} ions have optical fluorescent properties that are particularly useful in laser applications [233]. Composite polidimetilsiloxano (PMDS) and $BaTiO_3$ nanoparticles are used to fabricate devices of mechanically flexible 3D terahertz photonic crystals (3D-TPCs) to Terahertz technology (frequencies ranging from 100 GHz to 30 THz) [234]. The modulation of refractive indices is from the concentrations of $BaTiO_3$ nanoparticles and structure geometry, and both could affect the terahertz properties of these 3D-TPCs. Besides, the THz properties of these 3D-TPCs could be tuned by the in situ application of external force fields on them. Changes of 3D-TPC sample under elongation and torsion as examples of external force fields. The measured THz wave transmission spectra in 3D-TPC sample with 40 wt.% $BaTiO_3$ under elongation (10, 20, 30, 40, and 50%) showed a shift between 0.5 and 0.6 THz, while 0°, 20°, 40°, and 60° torsions the measured THz photonic band gap peaks had a refined shift. Such a result has found novel applications in Terahertz technology.

Ferroelectric material type Sm-doped $(1-x)Pb(Mg_{1/3}Nb_{2/3})O_3$-$xPbTiO_3$ (PMN-PT) is highly transparent relaxor ceramics due to their high transparency of 69.6%, large electro-optic coefficient of $35 \times 10^{-16}\ m^2\ V^{-2}$, decent extinction ratio of 32 dB, and low half-wave voltage (113 V at $d = L = 1$ mm). Figure 32a, b [235] shows the design of its application as electro-optic modulator. The response of this material for application in optical communication is influenced by the scattering loss, which

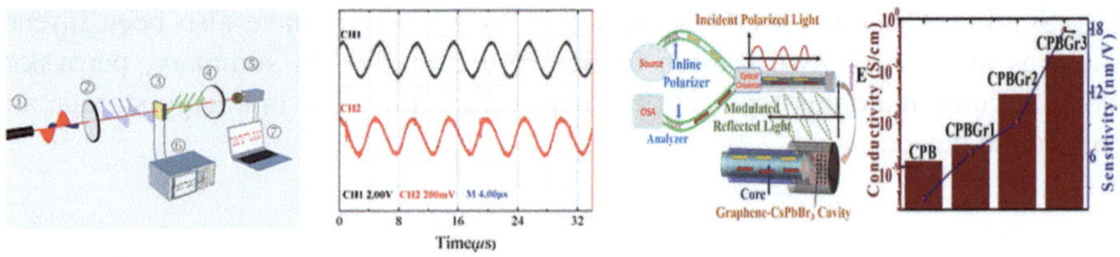

Fig. 32 Application of PMN-PT transparent ceramics in optical communication. **a** Schematic diagram of the electro-optical modulation system, **b** The modulation of sinusoidal signals at 200 kHz. **a, b** reprinted from Fang et al. [235], with permission from John Wiley and Sons. Copyright (2021). **c** Schematic of a nanocomposite-integrated high-speed electro-optic modulator. Reprinted from Aepuru et al. [236], with permission from American Chemical Society. Copyright (2020)

is the main factor affecting the transmittance, and it may be caused by the grain boundary scattering. The presence of Sm ions can influence ceramic density, transparency, and other properties. Likewise, Sm ions accumulation at the grain boundaries will lead to Rayleigh scattering loss due to the different refractive indices between grain boundaries and matrix. However, the optimization of PMN-PT ceramics has reached 0.5 mol.% Sm-doped PMN-PT ceramics. The driving voltage and modulation frequency can be tuned by ceramic width and thickness. Due to the electro-optic effect of PMN-PT ceramics, the varying electric signals can be modulated into optical signals and then received and demodulated by the silicon photodetector.

The interaction of organic–inorganic materials such as graphene and metal-halide perovskite (graphene–$CsPbBr_3$) has also been explored in new optical devices [236]. Different quantities of reduced graphene oxide, 10 mg (CPBGr1), 25 mg (CPBGr2), and 35 mg (CPBGr3) were added to $CsPbBr_3$ (CPB) nanocomposites in the powder. The presence of graphene in CPB induced dielectric polarization with strong electrical conduction and nonlinear dielectric behavior. The combination of organic–inorganic materials exhibits a significant enhancement in nanocomposite-integrated fiber-based electro-optic modulator design. The high optical and electrical response observed in the graphene–perovskite nanocomposites enable them to be further scrutinized as an active material for an electro-optic modulator. The optical fiber integrated design is scrutinized to explore light modulation and electro-optic sensitivity under an applied electrical field, as shown in Fig. 32c. The increase of graphene adjusts the electro-optic interference that leads to phase modulation of light by a change in the refractive index of the graphene–CPB nanocomposites, and that results in a higher electro-optic sensitivity of 18 nm/V than those of CPB nanocomposites, as shown the Fig. 32c.

7 Conclusion and Future Outlook

Perovskite-based ceramics are a significant class of innovative materials with fascinating properties, which are now receiving intensive research attention in energy conversion and storage, condensed matter physics, and optical device applications. In addition, inorganic perovskites exhibit many critical physical properties, such as ferroelectricity, magnetoresistance, and superconductivity. Perovskite ceramics are also categorized as the most important energy-focusing materials in batteries, fuel cells, photocatalysts, catalysis, thermoelectric, and solar thermal conversion.

A future outlook needs to be focused on the fundamental understanding of the perovskite mechanism, the design and preparation of perovskite nanoparticles with appropriate particle size, distribution, and crystal structures to stabilize large amounts of oxygen vacancies. Other than that, surface engineering and photocatalytic mechanisms are the approaches that can be further explored to enhance the perovskite ceramics performance. In materials development, it is essential to identify new materials, which may include a wide range of perovskite compounds-related compositions. Other than that, the development toward future industrial manufacturing and the scalability and impacts of ambient conditions for device fabrication should be well addressed.

Few advances have been made in optimizing the performance of perovskite oxide materials. Focus on interface engineering, which involves surface modification to produce uniformly dispersed active nanoparticles, can largely improve the performance of catalysts. Control of the composition and nonstoichiometry is essential in the growth of nanoparticles, which can eventually influence their chemical performance and stability. Furthermore, in defect engineering emphasis, strain modulation and dimensional confinement must be considered to optimize the electron filling state and improve the electronic structure of perovskite oxides. Thus, advanced characterization techniques that provide in situ and real-time information about perovskite materials in a working environment are essential for developing their applications. The outlook in this interesting field requires a multidisciplinary approach, including electrochemistry, optical, experimental solid-state chemistry, and physics, and advanced characterization with multiscale computational modeling.

References

1. S. V. Kailas, Mater, Sci. chrome-extension://efaidnbmnnnibpcajpcglclefindmkaj/https://gechstudentszone.files.wordpress.com/2015/07/material-science-iisc.pdf accessed at 7 June 2022
2. T. O. Mason, Encyclopaedia Britannica, *Ceramic Composition and Properties* (2011). https://www.britannica.com/technology/ceramic-composition-and-properties. Accessed at 7 June 2022
3. N.-G. Park, Mater. Today **18**, 65 (2015)
4. R.J.D. Tilley, *Perovskites Structure-Property Relationships*, 1st edn. (Wiley, United Kingdom, 2016), p.4

5. A.S. Bhalla, R. Guo, R. Roy, Mater. Res. Innov.Innov. **4**, 3 (2000)
6. L.-Q. Cheng, J.-F. Li, J. Mater. **2**, 25 (2016)
7. H. Wang, W. Su, J. Liu, C. Wang, J. Mater. **2**, 225 (2016)
8. Q. Ji, L. Bi, J. Zhang, H. Cao, X.S. Zhao, Energy Environ. Sci. **13**, 1408 (2020)
9. M.R. Levy, *Crystal Structure and Defect Property Predictions in Ceramic Materials*, Imperial College of Science, Technology and Medicine (2005)
10. T. Vinodkumar, B.G. Rao, B.M. Reddy, Catal. Today. Today **253**, 57 (2015)
11. Y. Cong, Z. Geng, Y. Sun, L. Yuan, X. Wang, X. Zhang, L. Wang, W. Zhang, K. Huang, S. Feng, A.C.S. Appl, Mater. Interfaces **10**, 25465 (2018)
12. X. Li, L. He, X. Zhong, J. Zhang, S. Luo, W. Yi, L. Zhang, M. Hu, J. Tang, X. Zhou, X. Zhao, Scanning, 1341608 (2018)
13. W. Tang, H. Ding, W. Bian, W. Wu, W. Li, X. Liu, J. Gomez, C. Regalado Vera, M. Zhou, and D. Ding, J. Mater. Chem. A Manuscr. **8**, 14600 (2020)
14. N.H. Hadi, M.R. Somalu, A.A. Samat, A. Muchtar, N.A. Baharuddin, M. Anwar, Int. J. Energy Res. **45**, 14357 (2021)
15. D.D. Athayde, D.F. Souza, A.M.A. Silva, D.V.E.H.M. Nunes, J.C. Diniz da Costa, W.L.Vasconcelos, Ceram. Int. **42**, 6555 (2016)
16. G. Yang, S.-J. Park, Materials **12**, 7 (2019)
17. A. Houbi, Z.A. Aldashevich, Y. Atassi, Z. Bagasharova Telmanovna, M. Saule, K. Kubanych, J. Magn. Magn. Mater. **529**, 167839 (2021)
18. A.A. Samat, M.S.A.A. Pisor, M.A. Faris, M.A. Tajudin, M.A.M. Nawi, M. Darus, N. Osman, N.A. Baharuddin, in *IOP Conference Series: Materials Science and Engineering*, vol. 957 (2020), p. 012055
19. A.V. Kuzmin, A. Yu Stroeva, V.P. Gorelov, Yu.V. Novikova, A.S. Lesnichyova, A.S. Farlenkov, A.V. Khodimchuk, Int. J. Hydrog. Energy **44**, 1130 (2019)
20. H. Ding, N.P. Sullivan, S. Ricote, in *18th International Conference on Solid State Protonic Conductors*, vol. 306 (2017), p. 97
21. Y. Kong, Y. Li, J. Li, C. Hu, X. Wang, J. Lu, Ceram. Int. **44**, 3947 (2018)
22. E. Rosa Silva, M. Curi, J.G. Furtado, H.C. Ferraz, A.R. Secchi, Ceram. Int. **45**, 9761 (2019)
23. A.S. Habiballah, A.M.M. Jani, A.H. Mahmud, N. Osman, S. Radiman, Mater. Chem. Phys. **177**, 371 (2016)
24. Z. Zhu, Y. Xu, J. Li, G. Cao, A.C.S. Appl, Nano Mater. **4**, 1294 (2021)
25. J. Xu, Y. Guo, K. Zhang, S. Liu, J. Zhao, E. Pawlikowska, M. Szafran, F. Gao, Appl. Surf. Sci. **531**, 147315 (2020)
26. B. Yue, Q. Hu, L. Ji, Y. Wang, J. Liu, RSC Adv. **9**, 38271 (2019)
27. H.-C. Song, D. Maurya, M. Sanghadasa, W.T. Reynolds, S. Priya, J. Phys. Chem. C **121**, 27191 (2017)
28. Attiq-ur-rehman, M.W. Ashraf, S. Tayyaba, S.M. Ali, S.M. Ramay, M. Saleem, Mater. Lett. **244**, 115 (2019)
29. Adelina-Carmen Ianculescu, C.-A. Vasilescu, L. Trupina, B.S. Vasile, R. Trusca, M. Cernea, L. Pintilie, A. Nicoară, J. Eur. Ceram. Soc. **36**, 1633 (2016)
30. Y. Xia, J. Wang, R. Chen, D. Zhou, L. Xiang, Crystals **6**, 148 (2016)
31. T. Liu, J. Ding, Z. Su, G. Wei, Mater. Today Energy **6**, 79 (2017)
32. J. Singh, A. Kumar, Ceram. Int. **46**, 11041 (2020)
33. V. Kumar, C.R. Mariappan, R. Azmi, D. Moock, S. Indris, M. Bruns, H. Ehrenberg, G.V. Prakash, ACS Omega, **2**, 6003 (2017)
34. K. Aisu, M. Osada, Y. Suzuki, J Mater Sci Nanotech. **1**, S101 (2018)
35. E.A. Reshetnikova, I.V. Lisnevskaya, E.A. Zalyubovskaya, V.V. Butova, A.V. Soldatov, Comments Inorg. Chem.Inorg. Chem. **40**, 314 (2020)
36. V.S. Puli, S. Adireddy, R. Elupula, S. Molugu, J. Shipman, D.B. Chrisey, in *AIP Conference Proceedings*, vol. 1837 (2017), p. 020001
37. S. Manafi, S. Joughehdoust, J. Nanoanalysis **7**, 117 (2020)
38. Z. Hanani, E.-H. Ablouh, M. 'barek Amjoud, D. Mezzane, S. Fourcade, M. Gouné, Ceram. Int. **44**, 10997 (2018)

39. A.M. Mendes, A.E. Martinelli, Mater. Lett. **256**, 126597 (2019)
40. K. Nakashima, Y. Toshima, Y. Kobayashi, M. Kakihana, J. Asian Ceram. Soc. **7**, 36 (2019)
41. M. Urbain, F. Riporto, S. Beauquis, V. Monnier, J.-C. Marty, C. Galez, C. Durand, Y. Chevolot, R. Le Dantec, Y. Mugnier, Nanomaterials **11**, 154 (2021)
42. K. Nakashima, Y. Toshima, Y. Kobayashi, Y. Ishikawa, M. Kakihana, J. Asian Ceram. Soc. **7**, 544 (2019)
43. M. François, F. Demoisson, M. Sennour, G. Caboche, Ceram. Int. **47**, 17799 (2021)
44. V.P. Junior, P.L. Rachadel, M.N. Quadri, D. Hotza, S.Y.G. González, Ceram. Int. **44**, 20671 (2018)
45. Z. Hanani, Mater. Sci. Green Energy Ifran City **127**, 109 (2019)
46. L. Liu, H. Fan, P. Fang, X. Chen, Mater. Res. Bull. **43**, 1800 (2008)
47. A. Abdul Samat, A.A. Jais, M.R. Somalu, N. Osman, A. Muchtar, K.L. Lim, J. Sol-Gel Sci. Technol. **86**, 617 (2018)
48. M.R. Somalu, A. Abdul Samat, A. Muchtar, N. Osman, Acad. J. Polym. Sci. **1**, 555571 (2018)
49. N. Osman, A. Abdul Samat, A.N. Che Mat, N.I. Abd Malek, J. Andas, Ceram. Int. **48**, 2289 (2022)
50. N.A. Mazlan, *A Study on the Role of Surfactant in the Synthesis of Cerate-Zirconate Ceramic Powder*, Universiti Teknologi MARA, Faculty of Applied Sciences (2017)
51. A.A. Samat, M.R. Somalu, A. Muchtar, H. Abd Rahman, N. Osman, Sains Malays. **49**, 3197 (2020)
52. A.A. Samat, S.H. Alias, M. Darus, M.R. Somalu, N.A. Baharuddin, N. Osman, in *IOP Conference Series: Earth and Environmental Science*, vol. 463 (2020), p. 012095
53. A. Abdul Samat, M. Rao Somalu, A. Muchtar, N. Akidah Baharuddin, N. Osman. *IOP Conference Series: Earth and Environmental Science*, vol. 268 (2019), p. 012137
54. L. Boudad, M. Taibi, A. Belayachi, M. Abd-Lefdil, Ceram. Int. **48**(5), 6087 (Nov 2022)
55. J.L. Clabel H., F.L. Zabotto, I.C. Nogueira, P. Schio, D. Garcia, O.F. de Lima, E.R. Leite, F.M.A. Moreira, C.A. Cardoso, J. Magn. Magn. Mater. **364**, 18 (2014)
56. J.L. Clabel H., F.A. Ferri, F.L. Zabotto, V.A.G. Rivera, I.C. Nogueira, D. Garcia, O.F. de Lima, E.R. Leite, M.A.Pereira-da-Silva, C.A. Cardoso, J. Magn. Magn. Mater. **407**, 160 (2016)
57. L.V. Yafarova, I.V. Chislova, I.A. Zvereva, T.A. Kryuchkova, V.V. Kost, T.F. Sheshko, J. Sol-Gel Sci. Technol. **92**, 264 (2019)
58. B.A. Josephine, A. Manikandan, V.M. Teresita, S.A. Antony, Korean J. Chem. Eng. **33**, 1590 (2016)
59. N. Singh, I.-W. Kim, S. Watanabe, T. K. Gundu Rao, V. Singh, Ceram. Int. **46**, 22108 (2020)
60. H. Liu, K. Zhu, Y. Liu, W. Li, L. Cai, X. Zhu, M. Cheng, W. Yang, Electrochim. Acta. **279**, 224 (2018)
61. F. Deganello, A.K. Tyagi, Prog. Cryst. Growth Charact. Mater. **64**, 23 (2018)
62. P.P. Khirade, V. Vinayak, P.B. Kharat, A.R. Chavan, Opt. Mater. **111**, 110664 (2021)
63. Y.-S. Yoo, Y. Namgung, A. Bhardwaj, S.-J. Song, J. Korean Ceram. Soc. **56**, 497 (2019)
64. A.T. Ravichandran, J. Srinivas, R. Karthick, A. Manikandan, A. Baykal, Ceram. Int. **44**, 13247 (2018)
65. M. Tenevich, A.P. Shevchik, V.I. Popkov, Res. Sq. **3**, 607249 (2021)
66. E. Novitskaya, J.P. Kelly, S. Bhaduri, O.A. Graeve, Int. Mater. Rev. **66**, 188 (2021)
67. M.A. Gabal, F. Al-Solami, Y.M. Al Angari, A.A. Ali, A.A. Al-Juaid, K.-W. Huang, M. Alsabban, Ceram. Int. **45**, 16530 (2019)
68. H. Aliasghari, A.M. Arabi, H. Haratizadeh, Ceram. Int. **46**, 403 (2020)
69. L. Ortega-San-Martín, A. Morán-Ruiz, A. Wain-Martin, K. Vidal, A. Larrañaga, M.A. Laguna-Bercero, M.I. Arriortua, Ceram. Int. **44**, 2240 (2018)
70. M. Musembi, F. Dejene, I. Ahemen, K. Tshabalala, Appl. Phys. A **126**, 1 (2020)
71. N. Asefi, S.M. Masoudpanah, M. Hasheminiasari, Mater. Chem. Phys. **228**, 168 (2019)
72. M.K. Musembi, F.B. Dejene, Heliyon **5**, e03028 (2019)
73. N. Sheng, C. Han, C. Zhu, T. Akiyama, Ceram. Int. **44**, 18279 (2018)
74. M.K. Hossain, E. Kecsenovity, A. Varga, M. Molnár, C. Janáky, K. Rajeshwar, Int. J. Self-Propagating High-Temp. Synth. **27**, 129 (2018)

75. C.O. Ehi-Eromosele, S.O. Ajayi, C.N. Onwucha, Mater. Chem. Phys. **259**, 124055 (2021)
76. G. Chandra Dhal, S. Dey, D. Mohan, R. Prasad, Mater. Today Proc. **4**, 10489 (2017)
77. N.A. Baharuddin, A. Muchtar, M.R. Somalu, M. Seyednezhad, Powder Technol. **313**, 382 (2017)
78. N.A. Baharuddin, A. Muchtar, M.R. Somalu, Mater. Lett. **194**, 197 (2017)
79. A. Han, Z. Wu, H. Zou, Ceram. Int. **43**, 16043 (2017)
80. M.K. Lee, S. Kang, Ceram. Int. **45**, 6665 (2019)
81. B.H. Devmunde, S.B. Somwanshi, P.B. Kharat, M.B. Solunke, J. Phys. Conf. Ser. **1644**, 012055 (2020)
82. S. Sasidharan, G. Jyothi, K.G. Gopchandran, J. Lumin.Lumin. **235**, 118048 (2021)
83. S.A. Kumar, P. Kuppusami, P. Vengatesh, Ceram. Int. **44**, 21188 (2018)
84. A.A. Ostroushko, O.V. Russkikh, T.Y. Maksimchuk, Ceram. Int. **47**, 21905 (2021)
85. Q.T.H. Vu, B. Bondzior, D. Stefańska, N. Miniajluk-Gaweł, M.J. Winiarski, P.J. Dereń, Sci. Rep. **11**, 22847 (2021)
86. T.H. Vu, B. Bondzior, D. Stefańska, N. Miniajluk, P.J. Dereń, Materials **13**, 1614 (2020)
87. T.A. Nguyen, V. Pham, T.L. Pham, L.T.T. Nguyen, I.Y. Mittova,V.O. Mittova, L.N. Vo, B.T.T. Nguyen, V.X. Bui, E.L. Viryutina, Crystals **10**, 219 (2020)
88. H. Hu, C.-E. Huang, J. Nan, H. Wang, C. Shen, Mater. Res. Express **7**, 016308 (2020)
89. Z. Li, C. Wang, Z. Wang, D. Zhang, Y. Qin, Q. Yang, Z. Wang, P. Zhao, X. Ma, M. Li, T. Ai, X. Yan, Y. Niu, B. Peng, S. Sun, D. Wang, Crystals **11**, 623 (2021)
90. X. Chen, X. Chao, Z. Yang, Mater. Res. Bull. **111**, 259 (2019)
91. A. Sharma, N. Sreenivasulu, T. Thomas, S.S. Bhattacharya, Materialia **9**, 100571 (2020)
92. M. Afqir, A. Tachafine, D. Fasquelle, M. Elaatmani, J.-C. Carru, A. Zegzouti, M. El Hammioui, Int. J. Miner. Metall. Mater. **27**, 1568 (2020)
93. R. Ao, L. Ma, Z. Guo, H. Liu, J. Yang, X. Yin, Q. Pan, Fuel **305**, 121617 (2021)
94. Y. Sim, J. Yoo, J.-M. Ha, J.C. Jung, J. Energy Chem. **35**, 1 (2019)
95. E.M. Kostyukhin, A.L. Kustov, L.M. Kustov, Ceram. Int. **45**, 14384 (2019)
96. X. Ji, C. Wang, T. Harumoto, S. Zhang, T. Rong, Q. Shen, J. Shi, Sci. Rep. **10**, 20352 (2020)
97. N. Asefi, S.M. Masoudpanah, M. Hasheminiasari, J. Sol-Gel Sci. Technol. **86**, 751 (2018)
98. F.V. Maziviero, R.L.B.A. Medeiros, D.M.A. Melo, H.P. Macedo, Â.A.S. Oliveira, T.R. Araújo, Mater. Chem. Phys. **264**, 124408 (2021)
99. W.D. Macedo, A.E. Souza, G.T.A. Santos, S.R. Teixeira, E. Longo, Ceram. Int. **44**, 953 (2018)
100. L.L. Gonzales, M. da Silva Hartwig, R.U. Fassbender, E.C. Moreira, M.B. Pereira, P.L.G. Jardim, C.W. Raubach, M.L. Moreira, S. da Silva Cava, Heliyon **7**, e06521 (2021)
101. R. Sahu, P. Kumar, Phase Transit. **93**, 91 (2020)
102. Deepa, N. Ahlawat, K. Rani, Aarti, Pooja, A. Rani, Mater. Today Proc. **54**(3), 895 (2022)
103. R. Kumar, M. Bag, J. Phys. Chem. C **125**, 16946 (2021)
104. V. Bashan, Y. Ust, Int. J. Energy Res. **43**, 7755 (2019)
105. J.L. Clabel H., J. Chacaliaza-Ricaldi, E. Marega Jr., Front. Nanotechnol. **4**, 827925 (2022)
106. C. Balamurugan, S.-J. Song, D.-W. Lee, Sens. Actuators B Chem. **272**, 400 (2018)
107. W.Y. Hernández, D. Lopez-Gonzalez, S. Ntais, C. Zhao, A. Boréave, P. Vernoux, Appl. Catal. B Environ. **226**, 202 (2018)
108. A. Hernández-Granados, A.N. Corpus-Mendoza, P.M. Moreno-Romero, C.A. Rodríguez-Castañeda, J.E. Pascoe-Sussoni, O.A. Castelo-González, E. Carmina Menchaca-Campos, J. Escorcia-García, H. Hu, Opt. Mater. **88**, 695 (2019)
109. J. Yu, R. Ran, Y. Zhong, W. Zhou, M. Ni, Z. Shao, Energy Environ. Mater. **3**, 121 (2020)
110. H. Gong, H. Tao Wang, X.F. Guo, X. Liu, L. Song, W. Xia, B. Gao, X. Huang, J. He, J. Mater. Chem. A. **6**, 16943 (2018)
111. Y. Dai, J. Yu, C. Cheng, P. Tan, M. Ni, Chem. Eng. J. **397**, 125516 (2020)
112. Y. Niu, C. Sun, X. Song, X. Yin, N. Zhang, J. Alloys Compd. **828**, 154291 (2020)
113. M. Mohsin, A. Yousaf, R. Raza, R. Zia, J. Alloys Compd. **791**, 248 (2019)
114. Q. Liu, Z. Xiao, H. Xie, J. Gao, M. Yuan, W. Dong, in *4th Conference on Surface,Coatings Nanostructured Materials 11–13 Oct. 2019 Xian China*, vol. 46 (2021), p. 9818

115. K. Ishii, M. Shimizu, H. Sameshima, S. Samitsu, T. Ishigaki, T. Uchikoshi, Ceram. Int. **46**, 13047 (2020)
116. F. Xiang, X. Chen, J. Yu, W. Ma, Y. Li, N. Yang, J. Mater. Sci. Technol. **34**, 1532 (2018)
117. C. Witt, A. Schmid, N. Leupold, M. Schultz, J. Höcker, A. Baumann, R. Moos, F. Panzer, ACS Appl. Electron. Mater. **2**, 2619 (2020)
118. N. Danilov, E. Pikalova, J. Lyagaeva, B. Antonov, D. Medvedev, A. Demin, P. Tsiakaras, J. Power. Sources **366**, 161 (2017)
119. M.M. Bazin, N. Ahmad, in *Advanced Maritime Technologies and Applications*, ed. by A. Ismail, W.M. Dahalan, A. Öchsner. Advanced Structured Materials, vol. 166 (Springer, Cham, 2022), pp. 325–333
120. M. Mohamed Bazin, N. Ahmad, Y. Nakamura, J. Asian Ceram. Soc. **7**, 417 (2019)
121. T.V. Barinova, I.P. Borovinskaya, V.Y. Barinov, I.D. Kovalev, K.B. Podbolotov, Int. J. Self-Propagating High-Temp. Synth. **26**, 119 (2017)
122. A. Agrawal, S.A. Siddiqui, A. Soni, K. Khandelwal, G.D. Sharma, Sol. Energy **226**, 9 (2021)
123. P. Huo, J. Wang, L. Fan, Z. Liu, B. Wang, A. Chang, J. Yao, Sens. Actuators Phys. **332**, 113156 (2021)
124. N. A. Baharuddin, N. F. Abdul Rahman, H. Abd. Rahman, M. R. Somalu, M. A. Azmi, and J. Raharjo, *Int. J. Energy Res.* 44, 8296 (2020)
125. M.R. Somalu, A. Muchtar, W.R.W. Daud, N.P. Brandon, Renew. Sustain. Energy Rev. **75**, 426 (2017)
126. M.R. Somalu, A. Muchtar, N.P. Brandon, Meet. Abstr. MA2015-03 **68**, 1323 (2015)
127. M.R. Somalu, A. Muchtar, N.P. Brandon, J. Mater. Sci. **52**, 7175 (2017)
128. N.A. Baharuddin, A. Muchtar, M.R. Somalu, M. Anwar, M.A. Shaikh Abdul Kader Abdul Hameed, J. Mah, Int. J. Appl. Ceram. Technol. **15**, 338 (2018)
129. E.S.M. Mouele, S. Ngqoloda, S. Pescetelli, A.D. Carlo, M. Dinu, A. Vladescu, A.C. Parau, A. Agresti, M. Braic, C.J. Arendse, L.F. Petrik, Coatings **10**, 1029 (2020)
130. N.A. Baharuddin, N.A. Mohd Nazrul Aman, A. Muchtar, M.R. Somalu, A. Abdul Samat, M.I. Aznam, Ceram. Int. **45**, 12903 (2019)
131. M.W. Lufaso, P.M. Woodward, Acta Crystallographica B. **60**, 10 (2004)
132. J.L. Clabel H., V.A.G. Rivera, Perovskite Ceramics: Recent Advances and Emerging Applications, 1st ed., (Elsevier, 2023), pp.1–27
133. S.W. Kwon, D.H. Yoon, Ceram. Int. **33**, 1357 (2007)
134. J. Song, T. Zhu, X. Chen, W. Ni, Q. Zhong, J. Mater. **6**, 377 (2020)
135. V.B. Jayakrishnan, S.K. Mishra, P.U. Sastry, Mater. Res. Bull. **144**, 111499 (2021)
136. T. Oku, in *Solar Cells-New Approaches and Reviews,* ed by L.A. Kosyachenko (IntechOpen, United Kingdom, 2015), p. 77
137. R.F. Egerton, *Physical Principles of Electron Microscopy: An Introduction to TEM, SEM, and AEM*, 2nd edn. (Springer Science+Business Media, New York, USA, 2005), pp. 155–175
138. M.U. Rothmann, W. Li, J. Etheridge, Y.B. Cheng, Adv. Energy Mater. **7**, 1 (2017)
139. Y. Zhou, H. Sternlicht, N.P. Padture, Joule **3**, 641 (2019)
140. C. De Caro, C. Haller, C.A. *UV/VIS Spectrophotometry—Fundamentals and Applications* (Mettler-Toledo Publication: Greifensee, Switzerland, 2015), pp. 4–19
141. D.L. Wood, J. Tauc, Phys. Rev. B **5**, 3144 (1972)
142. D. Di, K.P. Musselman, G. Li, A. Sadhanala, Y. Ievskaya, Q. Song, Z.K. Tan, M.L. Lai, J.L. MacManus-Driscoll, N.C. Greenham, R.H. Friend, J. Phys. Chem. Lett. **6**, 446 (2015)
143. J.L. Clabel H., G. Nicolodelli, G. Lozano C., V.A.G. Rivera, S.O. Ferreira, A.H. Pinto, M.S. Li, E. Marega, Phys. Chem. Chem. Phys. **23**, 18694 (2021)
144. W.Y. Pan, Y.C. Tang, Y. Yin, A.Z. Song, J.R. Yu, S. Ye, B.P. Zhang, J.F. Li, Ceram. Int. **47**, 23453 (2021)
145. L. Brus, J. Phys. Chem. **90**, 2555 (1986)
146. J.L. Clabel H., I.T. Awan, A.H. Pinto, I.C. Nogueira, V.D.N. Bezzon, E.R. Leite, D.T. Balogh, V.R. Mastelaro, S.O. Ferreira, E. Marega, Ceram. Int. **46**, 2987 (2020)
147. S. Si, H. Deng, W. Zhou, T. Wang, P. Yang, J. Chu, Ceram. Int. **44**, 14638 (2018)

148. M. Sendova, B.D. Hosterman, R. Raud, D. Koury, J. Raman Spectrosc.Spectrosc. **46**, 25 (2015)
149. J.L. Clabel H., V.A.G. Rivera, I.C. Nogueira, E.R. Leite, M.A. Pereira-da-Silva, M.S. Li, E. Marega, J. Lumin. **192**, 969 (2017)
150. W. Wang, W. Widiyastuti, T. Ogi, I.W. Lenggoro, Chem. Mater. **19**, 1723 (2007)
151. W.F. Zhang, Z. Yin, M.S. Zhang, Z.L. Du, W.C. Chen, J. Phys. Condens. Matter **11**, 5655 (1999)
152. J.L. Clabel H., I.T. Awan, V.A.G. Rivera, I.C. Nogueira, M.A. Pereira-da-silva, M.S. Li, S.O. Ferreira, E. Marega Jr., Appl. Surf. Sci. **493**, 982 (2019)
153. G.S. Nikolic, *Fourier Transforms-New Analytical Approaches and FTIR Strategies*, (IntechOpen, 2012), pp. 253–282
154. B.N. Parida, D.K. Pattanayak, R.K. Parida, B. Mohanty, N.C. Nayak, J. Mol. Struct. **1189**, 288 (2019)
155. Q. Jin, M. Jiang, S. Han, Y. Yan, J. Mater. Sci. Mater. Electron. **29**, 13407 (2018)
156. Y. Zhang, L. Sinatra, E. Alarousu, J. Yin, A.M. El-Zohry, O.M. Bakr, O.F. Mohammed, J. Phys. Chem. C **122**, 6493 (2018)
157. R.W. Collins, Measurement technique of ellipsometry, in *Spectroscopic Ellipsometry for Photovoltaics, Fundamental Principles and Solar Cell Characterization,* ed. by H. Fujiwara, R.W. Collins. Springer Series in Optical Sciences, vol. 212 (Springer Nature, Cham, 2018), pp. 19–58
158. H. Li, C. Cui, X. Xu, S. Bian, C. Ngaojampa, P. Ruankham, A.P. Jaroenjittchai, Sol. Energy **212**, 48 (2020)
159. M. Čebela, D. Zagorac, K. Batalović, J. Radaković, B. Stojadinović, V. Spasojević, R. Hercigonja, Ceram. Int. **43**, 1256 (2017)
160. M. Radović, Z. Dohcevic-Mitrovic, A. Golubović, V. Fruth, S. Preda, M. Šćepanović, Z.V. Popović, Ceram. Int. **39**, 4929 (2013)
161. F. Bourguiba, A. Dhahri, T. Tahri, K. Taibi, J. Dhahri, E.K. Hlil, Bull. Mater. Sci. **39**, 1765 (2016)
162. V. Khopkar, B. Sahoo, Phys. Chem. Chem. Phys. **22**, 2986 (2020)
163. M. Jain, L. Botter-Jensen, *Luminescence-Basic Concepts, Applications and Instrumentation*, ed. by H.S. Virk. Defect and Diffusion Forum, Vol. 357 Series (Trans Tech Publications Limited, 2014), pp. 245–260
164. J.L. Clabel H., S.O. Ferreira, E. Marega, *Latin America Optics and Photonics Conference OSA Technical Digest* (Optical Society of America, 2018), p. Tu5A.4.
165. J.L. Clabel H., I.T. Awan, G. Lozano, M.A. Pereira-Da-Silva, R.A. Romano, V.A.G. Rivera, S.O. Ferreira, E. Marega, Phys. Chem. Chem. Phys. *22*, 15022 (2020)
166. N.K. Kumawat, M.N. Tripathi, U. Waghmare, D. Kabra, J. Phys. Chem. A **120**, 3917 (2016)
167. W. Xie, G. Bai, Y. Cai, M. Cai, Y. Tian, F. Huang, S. Xu, J. Zhang, Ceram. Int. **45**, 5392 (2019)
168. Y. Lou, Y. Chen, C. Liu, P. Chen, R. Jia, X. Liu, L. Yang, J. Li, N. Dai, J. Lumin.Lumin. **223**, 117220 (2020)
169. Z. Abdelkafi, O. Bidault, I. Kriaa, H. Abdmouleh, Mater. Chem. Phys. **234**, 196 (2019)
170. J. Wu, Z. Wu, W. Qian, Y. Jia, Y. Wang, H. Luo, Mater. Lett. **184**, 131 (2016)
171. Z. Liu, H. Fan, Y. Zhao, G. Dong, J. Am. Ceram. Soc. **99**, 146 (2016)
172. J.L. Clabel H., S.N. Nazrin, G. Lozano C., M. Pereira da Silva, M. Siu Li, E. Marega, Vacuum **194**, 110562 (2021)
173. K. Li, L. Luo, Y. Zhang, W. Li, Y. Hou, A.C.S. Appl, Mater. Interfaces **10**(48), 41525 (2018)
174. H. Sun, Q. Zhang, X. Wang, C. Bulin, J. Am. Ceram. Soc. **98**, 601 (2014)
175. X. Zhang, G. Li, F. Mao, D. Han, C. Jing, Y. Chen, P. Yang, J. Chu, F. Yue, Ceram. Int. **46**, 18026 (2020)
176. Y. Huang, L. Luo, J. Alloys Compd. **706**, 312 (2017)
177. L. Li, V. Castaing, D. Rytz, A.D. Sontakke, Y. Katayama, S. Tanabe, M. Peng, B. Viana, J. Am. Ceram. Soc. **102**, 2629 (2019)

178. C. Ma, X. Wang, W. Tan, W. Zhou, X. Wang, Z. Cheng, G. Chen, Z. Zhai, Dalt. Trans. **49**, 5581 (2020)
179. N. Ding, W. Xu, D. Zhou, Y. Ji, Y. Wang, R. Sun, X. Bai, J. Zhou, H. Song, Nano Energy **78**, 105278 (2020)
180. Y. Duan, P. Li, Y. Lu, S. Xu, J. Zhang, Ceram. Int. **47**, 15219 (2021)
181. V. Jella, S. Ippili, J.-H. Eom, S.V.N. Pammi, J.-S. Jung, V.-D. Tran, V.H. Nguyen, A. Kirakosyan, S. Yun, D. Kim, M.R. Sihn, J. Choi, Y.-J. Kim, H.-J. Kim, S.-G. Yoon, Nano Energy **57**, 74 (2019)
182. R.J.D. Tilley, *Perovskites Structure-Property Relationships*, 1st edn. (Wiley, United Kingdom, 2016), pp. 1–40
183. R. Su, Z. Xu, J. Wu, D. Luo, Q. Hu, W. Yang, X. Yang, R. Zhang, H. Yu, T.P. Russell, Q. Gong, W. Zhang, R. Zhu, Nat. Commun.Commun. **12**, 2479 (2021)
184. F. Han, S. Ren, J. Deng, T. Yan, X. Ma, B. Peng, L. Liu, J. Mater. Sci. Mater. Electron. **28**, 17378 (2017)
185. K.S. Cole, R.H. Cole, J. Chem. Phys. **9**, 341 (1941)
186. D.H. Yoon, J. Ceram. Process. Res. **7**, 343 (2006)
187. J. Deng, L. Liu, X. Sun, S. Liu, T. Yan, L. Fang, B. Elouadi, Mater. Res. Bull. **88**, 320 (2017)
188. A. Ray, A. Roy, S. De, S. Chatterjee, S. Das, J. Appl. Phys. **123** (2018)
189. S.I. Shkuratov, C.S. Lynch, J. Mater. **8**, 739 (2022)
190. Y. Yang, Y. Zhou, J. Ren, Q. Zheng, K.H. Lam, D. Lin, J. Am. Ceram. Soc. **101**, 2594 (2018)
191. K. Tewatia, A. Sharma, M. Sharma, A. Kumar, Mater. Today: Proc. **44**, 4548 (2020)
192. X.K. Wei, Y.T. Su, Y. Sui, Q.H. Zhang, Y. Yao, C.Q. Jin, R.C. Yu, J. Appl. Phys. **110**, 114112 (2011)
193. D.N. Bhoyar, S.B. Somvanshi, P.B. Kharat, A.A. Pandit, K.M. Jadhav, Phys. B Condens. Matter **581**, 411944 (2020)
194. T. Wang, H. Deng, X. Meng, H. Cao, W. Zhou, P. Shen, Y. Zhang, P. Yang, J. Chua, Ceram. Int. **43**, 8792 (2017)
195. Y. Li, W. Chen, J. Zhou, Q. Xu, H. Sun, R. Xu, Mater. Sci. Eng. B **112**, 5 (2004)
196. M. Xu, B. Peng, J. Zhu, L. Liu, W. Sun, G.J.T. Leighton, C. Shaw, N. Luo, Q. Zhang, J. Alloy. Compd. **789**, 303 (2019)
197. C. Sun, J.A. Alonso, J. Bian, Adv. Energy Mater. **11**, 2000459 (2021)
198. J. Deng, F. Han, B. Schwarz, M. Knapp, H. Ehrenberg, W. Hua, M. Hinterstein, G. Li, Y. He, J. Wang, Y. Yuan, L. Liu, Inorg. Chem.. Chem. **60**, 6999 (2021)
199. Y. Zhu, J. Zhu, H. Song, J. Huang, Z. Lu, G. Pan, J. Rare Earths **39**(4), 374 (2021)
200. X. Zheng, L. Zhang, X. Wang, Y. Qing, J. Chen, Y. Wu, S. Deng, L. He, F. Liao, Y. Wang, J. Geng, J. Sun, G. Li, L. Liu, J. Lin, Inorg. Chem. Front. **7**, 3561 (2020)
201. V. Markovich, A. Wisniewski, H. Szymczak, in *Magnetic Properties of Perovskite Manganites and Their Modifications*. ed. by K.H.J. Buschow (Elsevier, UK, 2014), pp.1–9
202. H. Nakayama, H. Katayama-Yoshida, Jpn. J. Appl. Phys.. J. Appl. Phys. **40**, 1355 (2001)
203. M. Gromada, M. Biglar, T. Trzepiecinski, F. Stachowicz, in *Application of $BaTiO_3$ Perovskite Material for Piezoelectric Multilayer Actuators*, ed. by K. Ntalianis, A. Croitoru (Springer, Cham, 2018), pp. 29–35
204. T. Wang, L. Jin, C. Li, Q. Hu, X. Wei, J. Am. Ceram. Soc. **98**, 559 (2015)
205. J. Young, J.M. Rondinelli, Phys. Rev. B **89**, 174110 (2014)
206. S. Vasala, M. Karppinen, Prog. Solid State Chem. **43**, 1 (2015)
207. J.L. Clabel H., V.A.G. Rivera, A. Henrique P., E. Marega J., in *Multiferroic Perovskite Ceramics: Properties and Applications*, ed. by J.L. Clabel H., V.A.G. Rivera, (Elsevier, 2023), pp. 339–381.
208. A.M. Moustafa, S.A. Gad, G.M. Turky, L.M. Salah, ECS J. Solid State Sci. Technol. **11**, 033008 (2022)
209. Q. Lin, X. Yang, J. Lin, Z. Guo, Y. He, Int. J. Hydrog. Energy **43**(28), 12720 (2018)
210. J. Irvine, J.L.M. Rupp, G. Liu, X. Xu, S. Haile, X. Qian, A. Snyder, R. Freer, D. Ekren, S. Skinner, O. Celikbilek, S. Chen, S. Tao, T.H. Shin, R. O'Hayre, J. Huang, C. Duan, M. Papac, S. Li, V. Celorrio, A. Russell, B. Hayden, H. Nolan, X. Huang, G. Wang, I. Metcalfe, D. Neagu, S.G. Martín, J. Phys. Energy **3**, 031502 (2021)

211. C. Sun, J.A. Alonso, J. Bian, Adv. Energy Mater. **11**, 2000459 (2020)
212. C.E. Beall, E. Fabbri, T.J. Schmidt, ACS Catal.Catal. **11**, 3094 (2021)
213. J. Liu, H. Liu, H. Chen, X. Du, B. Zhang, Z. Hong, S. Sun, W. Wang, Adv. Sci. **7**, 191614 (2020)
214. M.T. Azcondo, M. Orfila, J. Marugan, R. Sanz, A. Muñoz-Noval, E. Salas-Colera, C. Ritter, F. García-Alvarado, U. Amador, Chemsuschem **12**(17), 4029 (2019)
215. B. Hua, M. Li, W.Y. Pang, W. Tang, S. Zhao, Z. Jin, Y. Zeng, B. Shalchi Amirkhiz, J.-L. Luo, Chem **4**, 2902 (2018)
216. W. Wang, M. Xu, X. Xu, W. Zhou, Z. Shao, Angew. Chem., Int. Ed. **59**, 136 (2020)
217. D. Kim, D.K. Lee, S.M. Kim, W. Park, U. Sim, Materials **13**(1), 210 (2020)
218. Q. Tao, T. Lu, Y. Sheng, L. Lia, W. Lu, M. Li, J. Energy Chem. **60**, 351 (2021)
219. M.Q. Ren, J.B. Zhang, M.M. Fan, P.M. Ajayan, J.M. Tour, Adv. Mater. Interfaces **6**(19), 1901035 (2019)
220. J. Lu, Y. Li, J. Mater. Sci. Mater. Electron. **32**, 9736 (2021)
221. L. Xu, S. Tang, Y. Cheng, K. Wang, J. Liang, C. Liu, Yuan-Cheng Cao, F. Wei, L. Mai, Joule **2**, 1991 (2018)
222. M.B. Hanif, S. Rauf, M. Motola, Z.U.D. Babar, C.-J. Lia, C.-X. Li, Mater. Res. Bull. **146**, 111612 (2022)
223. C.W. Sun, R. Hui, J. Roller, J. Solid State Electrochem.Electrochem. **14**, 1125 (2010)
224. I.R. Larramendi, N.O. Vitoriano, I.B. Dzul-Bautista, T. Rojo, in *Perovskite Materials*, ed. by L. Pan, G. Zhu (Intech Open, 2016), pp. 589–617
225. K. Zhang, J. Sunarso, Z. Shao, W. Zhou, C. Sun, S. Wanga, S. Liu, RSC Adv. **1**, 1661 (2011)
226. C. Zhu, X. Niu, Y. Fu, Nat. Commun.Commun. **10**, 1 (2019)
227. F.S. Silva, T.M. Souza, Int. J. Hydrog. Energy **42**, 26020 (2017)
228. J.C. Ruiz-Morales, J. Canales-Vazqzez, C. Savaniu, D. Marrero-López, W. Zhou, J.T.S. Irvine, Nature **439**, 568 (2006)
229. S.W. Tao, J.T.S. Irvine, Nat. Mater. **2**, 320 (2003)
230. V. Hornebecq, C. Huber, M. Maglione, M. Antonietti, C. Elissalde, Adv. Funct. Mater. **14**, 899 (2004)
231. J.L. Clabel H., J.C. Sczancoski, A. Henrique P., E. Marega J., in *Methods for the Synthesis of Ceramic Materials with Perovskite Structure*, ed. by J.L. Clabel H., V.A.G. Rivera (Elsevier, 2023), pp. 31–75
232. J.F. Ihlefeld, D.T. Harris, R. Keech, J.L. Jones, J.P. Maria, S. Trolier-McKinstry, J. Am. Ceram. Soc. **99**, 2537 (2016)
233. L.V. Maneeshya, P.V. Thomas, K. Joy, Opt. Mater. (Amst). **46**, 304 (2015)
234. P. Zhu, W. Yang, R. Wang, S. Gao, B. Li, Q. Li, Adv. Opt. Mater. **5**, 1 (2017)
235. Z. Fang, X. Jiang, X. Tian, F. Zheng, M. Cheng, E. Zhao, W. Ye, Y. Qin, Y. Zhang, Adv. Opt. Mater. **9**, 2002139 (2021)
236. R. Aepuru, C.N. Rao, R. Udayabhaskar, D. Vennu, V.M. Gaikwad, V. Kumar, H.S. Panda, P. Aqueveque, P. Cao, R.V. Mangalaraja, J. Phys. Chem. C **124**, 26648 (2020)

Advanced Perovskite Solar Cells

Yuqin Tian, Can Zhu, Kun Hong, Kai Qiu, and Renhui Zhang

Abstract The energy crisis is a huge challenge facing the world today. Natural resources such as coal and oil are consumed in large quantities and their reserves are gradually decreasing. It is imperative to advocate energy conservation. Meantime, it is very important to develop green and clean energy. Solar energy has become one of the most promising green energy sources in recent years because of its sustainable and safe advantages. Solar energy can be converted into effective energy such as heat energy through photoelectric conversion because it doesn't produce harmful gases, solid waste, and other pollutants in the conversion process. In addition, the new solar cells have the advantages of low cost, cleanliness, and they are efficient. Since 2009, a new type of perovskite solar cell has developed rapidly. In order to further improve the photoelectric conversion efficiency of batteries, more researchers in recent years have tried to apply new ceramic materials (perovskite materials) to batteries, and have achieved remarkable results. Relevant research reports show exponential growth. Perovskite solar cells use crystals with perovskite structure as electron transfer materials to improve the light absorption efficiency of the solar cells. Studies show that the structure and performance of the electron transfer layer directly affect the stability and life of the battery, which proves that the appropriate electron transfer layer is very important. New perovskite ceramic materials have been widely used in solar cell devices. This chapter mainly introduces the most common perovskite thin films and their preparation methods, organic–inorganic perovskite solar cells, etc., focusing on their development status, and the main factors affecting their stability. Finally, the current problems and development prospects in the research and application of perovskite solar cells are introduced, which will lay a solid foundation for the deeper understanding of perovskite solar cells and the preparation of new and efficient ones.

Keywords Advanced ceramics · Perovskite solar cell · Electronic transmission material

Y. Tian · C. Zhu · K. Hong · K. Qiu · R. Zhang (✉)
East China Jiaotong University, Nanchang 330013, China
e-mail: 3067@ecjtu.edu.cn

S. J. Ikhmayies (ed.), *Advanced Ceramics*, Advances in Material Research and Technology, https://doi.org/10.1007/978-3-031-43918-6_3

1 Introduction

The continuous exploitation of non-renewable resources, such as petroleum and mineral resources, not only brings energy crisis but also causes serious environmental pollution. Therefore, it is an effective solution to solve the soaring world energy demand and global climate crisis to find corresponding alternative energy resources. As the cleanest and most abundant renewable energy, solar energy has attracted the attention of many researchers. Solar energy mainly converts light energy into electric energy and heat energy through photoelectric conversion and photothermal conversion. In recent years, solar cells have experienced the first generation (crystalline silicon solar cells, the most commonly used monocrystalline silicon thin film solar cells, with a maximum conversion efficiency of 24.7%) and the second generation (thin film solar cells, with a maximum conversion rate of 20.3%). Despite their high conversion efficiency, both the first- and second-generation solar cells share a common disadvantage that cannot be ignored, namely, on the one hand, they cause serious harm to the environment, and on the other hand, silicon processing and raw material costs are high. Based on this, the third-generation solar cell–new solar cell, while maintaining high conversion efficiency, has gradually become a research hotspot with the standard of low cost, rich raw materials, non-toxic, and eco-friendly, especially the new perovskite solar cell.

So far, different types of solar cells, such as polysilicon solar cells, single-crystal silicon solar cells, copper indium gallium selenide (CIGS) solar cells, cadmium telluride (CdTe)-based solar cells, quantum dot (QD) solar cells, organic photovoltaic, and perovskite solar cells have been reported. Specifically, energy conversion efficiency and material cost plays a crucial role in the commercialization of solar cells. At present, crystalline silicon solar cells occupy a dominant position in the market. However, expensive manufacturing process requirements and expensive raw materials have forced researchers to try to find an efficient and low-cost new photovoltaic technology. Perovskite materials have attracted much attention because of their excellent photoelectric properties, low material consumption, and considerable production cost [1]. The steady progress of research and development in the perovskite field has increased the possibility that it can replace existing solar cell technology, and it has good prospects [2].

Perovskite is named after the Russian mineralogist L.A. Perovski. The molecular formula of the perovskite structure material is ABX_3, which is generally a cubic or an octahedral structure, and is shown in Fig. 1 [3]. As shown in the structure, the larger A ion occupies an octahedral position shared by 12 X ions, while the smaller B ion is stable in an octahedral position shared by 6 X ions [3, 4]. The crystallographic stability and probable structure of perovskites can be deduced by considering a tolerance factor t and an octahedral factor μ, As shown in formula (1), and (2) [5], Where, t is defined as the ratio of the distance between A–X and B–X in the ideal solid sphere model, which is defined by formula (1), where RA, RB, and RX are the ionic radii of the corresponding ions, μ is defined by formula (2). For halide perovskite (X = F, Cl, Br, and I), and t is usually in the range $0.81 < t < 1.11$.

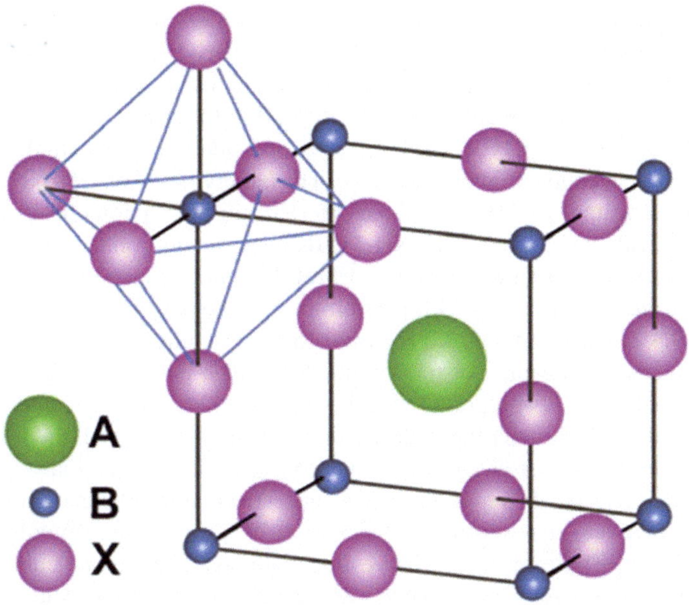

Fig. 1 Perovskite crystal structure. Reprinted from Park [3], with permission from Elsevier. Copyright (2015)

while μ is in the range $0.44 < \mu < 0.90$. For a narrow range of $0.89 < t < 1.0$, the cubic structure in Fig. 1 is likely to exist, and for a lower t value the less symmetrical tetragonal or orthogonal structures will exist [6, 7]. For organic–inorganic halide perovskite, the larger cation A is organic (methylammonium), and the anion X is generally a halogen (I, Br, and Cl), and for high-efficiency solar cells, the B ion is generally Pb [8].

$$t = (r_A + r_X)/(\sqrt{2}[r_B + r_X]) \tag{1}$$

$$\mu = R_B/R_X \tag{2}$$

2 Research Status of Perovskite Ceramic Materials

According to the survey, the PCE (power conversion efficiency) of perovskite solar cells increased rapidly from 3.8% to 22.1% in just 7 years from 2009 to 2016 [9]. As a new all solid-state planar solar cell, perovskite solar cell has developed rapidly because of its advantages of simple preparation process, low cost, and high efficiency.

In 2009, Kojima et al. [10] first studied the photovoltaic function of organic–inorganic lead halide perovskite compounds $CH_3NH_3PbBr_3$ and $CH_3NH_3PbI_3$ as visible light sensitizers in photoelectric chemical cells. The nanocentricity self-assembled in the mesoporous titanium dioxide (TiO_2) film exhibits a strong band gap absorption characteristic. Based on $CH_3NH_3PBI_3$, the spectral sensitivity is as high as 800 nm, and the solar conversion efficiency is 3.8%. The battery based on $CH_3NH_3PBB_3$ has a high optical voltage of 0.96 V, and the external sub-conversion efficiency is 65%. In 2011, Im et al. [11] formed (CH_3NH_3) PBI_3 quantum dots on the surface of nano-TiO_2. Kim et al. [12] permeated the spiro-MeOTAD hole transport material into the perovskite-sensitized nanorods' film, and improved the power conversion efficiency (PCE) to 9.4%. Christians et al. [13] used to successfully realize a predecessor's power conversion efficiency of 6.0%, and they had a good photocurrent stability solar cell. Hou et al. [14] reported that the interface between SnO_2 and perovskite was modified by dopamine, the hydrophilicity of SnO_2 increased, the high-quality transformation of the perovskite film was promoted, the prepared absorbing interface has achieved 16.65% high-optoelectronic conversion efficiency, the stability of the interface and has improved. Tavakoli et al. [15] reported modifications to suppress defects in perovskite light harvesting materials. These modifications include engineering of the perovskite interface with the electron and hole selective contact materials. They introduced a planar SnO_2/TiO_2 double layer to ensure fast electron extraction. To passivate surface trap states, they treated the surface of the perovskite facing the hole conductor with iodine dissolved in isopropanol. The modifications resulted in retardation of radiationless carrier recombination, increasing the photoelectric conversion efficiency of the interface to 21.65%, increasing the open circuit voltage to 1.24 V, and decreasing the opening voltage loss to only 0.37 V. At the same time, the device efficiency at 20, 50, and 65 °C remained 96%, 90%, and 85%, respectively, of the initial photoelectric conversion efficiency under continuous light, which is one of the most stable perovskite solar devices reported so far. Qi et al. [16] treated the mixed perovskite with phenylethyl ammonium iodide (PEAI). The existence of PEAI passivated and reduced the surface defects of the film. The prepared perovskite solar cell achieved a high photoelectric conversion efficiency of 23.6%, and the verified steady-state photoelectric conversion efficiency reached 23.32%. Tong et al. [17] reported in 2019 that the addition of thiocyanate (GuaSCN) can significantly improve the structure and photoelectric properties of Sn–Pb mixed low band gap perovskite films. The defect density is reduced by a factor of 10, resulting in a carrier lifetime of more than 1 microsecond and a diffusion length of 2.5 microns. A series of four-terminal solar cells with a high efficiency of 25%, and two-terminal full perovskite polycrystalline thin films of a high efficiency of 23.1% are realized. In 2020, the laminated photovoltaic device achieved a bigger breakthrough. Xiao et al. [18] added the ion-reducing sulfonic acid (FSA) drug, and zwitterionic antioxidant to inhibit Sn^{2+} oxidation, and passivating the mixed lead–tin perovskite film, making the efficiency of monochrome solar cells reach 21.7% (certified 20.7%). The entire perovskite series has a certification efficiency of 24.2% in an area of 1 cm^2. It is currently the highest certified photoelectric conversion efficiency of the global calcium titanium mineral cells.

Because of product instability and human health, researchers are working hard to achieve the Pb-free method. The most direct method to achieve a perovskite material is to use elements with similar properties (such as Sn, Ge) instead of Pb element. Jokar et al. [19] studied organic cation-doped diiodide ($EDAI_2$) and iodide (BAI) as defect passivating surface additives to improve device performance and durability of $FASnI_3$ perovskite solar cells. The addition of BAI significantly changed the orientation of crystal growth, and improved the connectivity of grains. The addition of $EDAI_2$ can cure the pinhole problem, passivate the surface defect states, prevent Sn^{2+}/Sn^{4+} oxidation and induce the slow relaxation of crystal structure. Nishimura et al. [20] prepared tin halide PSCs with an efficiency of more than 13% by adjusting the A-site cation, making its tolerance factor close to 1. The formamidinium cation was partially substituted with an ethylammonium cation, and a more stable tin calcium ore crystal was obtained, with inhibition of order 1 of trap density. Another way to reduce toxicity is to partially replace Pb in perovskite materials. A certain amount of Pb-doped tin perovskite can improve the stability and performance, so that Sn–Pb mixed perovskite has adjustable band gap, which makes it an ideal light absorbing layer for all perovskite laminated batteries [21]. However, the high defect density caused by Sn^{2+} oxidation limits carrier life and charge collection in tin-based titanium mine. Lin et al. [22] recently reported a strategy to reduce Sn vacancies in mixed Pb–Sn narrow-band gap perovskite in which Sn^{4+} (oxidation product of Sn^{2+}) is reduced to Sn^{2+} by proportional reaction with metal tin, A monolithic all perovskite series cell was fabricated, which obtained 24.8% PCCE certification in small area devices (0.049 cm^2) and 22.1% PCCE certification in large area devices (1.05 cm^2). The cell series can still maintain 90% performance after 463 h of operation at the maximum power point under full single solar illumination. From the perspective of the environment, the use of Pb in perovskite solar cells is still a threat. Partial substitution of lead has shown great advantages. In current studies, we must focus on the research of Pb device packages and appropriate recycling mechanisms, the comprehensive transformation of the perovskite solar cell will enable the photovoltaic field to reach a new height.

3 Perovskite Films and Their Preparation Methods

Perovskite is a natural calcium titanate mineral, and its chemical molecular formula is $CaTiO_3$. Typically, a material having the same crystal structure as $CaTiO_3$ is referred to as a perovskite material. In the past few decades, inorganic perovskite oxides and halides have been widely studied because of their wide applications in the fields of optics, magnetism, electronics, and superconductors [23]. Among them, the research of perovskite films in solar cells has attracted great attention. In the early devices, due to the poor quality of perovskite films, there are many defects, which not only reduce the light absorption performance and photocurrent, but also cause the contact between electron and hole transport layer, resulting in the decline of the overall performance. The prepared films used in perovskite batteries also have some problems, such as low viscosity of precursor solution, low crystallinity caused by fast crystallization, and poor contact between substrate and perovskite layer. Therefore, it is crucial for the preparation of high-efficiency solar cells to design a simpler, and more effective preparation method of perovskite films, adjust the coverage and uniformity of the films, and reduce the formation of defect sites [24].

With the advancement of technology, optimizing the preparation technology of perovskite thin films is a necessary condition for the fabrication of high-efficiency devices. Typically, the chemical measurement, crystal phase, and grain structure of fine control materials are needed to prepare an efficient perovskite solar cell. Since these properties can be controlled by the preparation method, a large number of studies have been carried out on the development of thin film deposition techniques. In the evolution process of perovskite solar cells, four main deposition methods are studied, as shown in Fig. 2, which are one-step dissolution process, two-step dissolution process, steam-assisted dissolution process, and thermal evaporation process [25–28]. In addition, in order to improve the quality of perovskite films and improve the efficiency of perovskite batteries, researchers have put forward more advanced film preparation technologies after continuous exploration.

3.1 One-Step Deposition Method

One-step deposition is the earliest technique for preparing perovskite solar cells. Because of its simplicity and low cost, it is the most widely used technique. Chien Huang Chiang et al. [30] developed a one-step hot solution spin coating method to prepare high-quality mixed halide perovskite films. High-quality thin films can be prepared by a simple one-step method with an efficiency of over 16% [31]. However, due to the low surface coverage of perovskite film prepared by one-step solution deposition method, and micropores will occur in the film during the preparation process, which will affect the quality of the film, this method has been gradually replaced by subsequent preparation technologies.

3.2 Two-Step Solution Deposition Method

The perovskite film prepared by one-step deposition method has poor surface coverage and inevitable non-uniformity. In order to overcome this problem, Burschka et al. [26] developed a two-step solution deposition method based on Mitzi et al.'s [10] pioneering work to prepare uniform $MAPBI_3$ films on the MP-TiO_2 layer. The gas-assisted solution method was developed into an improved two-step solution deposition method, and methylammonium iodide (MAI) was introduced on the basis of gas-phase deposition. Chen et al. [27] prepared homogeneous conformers methylammonium lead iodide $CH_3NH_3PbI_3$ ($MAPBI_3$) films with high phase purity by the two-step deposition technique. Heping Shen and his team used the improved

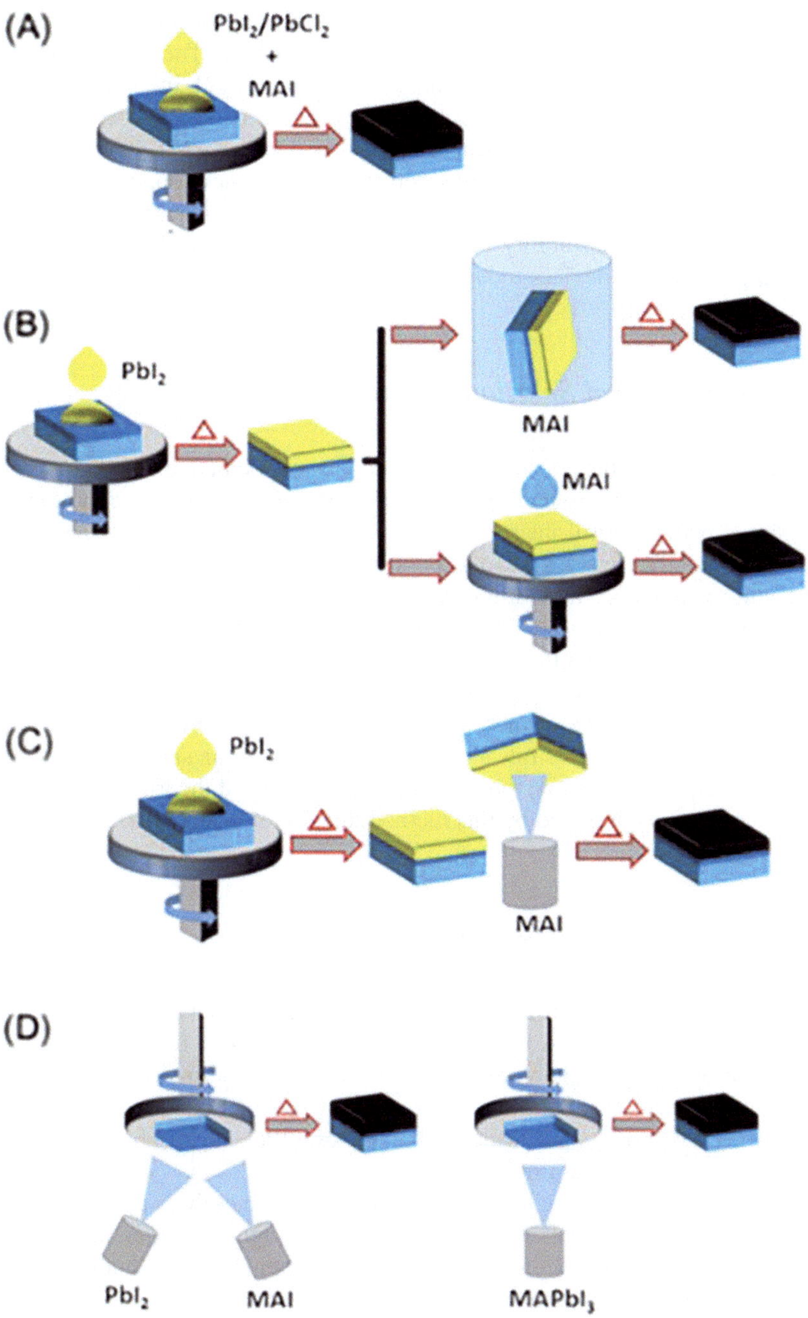

Fig. 2 The four main deposition methods to prepare perovskite (calcium titanium mine) film. **a** One-step dissolution process. **b** Two-step dissolution process. **c** Steam-assisted dissolution process. **d** Thermal evaporation process. Reprinted from Watthage et al. [29], with permission from Elsevier. Copyright (2018)

two-step deposition method to obtain a large area of high-quality perovskite film. The prepared film has less pinholes and good uniformity. The residue of PbI_2 in perovskite films prepared by this method is the minimum [32]. Julian Burschka et al. prepared perovskite thin films by two-step deposition method and achieved about 15% power conversion efficiency (measured under standard AM1.5G test conditions on solar energy). This two-step approach will provide new opportunities for the manufacture of processed photovoltaic cells. It has unprecedented energy conversion efficiency and high stability, which is equivalent to or even higher than the best thin-film photovoltaic devices today [26].

3.3 Vapor-Assisted Solution Deposition

In an improvement of the two-step solution deposition method, the successful preparation of perovskite films can be realized by vapor deposition technology. This method can better control the morphology and grain size through gas–solid crystallization, and effectively avoid the possible film delamination in the process of liquid–solid interaction. In addition, the perovskite films prepared by this preparation technology have the characteristics of uniform surface coverage, large grain size, and high conversion efficiency. Chen Qi et al. demonstrated the preparation of perovskite films and planar geometry photovoltaic devices by vapor-assisted solution method (VASP). The formed perovskite films have the characteristics of full surface coverage on the substrate and have a significant grain size on the microscale [27]. However, due to the long gas–solid reaction time and the low efficiency (only 10–12%) of devices prepared by this method, the use of this method is limited.

3.4 Low Pressure-Assisted Solution Method

Low pressure-assisted solution method is a recently developed method for the preparation of perovskite films. A high-quality double perovskite film $Cs_2AgBiBr_6$ can be prepared using this technology, which has excellent thermal stability and environmental stability. The optimum PCE of the $Cs_2AgBiBr_6$ film is 1.44% under AM1.5 (100 mW cm^{-2}) light, which is the highest efficiency of Pb-free high-stable perovskite solar cell at that time.

The preparation process of the perovskite film determines the use of the perovskite film. Different requirements for battery equipment require different preparation processes. With the advancement of commercialization and technology, a large number of more efficient preparation processes will emerge.

4 Factors Affecting the Stability of Perovskite Solar Cell

The efficiency of perovskite solar cells is increasing because of their excellent photoelectric performance, but the stability is the key factor for the commercialization of perovskite solar cells. The current research results show that the main factors affecting their stability include electron transport materials in solar devices; hole transport layer materials in devices; environment, such as humidity, temperature, etc.

4.1 Electronic Transmission Layer Material

Electronic transport layer (ETL) materials refer to materials that accept and transmit negatively charged electronic carriers, a semiconductor material (i.e., N-type semiconductor) has a higher electron affinity and is commonly used as ETL.

The ideal ETL should have the following factors: (1) The frontier orbits of molecules in ETL should match the conduction and valence bands of perovskite layers; (2) High Electron Mobility; (3) The solubility of ETL in the orthogonal solvent is much higher than the perovskite layer to prevent damage of the Perovskite layer after spinning; (4) Full film coverage; and (5) Ability to passivate defects in perovskite layers.

Electrons in the perovskite layer are transferred to ETL, which is usually collected by metal electrodes [33]. An important feature of ETL is that it can be aligned with the perovskite. For the perovskite layer to be able to absorb light effectively, its UV–visible transmittance must be high [34].

4.1.1 Metal Oxide Electron Transport Layer Material

The electron transport layer of perovskite solar cells studied at present mostly refers to the commonly used TiO_2 in dye-sensitized solar cells [26, 28]. In order to increase the growth reaction speed of perovskite materials, TiO_2 thin films with high specific surface area are usually prepared using nano-sized TiO_2 particles. Dense titanium dioxide can transfer electrons, which is beneficial to both enhancing the light absorption inside the solar cell and improving the conductivity of electrons. However, the porous TiO_2 is easily affected by ultraviolet light, which directly causes a decrease in solar cell efficiency [35]. To avoid the inherent defect of porous titanium dioxide, porous titanium dioxide can be replaced or modified with other materials [36].

Snaith et al. [37] Replaced porous TiO_2 with porous Al_2O_3, thus avoiding its inherent instability. There is still a stable photocurrent when the solar cell is irradiated with the whole solar spectrum for more than 1000 h. The prepared solar cell has good stability and can maintain the original efficiency of nearly 90% after being exposed to air for 500 h without packaging [38].

SnO_2 is another well-known ETL material because of its high transparency, electron mobility, and wide band gap [39]. As a perovskite ETL material, SnO_2 can be stable in ambient air for 30 days after low-temperature treatment [40]. Low-temperature atomic layer based on SnO_2 is relatively better than TiO_2 [41]. WO_3 has good stability and can resist acid corrosion, which has higher mobility than TiO_2 [42].

4.1.2 Organic Electron Transport Layer Materials

Organic electron transport layer materials can be divided into two categories: The first one is n-type organic small molecules, including naphthalene diimide (NDI) and fullerene derivatives: [6]-phenyl-C61-butyric acid methyl ester (PCBM), etc., and the second category is n-type conjugated polymers. Due to the advantages of easy synthesis and purification, adjustable frontier molecular orbital, high electron mobility, low cost, and good chemical/thermal stability, these materials have become a promising ETL material in inverted perovskite solar cells (p–i–n PSC) [43]. Due to the reliability of PCBM as ETL material, Snaith and colleagues demonstrated the experiment of PCBM as ETL material to explore the charge diffusion length in $MAPbI_3$ for the first time [37]. In addition, Lam and colleagues used PCBM as ETL to understand the origin of high performance of $MAPbI_3$ solar cells, and found that PCBM plays a decisive role in inhibiting the migration of halide ions from perovskite layer through halide-π non-covalent interaction [44].

4.1.3 Composite Electron Transport Layer Material

The material using more than one different material as an electron transport layer is classified as a composite material.

Taking TiO_2(compact)/Al_2O_3(mesoporous) electron transport layer as an example, trihalide permeates into the Al_2O_3 nanoparticles skeleton with a thickness of about 400 nm, which can improve the uniformity of perovskite film and inhibit the leakage caused by the appearance of pinholes [45]. Electrons are transported to the dense TiO_2 electron transport layer in the calcium titanium ore body along the surface of the Al_2O_3 nanoframework [25], Using Al_2O_3 instead of TiO_2 as nanoframework can reduce the sintering temperature to 150 °C and obtain devices with an efficiency of 10.9%.

Abrusci et al. [42] introduced benzoic acid substituted C_{60} self-assembled monolayer as a modified layer on the surface of TiO_2 framework, which significantly improved the electron collection efficiency, and increased the photoelectric conversion efficiency of the solar cell to 11.7%. The coating layer can act as a barrier to prevent or reduce exciton recombination, thus reducing the degradation of perovskite material on porous TiO_2 surface and making the device more stable. Ito et al. [46] prepared a layer of Sb_2S_3 on TiO_2 layer by chemical deposition method, which improved the efficiency and photostability of the solar cell in the form of interfacial barrier layer. Under the same conditions, the device covered with Sb_2S_3 barrier retained 65% of the original efficiency after 12 h [47].

Wang et al. [48] used graphene/TiO_2 nanoparticle composites as the electron transport layer. Because graphene has the characteristics of high conductivity and its work function (WF) lies between those of Fluorine-doped Tin Oxide) (FTO) and Titanium dioxide (TiO_2), the electron transport performance is greatly improved, and the photoelectric conversion efficiency of the solar cell is as high as 15.6%. In addition, Zhou et al. [49] doped yttrium in TiO_2 to obtain 19.3% planar heterojunction perovskite solar cells. In addition to the above-mentioned improvement of the performance of perovskite solar cells by doping to obtain composite electron transport layer materials, Table 1 lists some examples of metal ion doped ETL materials.

Table 1 Metal ion-doped ETL materials for perovskite solar cells. Reprinted from Rao et al. [50], with permission from Elsevier. Copyright (2021).

Metal doped ETL	Deposition technique	J_K (mA cm)	V_c (V)	PCE (%)
Mg–TiO, natiorod	Microwave-assisted hydrothermal method	10.4	0.S02	4.17
Nb–TiOj	Spray pyrolysis	18.08	0.880	10.26
Nb–TiOs nanoiod	Hydrothermal	16.50	0.87	7.5
Al–$Ti0_2$	Spin coating	20.0	1.07	13.8
Mg–$Ti0_2$	Spin coating	18.34	1.08	12.23
Zr–TIO	Spin coating	14.4	0.93	9.1
Mg–TiO_a	Dip coaling	21.48	1.09	17.65
Cs–TIO	Spin coating	14.4	0.64	5.26
Nd–TiOa	Spin coating	22.3	1.133	18.2
Li–$Ti0_2$	Spin coating	23	1.114	19.3
Ta–TIO	Spray pyrolysis	19.01	1.01	13.76
Nb–TiO_a		19.10	1.00	13.81
Sm–TiO_2	Spin coating	18.07	1.04	13.1
Nb–TiO_2	Spin coating	22.25	1.08	16.96
Z_a–TiO_j	Spin coating	22.3	0.95	14
Nb–TiO_z	Chemical bath deposition	22.86	1.10	19.23
P.ii–$Ti0_2$	Spray pyrolysis	21.91	0 994	15.7
Sn–TIO	Co-electrodeposition	20.84	1.072	16.83
Ce–TiO_2	Hot-bach deposition	21.95	1.07	15
Co–TIO	Spin coating	24.078	1.027	15.75

4.2 Hole Transport Layer Materials

Hole transport layer (HTL) is another important component of PSC. As a hole transport layer, HTL must meet the following conditions: (1) Highest occupied molecular orbital (HOMO) energy level should be higher than the maximum value of valence band of perovskite material, so as to transport holes from perovskite layer to metal electrode; (2) It has high conductivity, which can reduce the series resistance and improve the Fill factor (FF); (3) The HTL layer and the calcium–titanium layer need to be in close contact [51]. The selection of HTL has a significant impact on the stability of PSCs. It is reported that the presence of HTL promotes better coverage of perovskite compared with transparent conducting oxide (TCO)/perovskite/metal [52, 53]. The use of hole transport layer (HTL) material has preliminarily solved the problems of instability and difficult packaging of liquid electrolyte perovskite solar cell.

4.2.1 Organic Hole Transport Layer Materials

Generally speaking, organic materials not only have good film-forming properties, but also have excellent hydrophobic properties, which can effectively block the erosion of water vapor to perovskite materials. Therefore, in order to solve the stability problem of perovskite solar cells, Kim et al. introduced a solid-state hole transport layer material (spiro-OMeTAD [2,2′,7,7′-tetrakis(*N*,*N*-di-*p*-methoxyphenyl-amine)9,9′-spirobifluorene]) into perovskite solar cells and prepa.red the first all-solid-state perovskite solar cells in 2012, with an efficiency of 9.7%. The results show that even without packaging, the efficiency attenuation of the solar cell is still very small after 500 h of operation [54]. In addition, the development of hole transport system with non-lithium salt additives will help to improve the stability of devices [35]. As a hole transport layer, poly[2,5-bis(2-decyldodecyl)pyrrolo[3,4-*c*]pyrrole-1,4(2*H*,5*H*)-dione-(*E*)-1,2-di(2,2′-bithiophen-5-yl) ethene] (PDPPDBTE) polymer can almost maintain the initial efficiency after being exposed to air for 1000 h without packaging [55]. Organic materials such as PTAA (poly[bis(4-phenyl) (2,4,6-trimethylphenyl) amine]) and PEDOT: PSS (poly(3,4-ethylene dioxythiophene): polystyrene sulfonate) polymer can also effectively isolate water vapor infiltration and improve system stability and other properties.

Poly[bis(4-phenyl)(2,4,6-trimethylphenyl)amine (PTAA) is a widely used amorphous polymer. It has been reported that by using PTAA as HTL to manufacture PSC, the PCE of solar cell is more than 20%. Zhao et al. [56] used undoped PTAA as HTL, tried PTAA with different molecular weights, analyzed its influence on perovskite layer deposition, and obtained 18.11% PCE. Their work shows that undoped low molecular weight PTAA can easily develop efficient PSC, which will contribute to the development of commercial PSC. PEDOT:PSS is a water-soluble polymer HTL material with a HOMO energy level of −5.5 eV [50]. PEDOT: PSS as a hole conductor has been widely used in inverted PSCs to obtain high PCE. However, poly(3,4-ethylenedioxythiophene) polystyrene sulfonate (PEDOT:PSS) will degrade PSCs due to its low electron blocking ability and poor chemical stability caused by moisture absorption [57]. Wang et al. [48] modified the surface of PEDOT: PSS with sodium citrate ($Na_3C_6H_5O_7$), which removed part of polystyrene sulfonate (PSS) on the surface of PEDOT: PSS and helped perovskite crystallize in a larger area. In order to reduce the acidity of its PSS group, PEDOT: PSS was treated with ammonia, which not only improved the performance of the corresponding PSC but also enhanced the crystal structure of perovskite films. The lower hole transport efficiency of PEDOT: PSS will lead to the imbalance of charge transport in PSC and eventually lead to hysteresis. The surface treatment of PEDOT: PSS with ethylene glycol and methanol will help to reduce the influence of PSS weak ion conductor, enhance the size of perovskite crystal domain.

4.2.2 Inorganic Hole Transport Layer Materials

Inorganic hole transport layer (HTL) materials have become popular selective contact materials because of their inherent chemical stability and low cost. Inorganic HTL materials have the characteristics of appropriate energy levels and high carrier mobility. They can not only assist charge transport but also improve the stability of PSCs under environmental conditions [57].

The application of copper (I) oxide (Cu_2O) as inorganic HTL in PSCs has been proved to improve the stability of devices stored in glove boxes [58, 59]. In 2016, it was reported that n–i–p heterojunction solar cells composed of FTO/TiO_2/$CH_3NH_3PbI_{3-x}Cl_x$/Cu_2O/Au existed stably at room temperature for 2 months, and the initial efficiency of the device was 90% [60], Fig. 3 shows the structure of an n–i–p solar cell.

Copper(I) iodide (CuI) has high hole mobility and good solubility in organic solvents. It is a good candidate material for inorganic HTL materials [62]. The device made of inorganic compound CuI as the hole transport layer is exposed to continuous light in the air for 2 h without package, and the current basically remained unchanged [63]. Moreover, when CuI is used as the hole transport layer, its efficiency is reduced and it can be recovered. After being placed in the dark for a period of time, its efficiency can basically return to the initial value. Nickel oxide (NiO_x) HTL, which is widely used in PSCs, also shows good stability. In 2015, Nejand et al. [64] found that the long-term durability of NiO_x storage in the environment was proved to be 2 months when applied to equipment as HTLs.

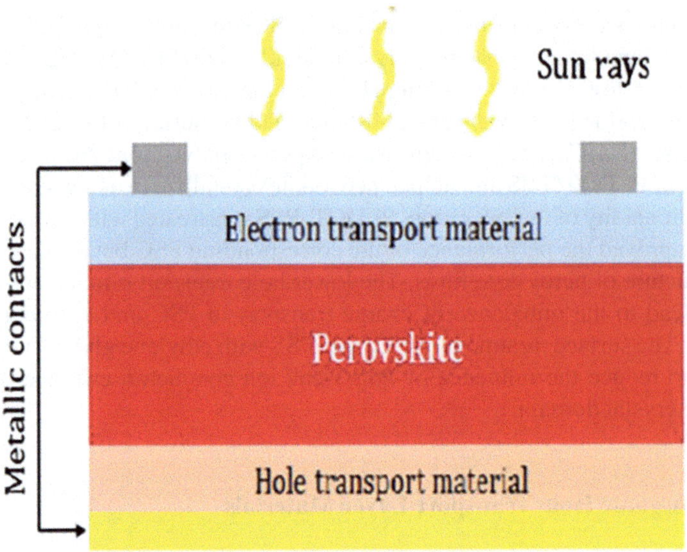

Fig. 3 The n–i–p perovskite Solar cell structure. Reprinted from Azri et al. [61], with permission from Elsevier. Copyright (2019)

4.3 Environmental Factors

In addition to the influence of electron transport layer (ETL) and hole transport layer (HTL) on the stability of PSCs, there are also environmental factors, such as humidity and temperature, which can not be ignored [65].

4.3.1 Humidity

The degradation of perovskite layer is a chemical process, because perovskite material is very sensitive to water and has polarity, so it is easy to cause hydrolysis in humid environment. Studies have shown that when $CH_3NH_3PbI_3$ is exposed to humid environment, PbI_2, $CH_3NH_2I_3$, and HI will be formed. In the process of perovskite degradation, Walsh et al. [66] combined water molecules (as Lewis base) with perovskite chemistry, received an H^+ from ammonium and degraded it through an intermediate step. The intermediate is degraded to CH_3NH_2, HI, and finally, PbI_2, in which HI and solid PbI_2 are dissolved in water. The decomposition of PbI_2 leads to further toxicity. CH_3NH_2 is a polar organic compound, so it is highly volatile and soluble in water [26, 66, 67].

The greater the air humidity, the worse the stability of perovskite solar cells. As shown in Fig. 4, an experimental device designed by researchers is used to control and measure the relative humidity of the film. The facility is considered essential for effective degradation studies.

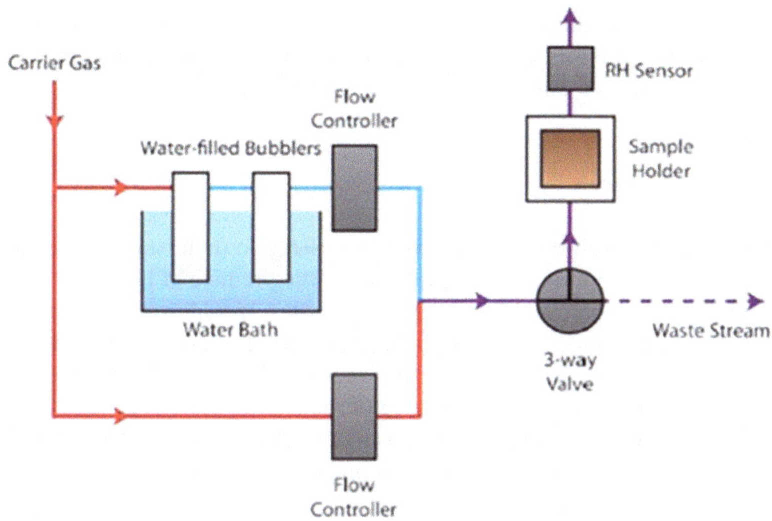

Fig. 4 Schematic representation of the system used to control relative humidity. Reprinted from Wang et al. [68], with permission from Elsevier. Copyright (2016)

4.3.2 Temperature

According to international standards, PSCs need to be stable at 85 °C for a long time in order to compete with silicon solar cells. Temperature is one of the important parameters that determine the overall efficiency of PSCs. In general, solar cells must be able to work in hot climate [50].

Leong et al. [69] studied the performance of m-TiO_2 based PSCs in the temperature range of 80–360 K to study the relationship between temperature and device performance. When $T < 330$ K, the efficiency shows an increasing trend, while when $T > 330$ K, the efficiency decreases significantly.

Under light and high temperature, the degradation of PSCs strongly depends on the composition of halide perovskite. For example, $MAPbI_3$ decomposed significantly when exposed to sample temperatures of 45–55 °C, while $MAPbBr_3$ did not. The stability of PSCs can be further improved by introducing transparent contact agents and strong perovskite components. After 160 h of continuous operation under nitrogen environment and light conditions at 85 °C, PSCs showed an initial efficiency of more than 90% [70].

Perovskite solar cells have developed rapidly in the past decade, and their efficiency has increased from 3.8% to 23.3%. In the practical application process of perovskite solar cells, in addition to improving the photoelectric conversion efficiency of the cells, the stability of the cells can be continuously improved by finding suitable materials and optimizing the preparation method of perovskite film. In addition, the preparation of large-area perovskite thin films is still an important factor restricting the large-scale and commercialization of perovskite solar cells. Therefore, the research and preparation of reliable, stable, and efficient perovskite solar cells are still a systematic, multidisciplinary, and complex project, which requires constant investment of funds and unswervant determination of researchers.

4.3.3 Other Factors

For perovskite solar cell devices, it is very necessary to understand device attenuation mechanism and optimize its stability under the premise of high efficiency. The influencing factors of perovskite stability are complex, In addition to temperature and humidity, light irradiation [71] (ultraviolet light), oxygen and the presence of electric field can cause degradation of perovskite solar cells, thus reducing their stability. In addition, the structure and performance of perovskite solar cells also affect their stability. Therefore, the selection of appropriate devices and reasonable packaging of perovskite solar cells are very critical to improve their stability [71, 72].

5 Industrialization and Challenges of Perovskite Solar Cells

In recent years, the energy conversion efficiency of perovskite solar cells has become comparable to that of conventional commercial solar cells. Due to their light weight and low production cost, perovskite solar cells have gradually replaced traditional solar cells, and many improvements are expected to break through higher efficiency. However, despite the promotion of further development of photovoltaic technology, the realization of large-scale production and commercialization of the perovskite solar cell, there are still some problems to be solved.

On the one hand, solar cells must have the ability to operate in harsh environments such as high humidity, high temperature, strong light, and changeable weather to adapt to different conditions. In order to successfully realize the commercialization of perovskite solar cell technology, scientists are committed to solving several fundamental problems: (1) Controllable growth and deposition of thin film materials; (2) High stability and long life; (3) Scale and repeatability; (4) Low toxicity; and (5) Low cost.

With low surface roughness and low density of structural defects, large-scale perovskite films are critical for industrial applications. A large number of researchers are committed to developing various film deposition technologies, including spin coating, dip coating, and spray deposition. The most widely used method for low-cost deposition of high-quality perovskite films is spin coating. However, this method could not achieve large-scale production because it is difficult to obtain large-area uniform films. The chemical vapor deposition can not only achieve a large area of film deposition but also eliminate the need for toxic organic solvents, which is a very developing film preparation process [73]. Wang et al. [74] introduced butylammonium cation into the mixed cation Pb mixed halide $FA_{0.83}CS_{0.17}Pb\ (Br_{1-y})_3$ to obtain a three-dimensional perovskite structure, which resulted in a considerable power conversion efficiency in the produced perovskite solar cell. It can maintain high stability in an analog illumination environment. Grancini [75] designed a 2D/3D perovskite structure, which not only achieved a high power conversion rate of 22% but also made an outstanding contribution to the improvement of the stability and conversion efficiency of perovskite solar cells.

The environment such as temperature, humidity, oxygen, etc., will cause the rapid aging of solar cells. Attempts have been made to adapt perovskite processing and solar cells to ambient humidity for mass production. Doping perovskite with halide ions, increasing hydrophobicity and interface modification are the main methods to make perovskite solar cells compatible with environmental humidity. In addition, proper packaging of the cell can effectively reduce the aging speed of the solar cell and improve its service life. Packaging technology can not only avoid the harm caused by Pb toxicity but also effectively improve the stability of the solar cell. Hashmi et al. [76] sealed the solar cell with epoxy glue and confirmed that the encapsulated solar cell still maintained good stability after the 1002 h aging test. This study shows that the commercial market prospect of perovskite solar cells is very considerable.

All customers, manufacturers, and investors are very concerned about the toxicity and safety of Pb-containing photovoltaic equipment. The toxicity of Pb has brought great limitations to the development of perovskite solar cells. In recent years, researchers have developed lead-free alternatives after extensive research. slavney et al. [77] combined non-toxic Bi^{3+} into the perovskite lattice of $CsAgBiBr_6$ using a double perovskite structure, which improved the thermal and wet stability of the material. However, we must admit that Pb-free materials are not perfect in the field of perovskite solar cells because the performance of these devices is still not enough to meet the needs of industrial commercialization.

In addition, since the perovskite layer plays a decisive role in the solar cell structure, measurements often were taken to improve the efficiency and stability of the solar cell include (1) Optimization of the preparation method of perovskite films; (2) Changing the chemical composition of the solar cell; and (3) Improving the electron transport layer and hole transport layer to better separate electrons and holes and smoothly reach the two poles to form current, so as to improve the stability of the solar cell and improve its service efficiency. 2D perovskite thin films have strong light absorption in the visible region and strong photoluminescence properties at room temperature. As a perovskite solar cell material, it can maintain its performance even under long-term exposure in high humidity environment [78]. Since the solar cell can be dismantled, timely recycling of the main components of the solar cell can not only save raw materials and production time, but also reduce the pollution of solid waste to the environment [79], According to the investigation, the Pb concentration in perovskite solar cells is as high as 0.55%, and the recovery of Pb solved the great problem caused by the toxicity of Pb components [80]. As shown in Fig. 5, Lichong et al. [81] recovered Pb from lead acid batteries for the preparation of perovskite films, which reduced lead pollution to the environment and greatly reduced resource waste.

The life cycle of a solar cell consists of four parts: raw materials, module manufacturing, module operation, and processing, which together constitute the entire

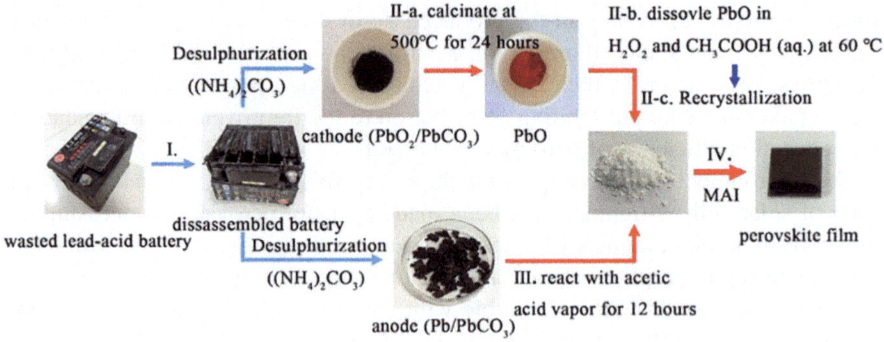

Fig. 5 Schematic diagram of lead recovery from a lead-acid battery. Reprinted from Li et al. [81], with permission from Elsevier. Copyright (2019)

life cycle of electronic equipment. When discussing the application and commercialization of equipment in reality, it is particularly important to comprehensively consider each step. Therefore, in addition to finding suitable materials as raw materials for batteries and solar cells, many factors such as battery/solar cell manufacturing cost, preparation process, and the impact of residues on the environment should be considered comprehensively.

6 Conclusion and Outlook

As a new generation of green cells, perovskite solar cells have gradually replaced crystalline silicon solar cells and occupy a certain position in the solar cell market. Since 2009, perovskite solar cells have developed rapidly because of their simple preparation process, rich raw materials, low cost, and environment-friendly. Organic–inorganic perovskite type light absorbers have made great progress in the past decade. The power conversion efficiency of single-junction solar cells using perovskite light absorbers has reached more than 25%. Some researchers even reported how to adjust the subsequent film thickness and performance to achieve an efficiency of more than 30%. During the development of perovskite solar cells, it has been found that the metal Pb in the solar cells is toxic and pollutes the environment, and the existence of Pb composition will also affect the stability of the cells. Therefore, improving the stability of solar cells and promoting the green and sustainable development of the solar cells are very important for the large-scale production and commercialization of the cells. Based on this, a large number of studies are devoted to the modification and interface modification of perovskite and the search for alternative materials.

Compared with other photovoltaic technologies, PSCs have a huge impact on global energy production. With their simple production process and low material cost, PSCs have made significant progress in conversion efficiency, achieved high efficiency in the laboratory, and will continue to mature with increased stability.

While achieving high efficiency, solving the problem of solar cell stability is very important to realize the practical application of high-efficiency PSCs. Although there are some limitations in the development of perovskite solar cells. Perovskite cells are still a kind of rapid development and great potential solar cells in the present and near future for a long time. We believe that with the continuous innovation of technology, perovskite solar cells will break through a series of problems currently facing, solve the energy crisis and environmental pollution problems, and achieve full coverage of new energy solar cells in the future.

References

1. P. Roy, N. Kumar Sinha, S. Tiwari, A. Khare, A review on perovskite solar cells: evolution of architecture, fabrication techniques, commercialization issues and status. Sol. Energy **198,** 665–688 (2020). https://doi.org/10.1016/j.solener.2020.01.080
2. A. ToshniwalV, Kheraj, development of organic-inorganic tin halide perovskites: a review. Sol. Energy **149**, 54–59 (2017). https://doi.org/10.1016/j.solener.2017.03.077
3. N.-G. Park, Perovskite solar cells: an emerging photovoltaic technology. Mater. Today **18**, 65–72 (2015). https://doi.org/10.1016/j.mattod.2014.07.007
4. Z. Song, S.C. Watthage, A.B. Phillips, M.J. Heben, Pathways toward high-performance perovskite solar cells: Review of recent advances in organo-metal halide perovskites for photovoltaic applications. J. Photon. Energy. **6** (2016). https://doi.org/10.1117/1.Jpe.6.022001
5. K.u.c. Zusammensetzung, Krystallbau und chemische zusammensetzung. Ber. Dtsch. Chem. Ges. **60**, 1263–1268 (1927)
6. M.A. Green, A. Ho-BaillieH, J. Snaith, The emergence of perovskite solar cells. Nat. Photonics **8**, 506–514 (2014). https://doi.org/10.1038/nphoton.2014.134
7. C. Li, X. Lu, W. Ding, L. Feng, Y. Gao, Z. Guo, Formability of abx3 (x = f, cl, br, i) halide perovskites. Acta Crystallogr. B **64**, 702–707 (2008). https://doi.org/10.1107/S01087681080 32734
8. N.K. McKinnon, D.C. ReevesM, H. Akabas, 5-ht3 receptor ion size selectivity is a property of the transmembrane channel, not the cytoplasmic vestibule portals. J. Gen. Physiol. **138**, 453–466 (2011). https://doi.org/10.1085/jgp.201110686
9. J.-P. Correa-Baena, A. Abate, M. Saliba, W. Tress, T. Jesper Jacobsson, M. Grätzel, A. Hagfeldt, The rapid evolution of highly efficient perovskite solar cells. Energ. Environ. Sci. **10**, 710–727 (2017). https://doi.org/10.1039/c6ee03397k
10. A. Kojima, K. Teshima, Y. Shirai, T. Miyasaka, Organometal halide perovskites as visible-light sensitizers for photovoltaic cells. J. Am. Chem. Soc. **131**, 6050–6051 (2009). https://doi.org/ 10.1021/ja809598r
11. J. H. Im, C. R. Lee, J. W. Lee, S. W. Park, N. G. Park, 6.5% efficient perovskite quantum-dot-sensitized solar cell. Nanoscale. **3**, 4088–93 (2011). https://doi.org/10.1039/c1nr10867k
12. H.S. Kim, J.W. Lee, N. Yantara, P.P. Boix, S.A. Kulkarni, S. Mhaisalkar, M. Gratzel, N.G. Park, High efficiency solid-state sensitized solar cell-based on submicrometer rutile tio2 nanorod and ch3nh3pbi3 perovskite sensitizer. Nano Lett. **13**, 2412–2417 (2013). https://doi.org/10.1021/ nl400286w
13. J. A. Christians, R. C. FungP. V. Kamat, An inorganic hole conductor for organo-lead halide perovskite solar cells. Improved hole conductivity with copper iodide. J. Am. Chem. Soc. **136,** 758–64 (2014). https://doi.org/10.1021/ja411014k
14. M. Hou, H. Zhang, Z. Wang, Y. Xia, Y. Chen, W. Huang, Enhancing efficiency and stability of perovskite solar cells via a self-assembled dopamine interfacial layer. ACS. Appl. Mater. Inter. **10**, 30607–30613 (2018). https://doi.org/10.1021/acsami.8b10332

15. M.M. Tavakoli, M. Saliba, P. Yadav, P. Holzhey, A. Hagfeldt, S.M. Zakeeruddin, M. Grätzel, Synergistic crystal and interface engineering for efficient and stable perovskite photovoltaics. Adv. Energy. Mater. **9** (2019). https://doi.org/10.1002/aenm.201802646
16. Q. Jiang, Y. Zhao, X. Zhang, X. Yang, Y. Chen, Z. Chu, Q. Ye, X. Li, Z. Yin, J. You, Surface passivation of perovskite film for efficient solar cells. Nat. Photonics **13**, 460–466 (2019). https://doi.org/10.1038/s41566-019-0398-2
17. J. Tong, Z. Song, D.H. Kim, X. Chen, C. Chen, A.F. Palmstrom, P.F. Ndione, M.O. Reese, S.P. Dunfield, O.G. Reid, J. Liu, F. Zhang, S.P. Harvey, Z. Li, S.T. Christensen, G. Teeter, D. Zhao, M.M. Al-Jassim, M. van Hest, M.C. Beard, S.E. Shaheen, J.J. Berry, Y. Yan, K. Zhu, Carrier lifetimes of >1 mus in sn-pb perovskites enable efficient all-perovskite tandem solar cells. Science **364**, 475–479 (2019). https://doi.org/10.1126/science.aav7911
18. K. Xiao, R. Lin, Q. Han, Y. Hou, Z. Qin, H. T. Nguyen, J. Wen, M. Wei, V. Yeddu, M. I. Saidaminov, Y. Gao, X. Luo, Y. Wang, H. Gao, C. Zhang, J. Xu, J. Zhu, E. H. Sargent, and H. Tan, All-perovskite tandem solar cells with 24.2% certified efficiency and area over 1 cm2 using surface-anchoring zwitterionic antioxidant. Nat. Energy. **5** 870–880 (2020). https://doi.org/10.1038/s41560-020-00705-5
19. E. Jokar, C.-H. Chien, A. Fathi, M. Rameez, Y.-H. Chang, E.W.-G. Diau, Slow surface passivation and crystal relaxation with additives to improve device performance and durability for tin-based perovskite solar cells. Energ. Environ. Sci. **11**, 2353–2362 (2018). https://doi.org/10.1039/c8ee00956b
20. K. Nishimura, M.A. Kamarudin, D. Hirotani, K. Hamada, Q. Shen, S. Iikubo, T. Minemoto, K. Yoshino, and S. Hayase, Lead-free tin-halide perovskite solar cells with 13% efficiency. Nano Energy. **74** (2020). https://doi.org/10.1016/j.nanoen.2020.104858
21. F. Hao, C.C. Stoumpos, R.P. Chang, M.G. Kanatzidis, Anomalous band gap behavior in mixed sn and pb perovskites enables broadening of absorption spectrum in solar cells. J. Am. Chem. Soc. **136**, 8094–8099 (2014). https://doi.org/10.1021/ja5033259
22. R. Lin, K. Xiao, Z. Qin, Q. Han, C. Zhang, M. Wei, M. I. Saidaminov, Y. Gao, J. Xu, M. Xiao, A. Li, J. Zhu, E. H. Sargent, and H. Tan, Monolithic all-perovskite tandem solar cells with 24.8% efficiency exploiting comproportionation to suppress sn(ii) oxidation in precursor ink. Nat. Energy. **4** 864–873 (2019). https://doi.org/10.1038/s41560-019-0466-3
23. J.G. BednorzK, A. Müller, Perovskite-type oxides—the new approach to high-tcsuperconductivity. Rev. Mod. Phys. **60**, 585–600 (1988). https://doi.org/10.1103/RevModPhys.60.585
24. T. Zhang, Y. Zhao, Recent progress of lead halide perovskite sensitized solar cells. Acta Chim. Sinica. **73** (2015). https://doi.org/10.6023/a14090656
25. M.M. Lee, J. Teuscher, T. Miyasaka, T.N. Murakami, H.J. Snaith, Efficient hybrid solar cells based on meso-superstructured organometal halide perovskites. Science **338**, 643–647 (2012). https://doi.org/10.1126/science.1228604
26. J. Burschka, N. Pellet, S.J. Moon, R. Humphry-Baker, P. Gao, M.K. Nazeeruddin, M. Gratzel, Sequential deposition as a route to high-performance perovskite-sensitized solar cells. Nature **499**, 316–319 (2013). https://doi.org/10.1038/nature12340
27. Q. Chen, H. Zhou, Z. Hong, S. Luo, H.S. Duan, H.H. Wang, Y. Liu, G. Li, Y. Yang, Planar heterojunction perovskite solar cells via vapor-assisted solution process. J. Am. Chem. Soc. **136**, 622–625 (2014). https://doi.org/10.1021/ja411509g
28. M. Liu, M.B. JohnstonH, J. Snaith, Efficient planar heterojunction perovskite solar cells by vapour deposition. Nature **501**, 395–398 (2013). https://doi.org/10.1038/nature12509
29. S. C. Watthage, Z. Song, A. B. Phillips, and M. J. Heben, *Evolution of perovskite solar cells*, in *Perovskite photovoltaics*. ed. by S.Thomas, and A. Thankappan (Academic Press, 2018),pp. 43–88.
30. C.-H. Chiang, J.-W. LinC.-G. Wu, One-step fabrication of a mixed-halide perovskite film for a high-efficiency inverted solar cell and module. J. Mater. Chem. A. **4** 13525–13533 (2016).https://doi.org/10.1039/c6ta05209f
31. H. Shen, Y. Wu, J. Peng, T. Duong, X. Fu, C. Barugkin, T.P. White, K. Weber, K.R. Catchpole, Improved reproducibility for perovskite solar cells with 1 cm(2) active area by a modified

two-step process. ACS. Appl. Mater. Inter. **9**, 5974–5981 (2017). https://doi.org/10.1021/acsami.6b13868
32. K. Wang, C. Liu, P. Du, J. Zheng, X. Gong, Bulk heterojunction perovskite hybrid solar cells with large fill factor. Energ. Environ. Sci. **8**, 1245–1255 (2015). https://doi.org/10.1039/c5ee00222b
33. K. Mahmood, S. SarwarM, T. Mehran, Current status of electron transport layers in perovskite solar cells: Materials and properties. RSC Adv. **7**, 17044–17062 (2017). https://doi.org/10.1039/c7ra00002b
34. T. Leijtens, G.E. Eperon, S. Pathak, A. Abate, M.M. Lee, H.J. Snaith, Overcoming ultraviolet light instability of sensitized tio(2) with meso-superstructured organometal tri-halide perovskite solar cells. Nat. Commun. **4**, 2885 (2013). https://doi.org/10.1038/ncomms3885
35. D.-F. Zhang, L.-L. Zheng, Y.-Z. Ma, S.-F. Wang, Z.-Q. Bian, C.-H. Huang, Q.-H. Gong, and L.-X. Xiao, Factors influencing the stability of perovskite solar cells. Acta Phys. Sin. **64,** (2015). https://doi.org/10.7498/aps.64.038803
36. G. Niu, X. GuoL, Wang, Review of recent progress in chemical stability of perovskite solar cells. J. Mater. Chem. A. **3**, 8970–8980 (2015). https://doi.org/10.1039/c4ta04994b
37. C. Wehrenfennig, G.E. Eperon, M.B. Johnston, H.J. Snaith, L.M. Herz, High charge carrier mobilities and lifetimes in organolead trihalide perovskites. Ad. Mater. **26**, 1584–1589 (2014). https://doi.org/10.1002/adma.201305172
38. G. Yang, H. Tao, P. Qin, W. Ke, G. Fang, Recent progress in electron transport layers for efficient perovskite solar cells. J. Mater. Chem. A. **4**, 3970–3990 (2016). https://doi.org/10.1039/c5ta09011c
39. J. Song, E. Zheng, J. Bian, X.-F. Wang, W. Tian, Y. Sanehira, T. Miyasaka, Low-temperature sno2-based electron selective contact for efficient and stable perovskite solar cells. J. Mater. Chem. A. **3**, 10837–10844 (2015). https://doi.org/10.1039/c5ta01207d
40. K. Mahmood, B.S. Swain, A.R. Kirmani, A. Amassian, Highly efficient perovskite solar cells based on a nanostructured wo3–tio2core–shell electron transporting material. J. Mater. Chem. A. **3**, 9051–9057 (2015). https://doi.org/10.1039/c4ta04883k
41. H. Zheng, Y. TachibanaK, Kalantar-Zadeh, Dye-sensitized solar cells based on wo3. Langmuir **26**, 19148–19152 (2010). https://doi.org/10.1021/la103692y
42. A. Abrusci, S.D. Stranks, P. Docampo, H.L. Yip, A.K. Jen, H.J. Snaith, High-performance perovskite-polymer hybrid solar cells via electronic coupling with fullerene monolayers. Nano Lett. **13**, 3124–3128 (2013). https://doi.org/10.1021/nl401044q
43. A.A. Said, J. XieQ, Zhang, Recent progress in organic electron transport materials in inverted perovskite solar cells. Small **15**, e1900854 (2019). https://doi.org/10.1002/smll.201900854
44. S. Sun, T. Salim, N. Mathews, M. Duchamp, C. Boothroyd, G. Xing, T.C. Sum, Y.M. Lam, The origin of high efficiency in low-temperature solution-processable bilayer organometal halide hybrid solar cells. Energy Environ. Sci. **7**, 399–407 (2014). https://doi.org/10.1039/c3ee43161d
45. Y. Ogomi, K. Kukihara, S. Qing, T. Toyoda, K. Yoshino, S. Pandey, H. Momose, S. Hayase, Control of charge dynamics through a charge-separation interface for all-solid perovskite-sensitized solar cells. ChemPhysChem **15**, 1062–1069 (2014). https://doi.org/10.1002/cphc.201301153
46. S. Ito, S. Tanaka, K. Manabe, H. Nishino, Effects of surface blocking layer of sb2s3 on nanocrystalline tio2 for ch3nh3pbi3 perovskite solar cells. J. Phys. Chem. C **118**, 16995–17000 (2014). https://doi.org/10.1021/jp500449z
47. H.-K. Ting, L. Ni, S.-B. Ma, Y.-Z. Ma, L.-X. Xiao, and Z.-J. Chen, Progress in electron-transport materials in application of perovskite solar cells. Acta Phys. Sin. **64,** (2015). https://doi.org/10.7498/aps.64.038802
48. Y. Wang, Y. Hu, D. Han, Q. Yuan, T. Cao, N. Chen, D. Zhou, H. Cong, L. Feng, Ammonia-treated graphene oxide and pedot: Pss as hole transport layer for high-performance perovskite solar cells with enhanced stability. Org. Electron. **70**, 63–70 (2019). https://doi.org/10.1016/j.orgel.2019.03.048
49. H. Zhou, Q. Chen, G. Li, S. Luo, T. B. Song, H. S. Duan, Z. Hong, J. You, Y. Liu, and Y. Yang, Photovoltaics. Interface engineering of highly efficient perovskite solar cells. Science. **345,** 542–6 (2014). https://doi.org/10.1126/science.1254050

50. M.K. Rao, D.N. Sangeetha, M. Selvakumar, Y.N. Sudhakar, M.G. Mahesha, Review on persistent challenges of perovskite solar cells' stability. Sol. Energy **218**, 469–491 (2021). https://doi.org/10.1016/j.solener.2021.03.005
51. X. Yao, Y.-L. Ding, X.-D. Zhang, and Y. Zhao, A review of the perovskite solar cells. Acta Phys. Sin. **64,** (2015). https://doi.org/10.7498/aps.64.038805
52. K. Domanski, J.P. Correa-Baena, N. Mine, M.K. Nazeeruddin, A. Abate, M. Saliba, W. Tress, A. Hagfeldt, M. Gratzel, Not all that glitters is gold: Metal-migration-induced degradation in perovskite solar cells. ACS Nano **10**, 6306–6314 (2016). https://doi.org/10.1021/acsnano.6b02613
53. Q. Wei, H. Bi, S. Yan, and S. Wang, Morphology and interface engineering for organic metal halide perovskite-based photovoltaic cells. Adv. Mater. Interfaces. **5,** (2018). https://doi.org/10.1002/admi.201800248
54. H.S. Kim, C.R. Lee, J.H. Im, K.B. Lee, T. Moehl, A. Marchioro, S.J. Moon, R. Humphry-Baker, J.H. Yum, J.E. Moser, M. Gratzel, N.G. Park, Lead iodide perovskite sensitized all-solid-state submicron thin film mesoscopic solar cell with efficiency exceeding 9%. Sci. Rep. **2**, 591 (2012). https://doi.org/10.1038/srep00591
55. Y. S. Kwon, J. Lim, H.-J. Yun, Y.-H. Kim, and T. Park, A diketopyrrolopyrrole-containing hole transporting conjugated polymer for use in efficient stable organic–inorganic hybrid solar cells based on a perovskite. Energ. Environ. Sci. **7,** (2014). https://doi.org/10.1039/c3ee44174a
56. Q. Zhao, R. Wu, Z. Zhang, J. Xiong, Z. He, B. Fan, Z. Dai, B. Yang, X. Xue, P. Cai, S. Zhan, X. Zhang, J. Zhang, Achieving efficient inverted planar perovskite solar cells with nondoped ptaa as a hole transport layer. Org. Electron. **71**, 106–112 (2019). https://doi.org/10.1016/j.orgel.2019.05.019
57. P. K. Kung, M. H. Li, P. Y. Lin, Y. H. Chiang, C. R. Chan, T. F. Guo, and P. Chen, A review of inorganic hole transport materials for perovskite solar cells. Adv. Mater. Interfaces. **5,** (2018). https://doi.org/10.1002/admi.201800882
58. K.M. Reza, A. Gurung, B. Bahrami, S. Mabrouk, H. Elbohy, R. Pathak, K. Chen, A.H. Chowdhury, M.T. Rahman, S. Letourneau, H.-C. Yang, G. Saianand, J.W. Elam, S.B. Darling, Q. Qiao, Tailored pedot: Pss hole transport layer for higher performance in perovskite solar cells: Enhancement of electrical and optical properties with improved morphology. J. Energy Chem. **44**, 41–50 (2020). https://doi.org/10.1016/j.jechem.2019.09.014
59. C. ZuoL, Ding, Solution-processed cu2o and cuo as hole transport materials for efficient perovskite solar cells. Small **11**, 5528–5532 (2015). https://doi.org/10.1002/smll.201501330
60. J. You, L. Meng, T.B. Song, T.F. Guo, Y.M. Yang, W.H. Chang, Z. Hong, H. Chen, H. Zhou, Q. Chen, Y. Liu, N. De Marco, Y. Yang, Improved air stability of perovskite solar cells via solution-processed metal oxide transport layers. Nat. Nanotechnol. **11**, 75–81 (2016). https://doi.org/10.1038/nnano.2015.230
61. F. Azri, A. Meftah, N. Sengouga, A. Meftah, Electron and hole transport layers optimization by numerical simulation of a perovskite solar cell. Sol. Energy **181**, 372–378 (2019). https://doi.org/10.1016/j.solener.2019.02.017
62. H. Lei, P. Qin, W. Ke, Y. Guo, X. Dai, Z. Chen, H. Wang, B. Li, Q. Zheng, G. Fang, Performance enhancement of polymer solar cells with high work function cus modified ito as anodes. Org. Electron. **22**, 173–179 (2015). https://doi.org/10.1016/j.orgel.2015.03.051
63. J. A Christians,, R. C. Fung, P. V. Kamat, An inorganic hole conductor for organo-lead halide perovskite solar cells. Improved hole conductivity with copper iodide. J Am. Chem. Soc. 136(2), 758-764.(2014). https://doi.org/10.1021/ja411014k.
64. B.A. Nejand, V. AhmadiH, R. Shahverdi, New physical deposition approach for low cost inorganic hole transport layer in normal architecture of durable perovskite solar cells. ACS. Appl. Mater. Inter. **7**, 21807–21818 (2015). https://doi.org/10.1021/acsami.5b05477
65. Q. Wali, F. J. Iftikhar, M. E. Khan, A. Ullah, Y. Iqbal, and R. Jose, Advances in stability of perovskite solar cells. Org. Electron. **78,** (2020). https://doi.org/10.1016/j.orgel.2019.105590
66. J.M. Frost, K.T. Butler, F. Brivio, C.H. Hendon, M. van Schilfgaarde, A. Walsh, Atomistic origins of high-performance in hybrid halide perovskite solar cells. Nano Lett. **14**, 2584–2590 (2014). https://doi.org/10.1021/nl500390f

67. N. Rajamanickam, S. Kumari, V.K. Vendra, B.W. Lavery, J. Spurgeon, T. Druffel, M.K. Sunkara, Stable and durable ch3nh3pbi3 perovskite solar cells at ambient conditions. Nanotechnology **27**, 235404 (2016). https://doi.org/10.1088/0957-4484/27/23/235404
68. D. Wang, M. Wright, N.K. Elumalai, A. Uddin, Stability of perovskite solar cells. Sol. Energ. Mat. Sol. C. **147**, 255–275 (2016). https://doi.org/10.1016/j.solmat.2015.12.025
69. W.L. Leong, Z.E. Ooi, D. Sabba, C. Yi, S.M. Zakeeruddin, M. Graetzel, J.M. Gordon, E.A. Katz, N. Mathews, Identifying fundamental limitations in halide perovskite solar cells. Ad. Mater. **28**, 2439–2445 (2016). https://doi.org/10.1002/adma.201505480
70. T. Duong, Y. Wu, H. Shen, J. Peng, S. Zhao, N. Wu, M. Lockrey, T. White, K. Weber, K. Catchpole, Light and elevated temperature induced degradation (letid) in perovskite solar cells and development of stable semi-transparent cells. Sol. Energ. Mat. Sol. C. **188**, 27–36 (2018). https://doi.org/10.1016/j.solmat.2018.08.017
71. J.-W. Lee, D.-H. Kim, H.-S. Kim, S.-W. Seo, S. M. Cho, N.-G. Park, Formamidinium and cesium hybridization for photo- and moisture-stable perovskite solar cell. Adv. Energy. Mater. **5** (2015). https://doi.org/10.1002/aenm.201501310
72. S. Guarnera, A. Abate, W. Zhang, J.M. Foster, G. Richardson, A. Petrozza, H.J. Snaith, Improving the long-term stability of perovskite solar cells with a porous al2o3 buffer layer. J. Phys. Chem. Lett. **6**, 432–437 (2015). https://doi.org/10.1021/jz502703p
73. L. Qiu, L.K. OnoY, Qi, Advances and challenges to the commercialization of organic–inorganic halide perovskite solar cell technology. Mater. Today Energy. **7**, 169–189 (2018). https://doi.org/10.1016/j.mtener.2017.09.008
74. Z. Wang, Q. Lin, F.P. Chmiel, N. Sakai, L.M. Herz, H. J. Snaith, Efficient ambient-air-stable solar cells with 2d–3d heterostructured butylammonium-caesium-formamidiniumd lead halide perovskites. Nat. Energy. **2** (2017). https://doi.org/10.1038/nenergy.2017.135
75. G. Grancini, C. Roldan-Carmona, I. Zimmermann, E. Mosconi, X. Lee, D. Martineau, S. Narbey, F. Oswald, F. De Angelis, M. Graetzel, M.K. Nazeeruddin, One-year stable perovskite solar cells by 2d/3d interface engineering. Nat. Commun. **8**, 15684 (2017). https://doi.org/10.1038/ncomms15684
76. S.G. Hashmi, A. Tiihonen, D. Martineau, M. Ozkan, P. Vivo, K. Kaunisto, V. Ulla, S.M. Zakeeruddin, M. Grätzel, Long term stability of air processed inkjet infiltrated carbon-based printed perovskite solar cells under intense ultra-violet light soaking. J. Mater. Chem. A. **5**, 4797–4802 (2017). https://doi.org/10.1039/c6ta10605f
77. A.H. Slavney, T. Hu, A.M. Lindenberg, H.I. Karunadasa, A bismuth-halide double perovskite with long carrier recombination lifetime for photovoltaic applications. J. Am. Chem. Soc. **138**, 2138–2141 (2016). https://doi.org/10.1021/jacs.5b13294
78. D.H. Cao, C.C. Stoumpos, O.K. Farha, J.T. Hupp, M.G. Kanatzidis, 2d homologous perovskites as light-absorbing materials for solar cell applications. J. Am. Chem. Soc. **137**, 7843–7850 (2015). https://doi.org/10.1021/jacs.5b03796
79. J.M. Kadro, N. Pellet, F. Giordano, A. Ulianov, O. Müntener, J. Maier, M. Grätzel, A. Hagfeldt, Proof-of-concept for facile perovskite solar cell recycling. Energ. Environ. Sci. **9**, 3172–3179 (2016). https://doi.org/10.1039/c6ee02013e
80. A. Binek, M.L. Petrus, N. Huber, H. Bristow, Y. Hu, T. Bein, P. Docampo, Recycling perovskite solar cells to avoid lead waste. ACS. Appl. Mater. Inter. **8**, 12881–12886 (2016). https://doi.org/10.1021/acsami.6b03767
81. C. Li, Z. Zhu, Y. Wang, Q. Guo, C. Wang, P. Zhong, Z. a. Tan, R. Yang, Lead acetate produced from lead-acid battery for efficient perovskite solar cells. Nano Energy. **69** (2020). https://doi.org/10.1016/j.nanoen.2019.104380

Advanced Ceramics (Self-healing Ceramic Coatings)

Ali Shanaghi, Paul K. Chu, Ali Reza Souri, and Babak Mehrjou

Abstract Advanced ceramics have many commercial applications due to the high corrosion resistance and mechanical properties, but their implementation is frequently compromised by inherent defects including holes, pores, and microcracks. The number of coating defects can be reduced by using corrosion inhibitors since the corrosion products help to repair the defects. Small cracks and crevices, of which the internal sources are indistinguishable, can proliferate leading to sudden failure. Therefore, it is essential to develop techniques to identify and monitor cracks so that preventive measures can be taken if the cracks are short and remediable, or a self-healing mechanism can be implemented to enable automatic repair. In this respect, non-automatic repair requires that the defects are first identified, but the process can be difficult and costly. From the perspective of thermodynamics, a system away from thermodynamic equilibrium can be combined with a thermodynamic recovery force such as penetration to return the system to equilibrium. In fact, the regenerative power reduces the entropy to assist the repair processes. Therefore, self-repairing coatings are of great importance and have been progressing gradually from the laboratory to industrial adoption. For example, ceramics-based self-healing coatings containing inhibitors such as organic benzotriazole (BTA) have attracted much interest. Here, different formats of ceramics-based self-healing coatings including titania, zirconia, titanium–alumina, and zirconia–alumina incorporated with benzotriazole (BTA) as an inhibitor are described together with the fabrication processes. By releasing the benzotriazole from titania–alumina–benzotriazole coating, the vulnerable locations for corrosion would be confined and a protective layer is fabricated to enhance the corrosion resistance of the Al 2024 alloy. The amount of BTA plays a vital factor regarding the self-healing efficiency and slow release of BTA is observed to produce the optimal self-healing ability. Titania–alumina coating with 3.6% benzotriazole, provides 76% protection efficiency and the

A. Shanaghi (✉) · A. R. Souri
Materials Engineering Department, Faculty of Engineering, Malayer University, Malayer, Iran
e-mail: alishanaghi@gmail.com

P. K. Chu · B. Mehrjou
Department of Physics, Department of Materials Science and Engineering, and Department of Biomedical Engineering, City University of Hong Kong, Tat Chee Avenue, Kowloon, Hong Kong, China

S. J. Ikhmayies (ed.), *Advanced Ceramics*, Advances in Material Research and Technology, https://doi.org/10.1007/978-3-031-43918-6_4

self-healing behavior of the coating is confirmed by investigating the impedance with immersion time. Moreover, in the ZrO_2–Al_2O_3–benzotriazole coating, by reaction of oxygen and benzotriazole and subsequently formation of corrosion byproducts, the corrosion resistance and the cathodic reactions would be enhanced and delayed, respectively, lead to an improvement in the properties of double layer. The hardness and elastic modules and plastic deformation of the ZrO_2–Al_2O_3–benzotriazole coating, decreased and increased, respectively, compared to the coating without the presence of benzotriazole and moreover, the adhesion is the dominant mechanism of wear.

Keywords Self-healing coatings · Inhibitors · Corrosion resistance · Nanomechanical properties

1 Definition of Self-healing

The term "self-healing" in materials science means spontaneous repair of the basic properties of materials after destruction in the external environment. Self-healing is possible in two ways, namely, (1) voluntary (automatic) or without any external interference or (2) involuntarily or with the help of external factors. In some cases, external factors such as the temperature, radiation, pH, pressure, and/or mechanical agents are needed to initiate and maintain the self-healing actions. Some materials have indirect self-healing properties and self-healing materials must be able to repair damage without external forces. At present, self-healing is considered only as mechanical recovery by means of crack repair, but it can be defined as the situation where the small holes and cavities can also be repaired alongside the cracks [1–7]. There are several general means to design self-healing materials: (1) Release of healing agents, (2) Reversible cross-linking, and (3) Exploiting the effects of electrohydrodynamics, conductivity, shape memory effects, nanoparticle migration, co-precipitation, and so on. At present, release of healing agents in the form of particles embedded in the matrix is the most common approach.

2 Release of Healing Agents

Corrosive agents like monomers and catalysts are embedded in the materials during production and if there are cracks, these agents in the reservoirs move to the cracks via capillary force. By the presence of catalysts, they solidify and repair the cracks. This process takes place spontaneously without the need for an external force because the driving force leads to automatic movement and propagation. One of the important methods to embed restoration agents in the matrix is the introduction of micro-reservoirs. Micro-reservoirs retain the solid particles, liquid, or gas droplets in the inner shells that separate and protect them from the surrounding environment [8–14].

3 Principles and Mechanisms of Self-healing in Ceramic Structures

Advanced ceramics such as titania, alumina, zirconia, and composites have many desirable characteristics such as the high-temperature stability, robust strength, high corrosion resistance, special and unique magnetic and electrical properties including piezoelectricity, superconductivity, insulating/semiconducting properties, and so on, which appeal to the industry. In manufacturing of industrial parts, properties such as strength, wear resistance, and anti-corrosion properties are very important. For example, in the chemical industry, strong resistance to acids and other corrosive substances is essential. In the aerospace industry, the heat resistance is critical and in the electronics and communication industry, optical and electrical properties are crucial. In recent years, the widespread use of ceramic materials in the electronics industry, medical technology, and automotive industry has spurred significant development of advanced ceramics as well as ceramic coatings.

Ceramic materials such as alumina, titania, and zirconia have many commercial and practical applications owing to the high corrosion resistance and mechanical properties, but failure can occur due to imperfections including holes, pores, and micro-cracks. Owing to micro-cracks, coatings may lose the protective properties with time. Micro-cracking is initiated by thermal, chemical, and mechanical stress or fatigue, and loss of protective properties exposes the underlying metals leading to corrosion. Repair and replacement of traditional coatings requires physical intervention and so the concept of self-healing has raised enormous interest. Ceramic structures are brittle and can break abruptly. In fact, the fracture toughness of ceramics is not an inherent strength and depends on the fracture toughness and crack geometry. Stress concentration initiated by an applied force can cause crack expansion and fracture at stress level below the fracture stress. In general, two types of surface defects are considered: (1) micro-cracks, cavities, and pores created during production and heat treatment and (2) micro-cracks and cavities caused by penetration of corrosive ions in the corrosion process.

If the small surface cracks, cavities, and pores can be repaired, the strength of the ceramic structure and corrosion resistance can be boosted significantly. According to the types of defects, there are different mechanisms pertaining to the heat treatment in order to repair cracks during production of ceramic materials [15, 16], namely re-sintering, relaxation of the tensile residual stress at the indentation site, and crack bonding by oxidation. The common failure mechanisms in ceramic coatings are (1) nucleation and instantaneous failure of the materials and (2) nucleation, crack development and propagation, and delayed failure. To repair cracks by penetration of corrosive ions, the use of healing agents and creation of corrosion products at defects and cavities are effective. In general, it is necessary to restore the strength of the area by the following means [17–23]: (1) repair done simultaneously with cracking, (2) complete filling of the volume between the crack walls and products, and (3) creating sufficient bonding strength between the products and crack walls.

In a system deviating from the thermodynamic equilibrium, a thermodynamic recovery force such as penetration can revert the system back to equilibrium. In fact, the recovery force raises the local entropy to help the healing process. For example, creation of a supersaturated solid solution leads to the establishment of semi-stable but non-equilibrium conditions. Failure and rupture due to defects such as cracks and so on destabilize the semi-stable conditions to move toward equilibrium. Therefore, coatings have the ability to self-heal when they are in the semi-stable state before surface defects, mechanical damage, or penetration of corrosive agents occur. Two mechanisms can be considered: (1) self-healing through supersaturated solid solutions of the active elements and (2) self-healing through the reaction of monomers and inhibitors in the coatings.

Surface defects, mechanical damage, and corrosive agents can move the coatings from the semi-stable state to the stable state, which is the same as performing a chemical reaction and creating chemical products to repair defects. Recent advances in surface engineering have revealed new avenues to construct active surfaces consisting of a host (inactive) and guest (active) structure. This can be done by inserting the active component into the layered structure so that the inactive component layers contain the active components (such as inhibitors) or by forming tanks (capsules) on the micro- or nano-scale in the coatings to store the active components.

4 Thermodynamics of Healing Processes in Coatings

Destructive processes such as wear, corrosion, fatigue, failure, and creep often involve reactions with different characteristics and different hierarchies on the macro-, micro-, and nano-scale that occur naturally in the system. Friction and abrasion (including contact of abrasive and coarse particles with the micro-, macro-, and nano-dimensions), capillary action, adhesion, chemical bonding [24], and corrosion processes involve different anodic and cathodic ones on the macro-, micro-, and nano-scale. In most cases, these reactions produce irreversible energy loss but in some cases, entropy production at one level such as macro-scale may offset that at another level. Since the processes are independent of the different levels and entropy is a function of the state, the net entropy can be expressed as the sum of the entropies of the system as shown by Eq. (1) [25]:

$$\Delta S_{net} = \Delta S_{macro} + \Delta S_{micro} + \Delta S_{nano} \tag{1}$$

For example, a homogeneous solid structure, in which the micro-/nanosurface corresponds to atomic vibrations in the crystal lattice, porous structure, grains, defects, and dislocations on the micro-scale, can be considered. A defect-free single crystal has lower entropy (ΔS_{micro}) than a defective structure. Larger defects such as cracks and voids are considered in macro (ΔS_{macro}) components and so materials with regular microstructures or surfaces with distinct textures have smaller entropy (ΔS_{macro}) than micro-structured materials which have irregularities or surfaces with different textures [25] for defect repair. In fact, it is assumed that macro-scale defects such as cracks and cavities have additional entropy (ΔS_{macro}). The healing process can be accomplished by releasing self-healing agents and breaking the microcapsules. Rupturing microcapsules decreases the microstructural order and increases ΔS_{macro}. When $|\Delta S_{macro}| < |\Delta S_{micro}|$, the healing process is gone with a decrease in entropy on the macro-scale and an increase in entropy on the micro-scale. Therefore, according to the macro-scale curing process, the net entropy is equal to the entropy on the macro-scale: $\Delta S_{net} = \Delta S_{macro}$. Increasing the order of the structure and decreasing the entropy on the macro-scale can be accompanied by increasing the entropy on the smaller micro- or nano-scale. The structure becomes more orderly, although increased loss and excess entropy on the nano-scale occur each time when grain boundaries propagate [25].

5 Advanced Self-healing Ceramic Coatings

Ceramics-based coatings such as titania, alumina, zirconia, and their compounds have good corrosion resistance, abrasion properties, and hardness but over time, penetration of corrosive ions reduces the ceramic coating adhesion strength to the substrate and consequently reducing the lifetime. The self-healing ability of ceramic coatings prevents penetration of corrosive ions leading to better lifetime and performance in corrosive and harsh environments. In order to create an active or inherent protection state in the coatings, corrosion inhibitors are placed into the coating matrix so that if the coating is damaged and the substrate is exposed to a corrosive environment, the inhibitors can neutralize the metal surface and prevent corrosion reactions. Inhibitors are chemical compounds that in the presence of small amounts in the corrosive environment of a metal, they can reduce the corrosion rate.

5.1 Types of Inhibitors

Inhibitors can be organic or inorganic or a combination. Inorganic inhibitors often work by regenerating damaged or anodic areas and forming insoluble deposits. These deposits precipitate and provide a barrier against penetration of electrolytes, water and oxygen. Common inhibitors include phosphates, nitrites, molybdates, tungstates, and chromates [26–29], which are described in the following:

– **Phosphates**

Phosphates and their derivatives are classical inhibitors and their derivatives such as zinc phosphate are well-known corrosion inhibitors that create corrosion products consisting of iron and phosphate compounds that prevent corrosion of steels and reduce cathodic delamination. Other phosphate compounds such as ammonium magnesium hydrogen phosphate are considered "greener" corrosion inhibitors for protection of epoxy coatings [30].

– **Nitrites**

Nitrites can prevent pitting corrosion and impose less detrimental effects for the environment than inhibitors containing heavy metals, for example, CrO_4^{2-}, WO_4^{2-}, MoO_4^{2-}, and $Cr_2O_7^{2}$. The electrostatic force between nitrite ions and coatings results in easy reactions with surface compounds, thereby increasing the corrosion resistance by creating corrosion products [31].

– **Molybdates**

Sodium molybdate, Na_2MoO_4, as a mineral inhibitor, is highly effective in increasing the corrosion resistance of metals such as steels and its derivatives are considered non-toxic and green inhibitors [32].

– **Tungstates**

Tungstates constitute one type of environmentally friendly corrosion inhibitors. Among them, Tungstate anions (WO_4^{2-}) have been extensively studied in different corrosive media for active protection [33]. In addition, sodium tungstate can protect metals in neutral, acidic, and alkaline solutions, and is considered environmentally friendly anodic corrosion inhibitors. The inhibition effects of tungstate ions are derived from the formation of iron–tungstate complexes on the surface. However, it is better to avoid using tungstate alone for corrosion protection applications on account of the low inhibition efficiency at low concentrations, in addition to the high cost and poor oxidation ability [33].

– **Chromates**

Chromates which are effective due to the low water solubility, strong oxidation behavior, and neutral nature of corrosion products are largely used in the aerospace industry and ester pretreatment. However, they pose carcinogenic issues and efforts are being made to replace chromates with more environmentally friendly and safer materials [34].

– **Vanadates**

Vanadium compounds are non-chromatic corrosion inhibitors. Although vanadate conversion coatings (VCC) are deposited on aluminum alloys, they are about half a micrometer thick and contain a network of cracks. Vanadium compounds work as pigments in the "super-primary" coating system and inhibition arise from the barrier oxide film formation and its deposition by the sol–gel process. This film is insoluble in a wide range of pH and environmental conditions [35].

Organic compounds also have significant applications in surface corrosion inhibition due to good bonding with the surface of materials leading to increased corrosion inhibition, cost-effectiveness, ease of use and synthesis, as well as the possibility of recycling [36]. Organic compounds including aliphatic and aromatic amines such as ethylene diamine, ammonia, cyclohexylamine, benzylamine, aniline, methyl aniline, dimethyl aniline, phenylhydrazine, phenylamine, and ortho- and para-tluatin prevent corrosion by the formation of an adsorbed layer on the metal surface [37]. The prerequisite for organic inhibitors to be effective is attachment to the metal surface. They also work by increasing the local pH and neutralizing the substrate. Organic inhibitors have been reported to reduce and minimize corrosion when the corrosive medium contains dissolved oxygen, salts, and weak acids. When the corrosive environment consists of strong acids such as hydrochloric acid and is subject to a high temperature and microbial activity, polar organic compounds and colloidal organic matters are better inhibitors against corrosion [38]. Among the various organic inhibitors, heterocyclic compounds are of considerable importance as they absorb heterocyclic compounds on the surface to prevent corrosion reactions. Heterocyclic compounds make use of their electron-rich centers such as polar functional groups including –NHMe, $-NH_2$, $-NMe_2$, –OH, $-NO_2$, $-OCH_3$, –O–, –CN, $-CONH_2$, $COOC_2H_5$, etc., as well as π-electrons of the hetero-species such as NC=O, NC=N–, NC=S, –C≡N, –N=O, and –N=S, and homo-species including NC=C, –N=N–, and –C≡C–. Compounds with a larger molecular size are better corrosion inhibitors than smaller compounds because they cover a larger surface [36]. In the following section, common organic inhibitors used in the industry are described.

- **Azoles and Thiazoles**

Azole and thiazole inhibitors contain nitrogen and sulfur donor atoms and form suitable compounds with metal cations with polymeric properties as surface oxidation barriers. The inhibitory properties of azoles stem from physical or chemical adsorption via formation of single molecular layers. Much work has been done on the application of 1,2,3-benzo-triazole (BTA), 2-mercapto-benzothiazole (MBT), and their derivatives as corrosion inhibitors [39].

- **Imidazole**

Imidazole-based imidazole carbonitriles which are heterocyclic compounds containing two nitrogen atoms (diazole) in the ring have biological and industrial applications. The imidazole ring is contained in histamine, histidine, antifungal, and antibacterial drugs, antibiotics, and midazolam. By introducing an appropriate substitute into the five heterogeneous rings of imidazole, its properties can be modified and its easy dissolution in water can be achieved. Imidazole interacts strongly with charged metal surfaces due to its flat geometry and high bipolar torque. When an imidazole contains a hydrophobic chain in its structure it's the most effective inhibitors and electrochemical studies show that increased hydrophobicity raises the conservation efficiency [36].

- **Pyridine**

Carbonitrile pyridine is a six-member heterocyclic compound that contains one nitrogen atom in the heterocyclic ring. It is a flammable, water-soluble, and basic liquid. This compound can be found in vitamins, pharmaceuticals, and agrochemicals and pyridine with high aqueous phase solubility shows the robust affinity to interact with a metallic surface [36].

- **Nitro Compounds**

Nitro compounds have at least one nitro group in the molecular structure to attract electrons, and their presence in organic inhibitors reduces the anti-corrosion effects. However, replacement of NO_2 with one nitrogen atom and two oxygen atoms in the smaller molecules increases the molecular size. $-NO_2$ being highly polarized reacts strongly with polar electrolytes and dissolves complex molecules that normally do not dissolve without $-NO_2$, consequently enhancing the inhibitory effects [36].

- **8-Hydroxyquinoline (8-HQ)**

8-Hydroxyquinoline (8-HQ) is one of the common corrosion inhibitors due to this fact that the small molecule 8-HQ has strong bidentate chelating ligand for metal ions. The structure consists of a nitrogen and oxygen atoms as donor atoms to complex of metal ions. Recently, 8-HQ has been incorporated as a corrosion inhibitor into self-healing and smart coatings [40].

Important factors regarding the long-term corrosion protection are the corrosion inhibitors concentration in the coating and the barrier properties of formed self-healing layer in the defect sites. Other factors such as solubility may have an impact on self-healing as a low solubility of the corrosion inhibitor reduces the amount of active agents in the damaged areas in the coatings. However, a high solubility causes rapid removal of active compounds from the coatings giving rise to only short-term protection [41, 42]. Inhibitors can react chemically with the components of coatings and in some cases, the barrier properties of the coatings are destroyed [43].

The release of inhibitor from the self-healing coatings should be fast enough to respond to changes in the environment or status of the coatings quickly. There are new strategies to produce self-healing coatings, for example, by combining nano-scale containers (carriers) with active barrier components in conventional coatings. This method produces new coating systems based on passive–active structures. In fact, nanoreservoirs which are evenly distributed contain active inhibitors to eliminate excessive deterrence and remove osmotic bubbles when the inhibitory salts are highly soluble or have a very small particle size.

When local changes in the environment or the corrosion process occur in the coating defects, nanoparticles respond to these signals and release active materials [44, 45]. The inhibitor loading mechanism in self-healing ceramics-based coatings is via incorporation of oxide nanoparticles as nanocarriers on which the inhibitors adsorb. Oxide nanoparticles are also considered coating enhancers and their addition improves the inhibitory properties [46–50]. In addition to increasing the corrosion resistance, adding a corrosion inhibitor to oxide nanoparticles creates the self-healing ability. For example, stabilization of an inhibitor such as Ce^{3+} on the surface of ZrO_2 nanoparticles during hydrolysis of the zirconia sol produces a hybrid nanocomposite coating containing oxide nanoparticles with cerium ions [51–53]. The small diameter of the oxide nanoparticles increases the carriers surface area and increases the loading capacity of the inhibitors. Consequently, long-term release of the inhibitor from the nanoparticle surface improves the corrosion behavior.

5.2 *Self-healing Ceramics-Based Coatings Preparation Methods*

In general, several processes can be used to produce self-healing coatings containing barriers, including (1) use of inhibitor containers, and (2) incorporation of an inhibitor directly into the coating. The use of inhibitory containers in polymer-based coatings is divided into multiple groups depending based on the organic or inorganic containers type [54].

5.2.1 Organic Containers

In this case, the inhibitors in micro or nano-organic containers are self-healing epoxy coatings. Researchers have tried to examine the impacts of various organic nanocontainers as well as their other properties such as size, shape, and location in epoxy coatings. These nanoenclosures contain anti-corrosion inhibitors improve the corrosion resistance of materials. In this process, the inhibitor is placed inside the chambers, followed by active ingredients released to the damaged area freely due to mechanical damage or pH change leading to improved corrosion resistance. Various parameters such as the diameter of the chambers and amount and distribution of the inhibitors in the coating affect the inhibitor release rate and the efficiency of the self-healing coating. Organic containers are cellulose nanoreservoirs and polymeric microcapsules. While cellulose nanoreservoirs are used, the epoxy matrix with a grid structure is preferred for better repair. Cellulose chambers improve the thermal mechanical properties of the epoxy matrix by creating lattice structures rendering them excellent reservoirs in self-healing coatings. Another effective and practical approach to protect the metal surface is using the polymer microcapsules loaded with inhibitors as they can release the inhibitors in a controlled manner. This method is quite common and the polymer capsules can be made from urea formaldehyde (UF) resins via polymerization techniques. The facile preparation steps with high degree of control on the microcapsule size and the shell thickness are the main advantages of the in situ micro-polymerization methods [54].

5.2.2 Inorganic Containers

Halloysite nanotubes (HNTs), titanium dioxide (TiO_2), and mesoporous silica are used to hold inhibitors for self-healing purposes. These fillers internally carry on the self-healing liquids can act as a reservoir. Due to the larger aspect ratio (L/D) ratio of TiO_2 and HNT compared to the polymeric nanocontainers, they can cover larger site on polymer matrix. Furthermore, this system significantly enhances the overall thermomechanical properties. Encapsulation of inhibitors in organic containers is an easy process and the layer-by-layer (LbL) approach is a suitable approach to achieve the self-healing epoxy coatings. The reservoirs regulate inhibitor loading and release, and additionally, the multilayer permeability of the system can be governed by factors like the temperature, pH, electromagnetic field, and ionic strength [54].

The aforementioned methods are applied to create self-healing polymer-based coatings containing inhibitors, but cannot be used for ceramics-based coatings or in the attempt to reduce adhesion of ceramic coatings. In the ceramic base coatings, direct use of a barrier by solution processes such as sol–gel can produce a homogeneous and uniform coating with the suitable distribution of the barrier in the ceramic-based coating.

5.2.3 Sol–Gel Coatings

There are various methods to deposit coatings on metals, for example, physical vapor deposition (PVD), chemical vapor deposition (CVD), electrical deposition, plasma spraying, and sol–gel process. The sol–gel process is one of the common techniques to synthesize homogeneous powders, coatings, and ceramic composites and has many advantages [55, 56]. It is a low-temperature process (close to room temperature) and so thermal evaporation and degradation of organic inhibitors are minimized. Since liquid precursors are utilized in this process, it is possible to form coatings with a complex shape and there is no need to melt or machine in order to produce the thin films. The sol–gel process is environmentally friendly and final rinsing is usually not required.

In the sol–gel method, an oxide network is prepared by continuing compression of the precursors in the liquid medium [55–57] and in general, sol–gel coatings can be prepared in two methods, namely, the inorganic or organic method. In the inorganic methods, a colloidal suspension (normally the oxide format) and the sol coagulating (suspension of colloids with particle at very small sizes, 1–100 nm) are formed to deform the network till a network can be shaped in the continuous liquid state. The organic method is more common which normally begins with a monomer metal–solution or metal-like alkoxy precursor $M(OR)_n$ in alcohol or other organic solvents with low molecular weight. Here, M represents the main element such as Si, Ti, Zr, Al, Fe, and B and R stands for the alkyl group ($C_xH_{(2x+1)}$).

The sol–gel process consists of four stages: (1) hydrolysis, (2) formation of chains and particles through densification and polymerization, (3) particle growth, and (4) polymer structure agglomeration and subsequent network formation that spreads throughout the liquid environment followed by its thickening to form the gel-like structure. Hydrolysis and compaction occur simultaneously at the beginning of the hydrolysis reaction and the products with low molecular weight such as alcohol and water can be formed after drying and contraction of the lattice causes more compaction [55–58].

Metal alkoxides (Ti, Zr, Ce, Al, Si, etc.) with the typical formula of $A(OR)_4$, where A, OR, and R stand for a metal atom, an alkoxy group, and alkyl ligand, respectively, are considered as the precursors for the sol–gel process. The typical reaction of alkoxides in contact with water is shown by Eq. (2) (hydrolysis) [59]:

$$A(OR)_4 + H_2O \rightarrow HO\text{-}A(OR)_3 + ROH \tag{2}$$

According to Eqs. (3) and (4), hydroxyl ions bind with atom A to form alcohol and partially hydrolyze the starting materials, which are then combined by condensation:

$$(RO)_3A\text{-}OH + HO\text{-}A(OR)_3 \rightarrow (RO)_3A\text{-}O\text{-}A(OR)_3 + H_2O \tag{3}$$

$$(RO)_3A\text{-}OR + HO\text{-}A(OR)_3 \rightarrow (RO)_3A\text{-}O\text{-}A(OR)_3 + ROH \tag{4}$$

Reactions (3) and (4) are the polymerization process in gelation in which the solid network contains the liquid phase in the cavities and continues to build the network containing the O-A-O connections.

The rate of hydrolysis is usually larger than that of condensation. The metal alkoxides reactivity is different and increases in the form of $Ce(OR)_4 > Zr(OR)_4 > Ti(OR)_4 \gg Si(OR)_4$. In some cases, due to very slow rate of alkoxides reaction with water, catalyst should be used to enhance the rate of process. Acidic catalysis is commonly used to form coatings. Some of the determining factors which make an impact on hydrolysis and condensation are the ratio of water to alkoxide, pH, temperature, concentration of reactants, and nature of the alkoxide groups. Changes in these parameters affect not only the kinetics of the sol–gel process but also the microstructure and nanostructure of the final product [50]. Among the sol–gel coatings, the hybrid one, is more usual compared to the inorganic ones for protection of metal surface from corrosion, based on two facts. Firstly, a thin crack-free micrometer layer can be formed with low oxide-layer sintering temperature (below 100 °C) and secondly, they can provide higher degree of flexibility when mixed with anti-corrosive agents such as inhibitors, pigments, etc.; therefore, it can be said that they can provide higher and better corrosion protection.

Spin coating and dip coating are the most two common techniques for applying the sol–gel coatings to the surface of metals. Spraying and electrical deposition have also been reported. Regardless of the method, after deposition of the coating, evaporation of a large amount of solvent and water leads to significant volume shrinkage and creates internal stress. The film fabrication conditions should be precisely adjusted, otherwise, the internal stress can produce cracks. Usually, curing and heat treatment steps are performed on sol–gel coatings depending on the microstructure, quality, and application [55–58].

5.3 *Titania-Based Self-healing Coatings*

Titanium (titanium oxide) has very good properties such as chemical stability, thermal resistance and low electron conductivity and TiO_2 is a very good anti-corrosion agent. However, there are not many reports on the application of TiO_2 films as a steel protective coating. CeO_2 has comparable properties, and it has been used in sol–gel fabricated coating with the role of corrosion inhibitor, although it has a wider range of applications in optics, catalyst chemistry, pigments, etc. The oxide layers with more than one component can perform better to protect steel-based materials from corrosive medium and consequently their range of application can be expanded [60–62].

The TiO_2–benzotriazole nanostructured hybrid sol–gel coating consisted of 1.4, 2.8, and 4.2% benzotriazole, applied on the 7075 aluminum alloys substrates by the sol–gel method. Alkoxide $Ti(OCH_2CH_2CH_3)_4$ precursor with long chain lengths and miscibility with organic solvents on the molecular level generates crack-free films and organic inhibitors such as benzotriazole has a more homogeneous reaction by alkoxide compounds resulting in better release of the inhibitor to defects for healing[60–62].

Upon usage of benzotriazole, robust complexes can be formed when it reacts with transition metals and consequently, it can enhance the quality of protection of copper-based alloys. Also, it has applications in other sectors such as deicing agents, antifreeze and brake fluids, lubricating oil, and within cooling systems. Furthermore, the corrosion behavior of Fe in brine solution can be improved by using benzotriazole and some of its compounds. In this case, inhibitory effectiveness subjects to its concentration as here insoluble complexes formation protects the surface of metal [60–63].

3-Glycidoxypropyltrimethoxysilane (GPTMS) with its two functional groups, glycidoxy (organic component) and silicon alkoxide (inorganic component), already showed its potential to establish better connectivity among the organic and inorganic compounds in the sol–gel method. Therefore, it can be used to make a network by reaction of glycidoxy group with the both organic and inorganic inhibitors, during the polymerization process. Moreover, a network can be made via hydrolysis for subsequent condensation reactions between SiO and titanium alkoxide [64–66]. Thus, the reaction can be seen between organic groups together compared to the organic and inorganic ones. There are studies regarding the inhibiting action of organic compounds like benzotriazole on aluminum alloys. Usually, it happens with polymeric coatings with the self-healing ability and the thickness between 10 and 100 μm, although it can be too much for some purposes. The thickness of ceramics-based coatings prepared by the sol–gel method is 1–10 μm with 1.2, 3.6, and 4.8% benzotriazole as the self-healing agent [60–62]. Structure and morphology of the TiO_2 nanostructured and its self-healing coating contained different benzotriazole concentrations are shown in Fig. 1, which indicates that adding benzotriazole formed a uniform and homogeneous structure with less micro-cracks compared to the titania nanostructured coating [60].

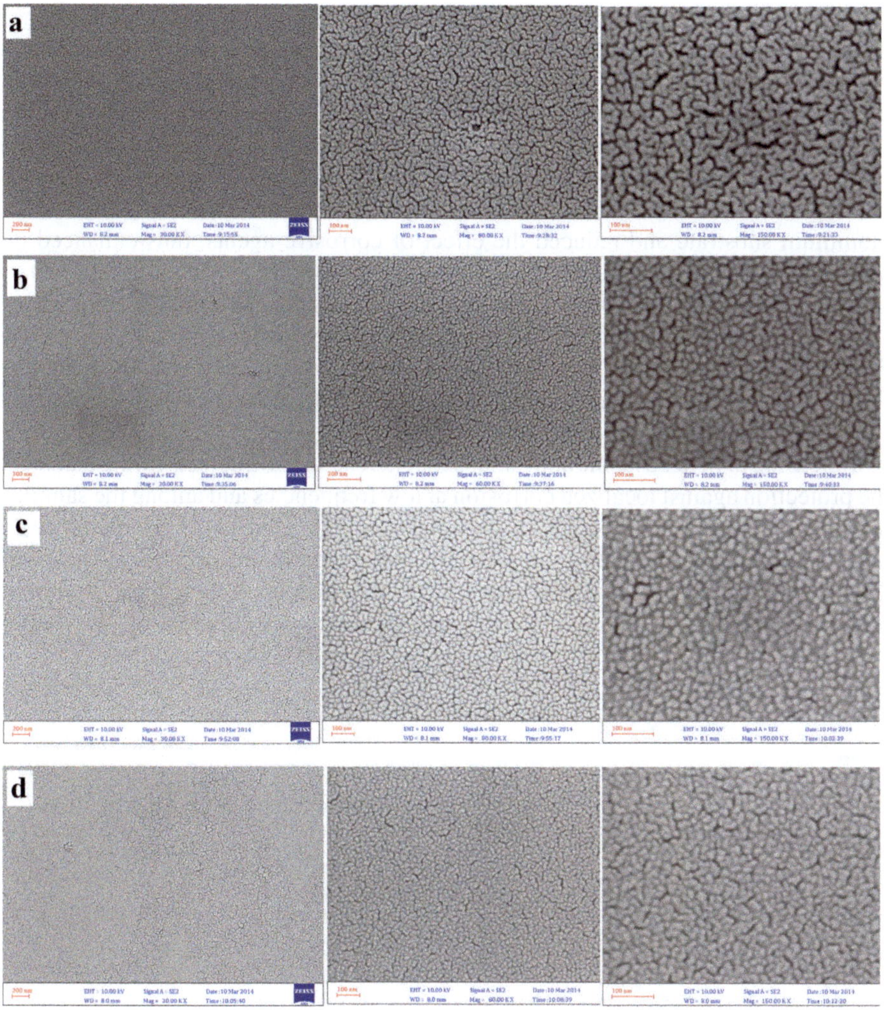

Fig. 1 FE-SEM images of **a** TiO_2 nanostructured, **b** TiO_2–1.4% *benzotriazole* self-healing, **c** TiO_2–2.8% benzotriazole self-healing, and **d** TiO_2–4.2% benzotriazole self-healing coatings, which are heat treated for 2 h at 150 °C of a heating rate of 1 °C/min. The magnifications of 30, 80, and 150 kX are shown in [60]

When the TiO_2 nanostructured coating was applied on the surface of 7075 Aluminum alloy, the corrosion behavior was deteriorated compared to the uncoated one, according to the results acquired from the Nyquist and bode-phase plots after samples were being immersed in 3.5% brine solution for 24 h (Fig. 2). However, the TiO_2 coating containing the benzotriazole showed higher corrosion resistance. The coating with 4.2% benzotriazole, showed better corrosion resistance compared to the ones with 1.4 and 2.8% due to its homogenous and uniform properties and also higher compactness of the coating as the defects can be filled out by releasing of benzotriazole. As it can be seen in Fig. 2, the sample containing 4.2% benzotriazole has two time-constant compared to the ones with 1.4 and 2.8%, showed that the corrosion reactions happened at the both coating and its interface with the substrate. It was shown that the inhibitor enhanced the corrosion resistance of TiO_2–4.2% benzotriazole upon immersion for 24 h. When the healing agent was released and subsequently reacted with the medium, a layer was formed between TiO_2 and the aluminum substrate and reduced the effect of corrosive agents, thus enhanced the corrosion resistance of the coating. According to the graphs, by increasing the benzotriazole content from 2.8 to 4.2%, protective layer was formed at higher frequencies. All in all, it can be said that, at both 2.8 and 4.2% benzotriazole, a protective layer was formed, due to releasing of benzotriazole at enough amount, however, the quality of the layer formed from the sample containing 4.2% benzotriazole was better, showed by its better stability and density [60–62].

Releasing of benzotriazole and its subsequent reaction and defects filling enhanced the protection against localized corrosion at low frequencies and among the samples, the coating with 2.8 and 4.2% of benzotriazole provided better corrosion resistance compared to the other groups, the uncoated aluminum substrate, coated with TiO_2 nanostructured and the one with 1.4% benzotriazole. Although the aluminum substrate, 7075, has good resistance to localized corrosion, however, by introducing

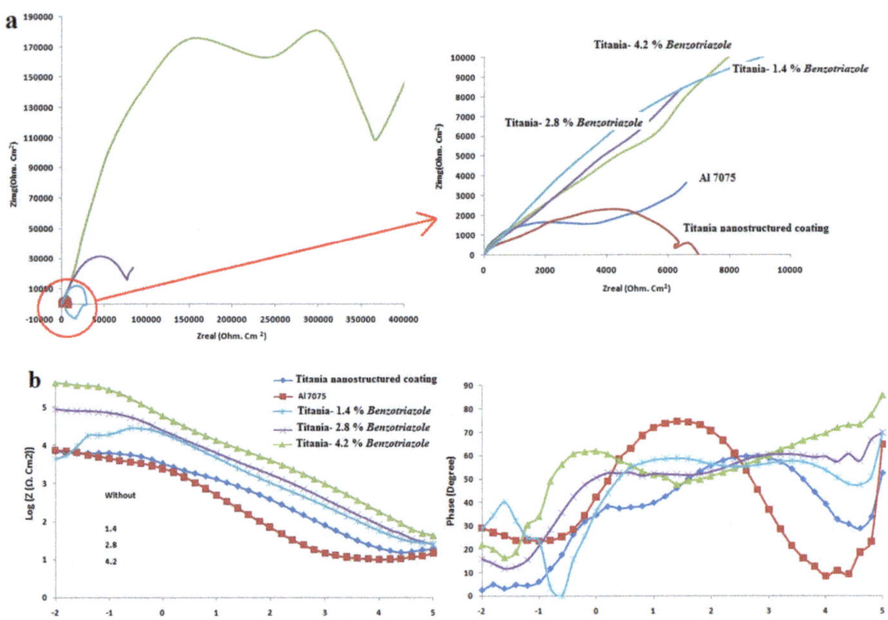

Fig. 2 **a** Nyquist plot and **b** Bode-phase plot of the Al 7075 alloy, titania nanostructure coating and titania nanostructured coatings containing 1.4, 2.8, and 4.2% benzotriazole after immersion for 24 h immersion in the 3.5 wt.% NaCl solution [60]

the TiO_2 nanostructured to the substrate, the corrosion resistance would be increased. Therefore, by coating the substrate with TiO_2 nanostructured and TiO_2–1.4% benzotriazole, the resistance to localized corrosion would be decreased as it can be understood from the Bode-phase plot, by the reduction of phase angle and its subsequent shifting toward the zero. But this deficiency can be improved by increasing the benzotriazole content from 1.4% to 2.8 and 4.2% [60–62]. Also, enhancement of localized corrosion resistance of the coating with TiO_2–4.2% benzotriazole can be achieved by increasing the immersion time to 120 h (Fig. 3) due to this fact that the absorption layer formed at the interface of coating and solution, has the higher stability. Extending the immersion time at 3.5% brine solution would decrease the corrosion and specifically localized corrosion resistance of TiO_2 nanostructured coating, however, by adding the healing agent the corrosion behavior, both locally and generally, improved significantly [60–62, 67]. Thermodynamically study of absorption and formation of protective layer between the medium and coating showed that by releasing benzotriazole from the coating with 4.2% concentration, and the subsequent reactions of nitrogen and oxygen molecules with water, an insulating layer was formed [60–62, 67].

Two mechanisms are involved in enhancing the corrosion resistance of the coating with 4.3% of benzotriazole:

(a) Blocking the surface defects via corrosion by-products, due to release of inhibitor and its subsequent interaction with the molecules and ions in the environment,

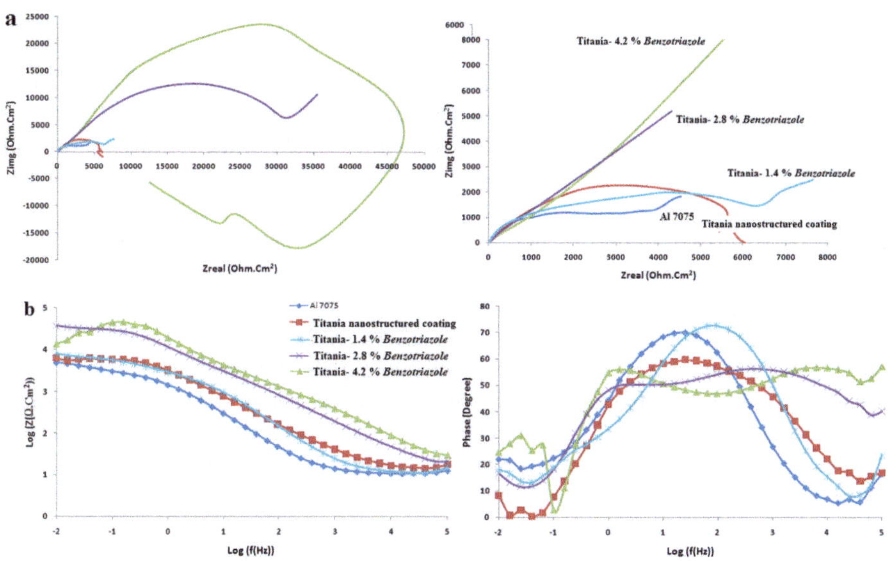

Fig. 3 **a** Nyquist plot, and **b** Bode-phase plot of the Al 7075 alloy, titania nanostructured coating, and titania nanostructured coatings containing 1.4, 2.8, and 4.8% benzotriazole after immersion for 120 h in the 3.5 wt.% NaCl solution [60]

(b) Formation of insulating layer between the coating and the medium due to the release of inhibitor and its surface adsorption.

Although it's been said that the benzotriazole released from the coating produces defects in the coating to ultimately reduce the resistance to the corrosion, both locally and generally, however, it is shown that the presence of 4.2% of benzotriazole in the coating can improve the corrosion resistance and healing reactions in the TiO_2 nanostructured coating.

5.4 Titania–Alumina Self-healing Ceramics-Based Coating

Titania and alumina coatings are protective coatings against corrosion and nanoparticles of titanium dioxide have anti-corrosion properties. Although titania is an active compound for the catalytic activity, the small surface area and change of the structure at elevated temperature (e.g., instability of the anatase structure) are limitations. Studies have been conducted to enhance the anti-corrosion properties by combining titanium oxide with other oxides [62, 68–70]. Alumina (produced by sol–gel technique) is structurally stable and has a large specific surface area [71, 72]. Alumina also has a good oxidation resistance to abrasion and provides excellent corrosion resistance [73].

After preparing the titania solution, the tri-butyl aluminum precursor is used as an alumina source to prepare the alumina solution. The solution containing 80% of the titanium solution and 20% of the alumina solution is added to tri-methoxy-silane as an encapsulate liquid to prepare benzotriazole solutions with three different concentrations [62]. According to FE-SEM image (Fig. 4), a more homogenous and smoother composite coating of titania-alumina containing benzotriazole can be achieved compared to the ones without it. benzotriazole reduces adhesion among particles. As shown in Fig. 4a, nanoparticles attached together with poor connection to each other can create cracks with small size and micro-channels. The stress caused by shrinkage upon the heat treatment and solvent removal can induce the crack formation. The corrosion rate can be increased due to more cracks formation, however, on the other hand, more benzotriazole can be stored in them. In this regard, by adding 1.2% benzotriazole, the cracks surface can be covered, and smoother and more homogenous structure can be achieved compared to the case without any benzotriazole. By further increasing the content of benzotriazole to 3.6%, as more cracks would be filled therefore, the homogeneity of surface structure would be enhanced. However, there is an optimal concentration regarding the application of benzotriazole to fill out the cracks, as increasing its content to 4.8%, a coating with higher porosity would be formed [62].

The coatings properties such as strength and adhesion are affected by the surface structure. For example, the amount of TiO_2 cracks deposited on Al 2024 alloy can be reduced by alumina [74, 75] and also higher surface roughness can induce more cracks [76–78]. The roughness and thickness of TiO_2–Al_2O_3 composite coatings are

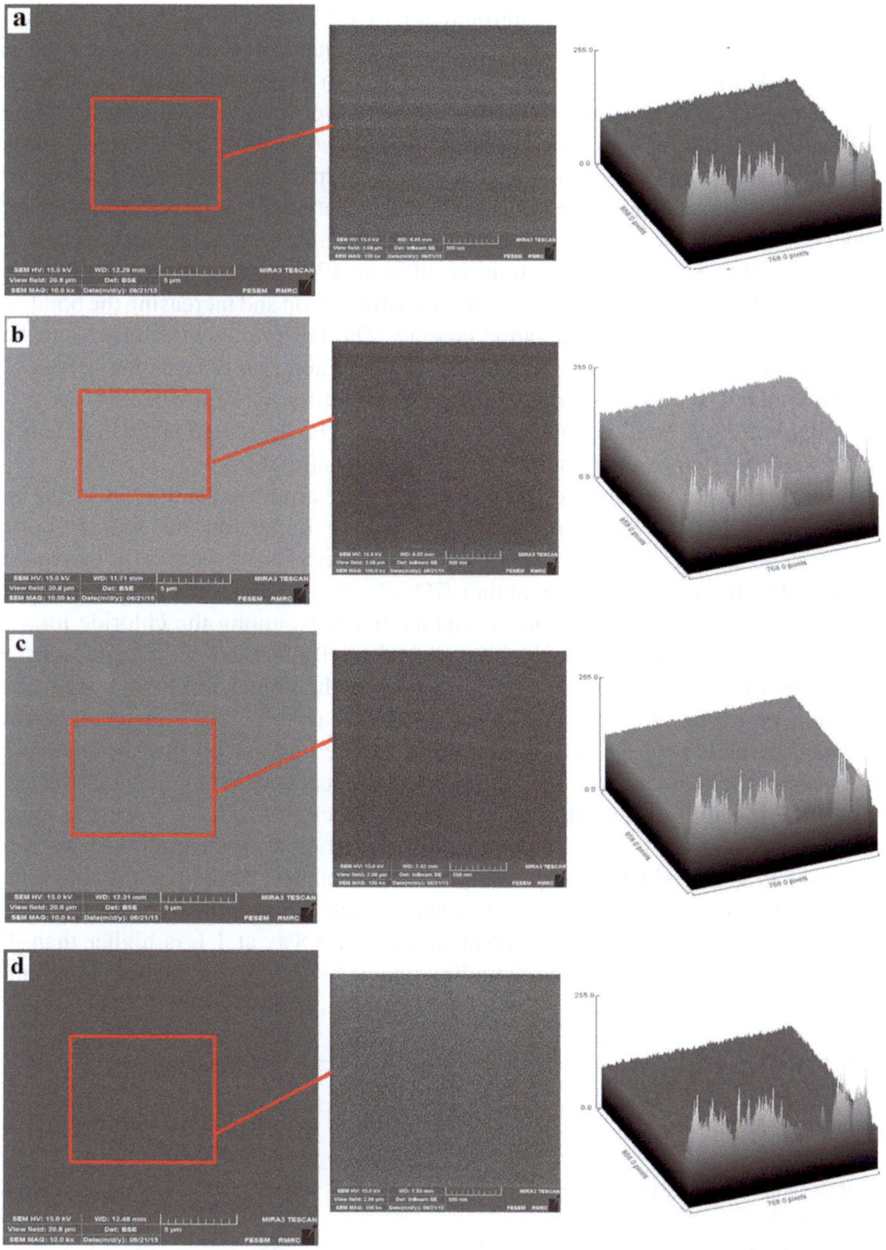

Fig. 4 FE-SEM images of the TiO_2–Al_2O_3 coatings: **a** Without benzotriazole; **b** 1.2% benzotriazole; **c** 3.6% benzotriazole; **d** 4.8% benzotriazole (10,000X and 100,000X magnification) [62]

summarized in Table 1. Upon the introduction of 1.2% of benzotriazole, the surface roughness (Ra) is increased, however, by further increasing the concentration of benzotriazole to 3.6 and 4.8%, better consistency can be attained. As you can see in Table 1, by adding more than 1.2% benzotriazole, lower surface roughness compared to the ones without the inhibitor can be achieved. The bonding quality between the Al 2024 substrate and the coating is affected by the mismatch of thermal expansion coefficient. The stress generated during the heat treatment can be accumulated and reduce the bonding strength of the coating as the thermal contraction and coefficient behavior of the coating and substrate is different [79]. In this regard, the role of alumina and benzotriazole is reducing the crack formation and increasing the bonding strength of the substrate and ceramic coating. The benzotriazole diffuses into the cracks upon its formation and probably reacts there and finally heals the defects. It should be noted that by further increasing the concentration of the benzotriazole to more than 4.8% the coating bonding strength would be decreased and delamination can be happened more easily in the ceramic coatings [60–62].

To study the coating corrosion behavior, they were immersed in the NaCl solution with 3.5 wt.% up to 96 h. It was found that when the immersion time was increased from 1 to 48 h, Fig. 5a–b, the polarization and corrosion resistance is increased. Compared to the coating only containing TiO_2–Al_2O_3, reactants attach to the surface, and due to redox reactions, the ion exchange happens among the chloride ions in the electrolyte and the ones in the coating and finally reduces the charge transfer resistance (1 h immersion (Fig. 5a)). By increasing the immersion to 48 h (Fig. 5b), corrosion produces filled the cracks and made a barrier to the transfer of ion to the metal surface. From the linear section, it can be understood that the system is governed by the movement of reaction products or reactants from the metal surface toward the medium. This is the case when the reaction products cover the metal surface [80]. Followed by 96 h immersion (Fig. 5c), a higher concentration of chloride ions and a gradual reduction in the passive film resistance, formed on the top of the corrosion byproducts can be seen. It is shown that the polarization resistance of the coatings with benzotriazole content of 1.2 and 4.8%, at 1 h is higher than 48 h. Trimethoxysilane ($C_3H_{10}H_3Si$, coupling agent) by forming a membrane around

Table 1 Surface roughness (Ra) and thickness of the TiO_2–Al_2O_3 coatings containing 1.2%, 3.6%, and 4.8% benzotriazole [62]

Sample	Ra (nm)	Thickness of one layer (nm)	Total thickness after five times immersion (μm)
Titania–alumina coating (no benzotriazole)	20.45	450	2.25
Titania–alumina coating (1.2% benzotriazole)	0.14	510	2.55
Titania–alumina coating (3.6% benzotriazole)	0.03	600	3.00
Titania–alumina coating (4.8% benzotriazole)	0.19	640	3.20

benzotriazole acts as a bridge between benzotriazole (an organic inhibitor) and TiO_2–Al_2O_3 (minerals). Benzotriazole can be released by variation of the pH or scratching and form a thin layer to protect the corroded metal from the diffusion of corrosive agents. The inhibition efficiency depends on two factors, adsorption of the inhibitor on the surface of metal and its inhibition efficacy at the metal–solution interface [62, 81].

As the polarization resistance is still low after 1 h of immersion, it can be concluded that the molecular absorption on the surface of Al substrate hasn't yet been completed and the equilibrium state between the molecules in the solution and absorbed on the surface is not yet reached. The inhibitory effect of benzotriazole is associated to the BTAH adsorption to the surface of Al substrate and formation of [BTA]ads–Al layer. This process happens by the removal of H from BTAH and subsequently formation of BTA^- and its linkage to the Al ions and the formation of [BTA]ads–Al layer on the surface. [81, 82]. This process is subjected to various factors and the physical and chemical properties of inhibitor molecules like functional groups, electron density, π orbitals, as well as charged metal surface [82–84].

Benzotriazole improves the coatings corrosion resistance compared to the ones without any inhibitor molecules after 1 h immersion (Fig. 5a). However, it should be considered that by adding 1.2% benzotriazole, no improvement in corrosion resistance can be seen after 48 h immersion (Fig. 5b) and at the inhibitor concentration below than the critical value, corrosion behaviors are worsen compared to the ones without any inhibitor at all. The pitting deteriorates the corrosion behavior by breaking the passivation and reduces the anodic area compared to the cathodic one. Adding below than optimal concentration of benzotriazole, enriches the porosity and consequently damages the protection behavior, thus a critical point for the amount of benzotriazole to protect the surface should be considered [81]. Here in this study, it is found that the optimal concentration of benzotriazole is 3.6% as it showed better polarization resistance compared to the one with 4.8%. The constructive or destructive impacts of benzotriazole on protective behavior of the prepared coatings with sol–gel techniques hinge on various factors including, composition of sol, deposition parameters and the benzotriazole concentration in the sol part. An important factor to provide the stability in the corrosive medium is the corrosion protection behavior for a long term and to provide the long-lasting corrosion protection, factors such as the corrosion inhibitors concentration, the silica film structure, and the self-healing layer barrier property formed in the flaws should be carefully considered. Moreover, there are some other factors which can impact the self-healing properties like the solubility behavior of corrosion inhibitors in the corrosive medium. For instance, low solubility of the corrosion inhibitors causes active agents depletion from the affected area, however, high solubility can lead to fast leaching [81]. The polarization curves of the titania–alumina composite coatings with 1.2, 3.6, and 4.8% benzotriazole after immersion in 3.5 wt.% NaCl solution for 96 h are shown in Fig. 6, and the results, such as corrosion current density (i_{corr}), anodic slop (βa), cathodic slop (β_c), corrosion potential (Ecorr), passivation current density (i_{pass}), and inhibition efficiency (ηp), are summarized in Table 2. This table shows that when the concentration of benzotriazole increased from 3.6 to 4.8%, the corrosion current

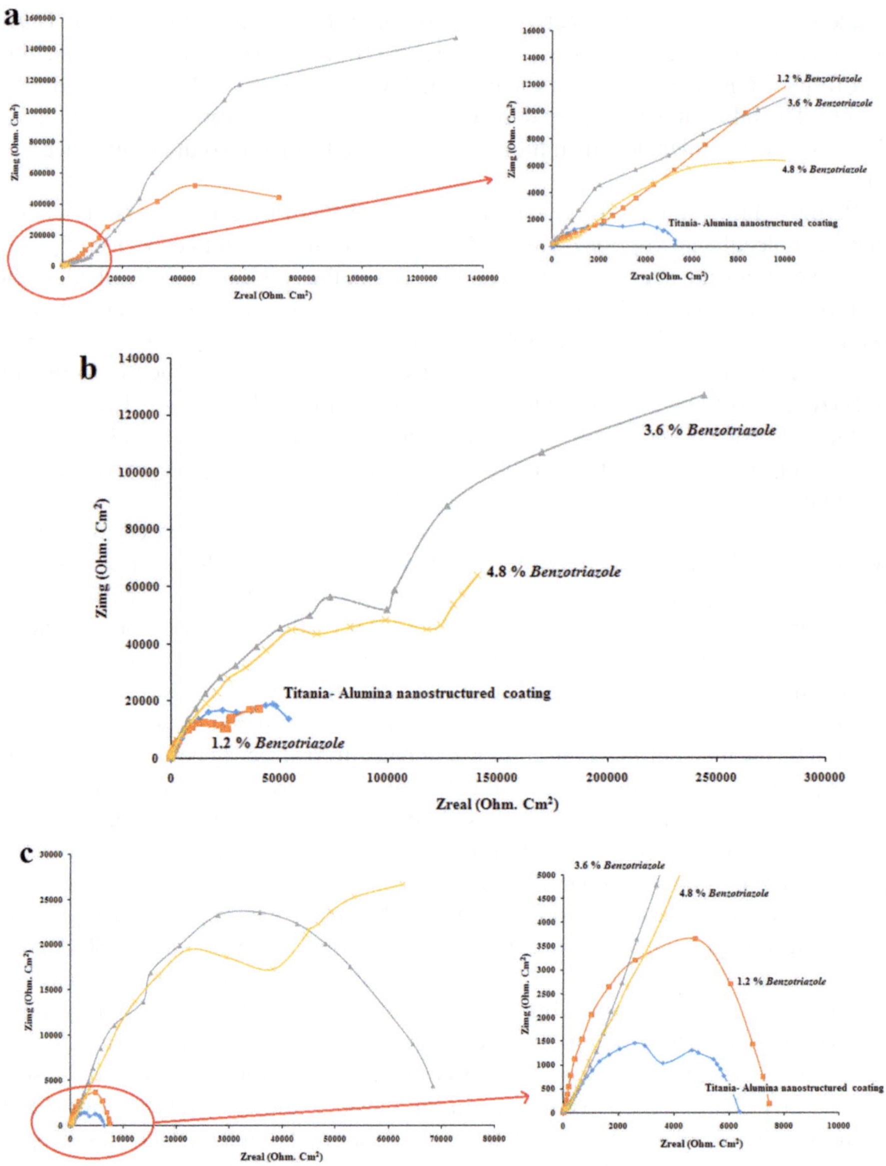

Fig. 5 Nyquist plots of Al 2024 alloy with the TiO_2–Al_2O_3 coatings: No benzotriazole, 1.2% benzotriazole, 3.6% benzotriazole, and 4.8% benzotriazole after immersion in 3.5 wt.% aqueous NaCl for **a** 1 h, **b** 48 h, and **c** 96 h [62]

density was decreased compared to the case of coating with 1.2% or even without any inhibitor, where the corrosion current is the current produced while electrochemical corrosion is occurring, and electrons move from the anode to the cathode. Probably, by increasing the benzotriazole content, more corrosion sites are blocked. The coatings with 3.6% benzotriazole, shows the self-healing ability. The coating and polarization resistance would be decreased when the corrosion reactions happen in the defects during the immersion. In the case of the sample with 3.6% benzotriazole, its slow release from the coatings when exposed to the solution and probably the passivation of defects, produces the self-healing property. However, passivation area of anodic branch experienced some oscillations. Casenave and Coll [85] by studying the aluminum corrosion in the presence of benzotriazole showed the benzotriazole suitability as a cathodic inhibitor. Other studies have named anodic and/or cathodic inhibitors [53, 64] and it is shown that the inhibitory mechanism of BTAH is related to its adsorption to the Al surface and subsequent formation of [BTA]ads:Al layer [62, 81].

Finally, it should be said that the benzotriazole concentration is a main factor regarding the self-healing properties of the coating as its steady release, specifically in the case of 3.6%, when exposed to an aqueous solution, render the self-healing ability. Also, elevated polarization resistance can be stemmed from the defects passivation. However, as there is a race between corrosion and healing process, oscillations can be seen in the cathodic branch of passivation area. The coating with 3.6% benzotriazole can provide the highest protection efficiency of 76% [62].

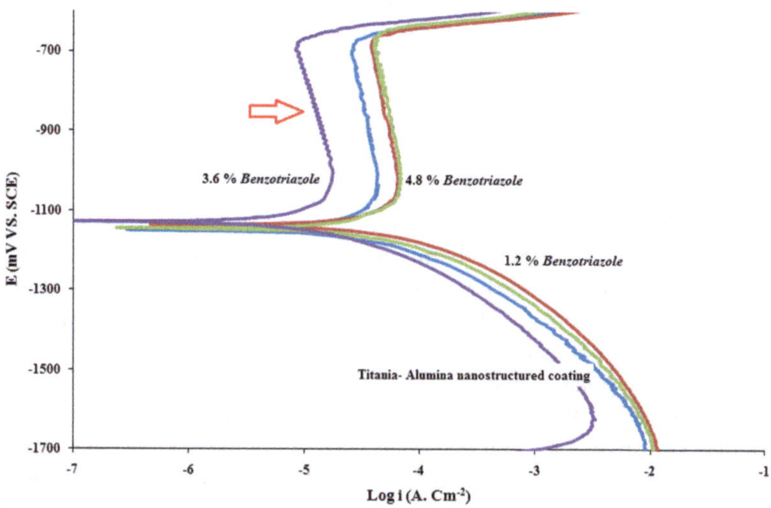

Fig. 6 Polarization curves (electrode potential versus the logarithm of the current density) of the TiO_2–Al_2O_3 coatings without benzotriazole, and with three various benzotriazole concentrations after being immersed in 3.5 wt. % NaCl solution for 96 h [62]

Table 2 Acquired data from polarization tests of the TiO_2–Al_2O_3 coatings containing benzotriazole at three concentrations [62]

Samples	i_{corr} (μA/Cm²)	βa (V/dec)	$β_c$ (V/dec)	Ecorr (mV)	i_{pass} (μA/Cm²)	ηp (%)
Titania–alumina (no benzotriazole)	16.6	0.668	0.277	−1149	36.3	–
Titania–alumina coating (1.2% benzotriazole)	20.4	0.710	0.332	−1135	56.2	−23
Titania–alumina coating (3.6% benzotriazole)	3.99	0.626	0.286	−1129	19.9	76
Titania–alumina coating (4.8% benzotriazole)	13.4	0.966	0.328	−1147	61.6	19

5.5 Zirconia Coating

Zirconia (ZrO_2) is one of the important ceramics with enhanced mechanical strength, thermal stability, wear, and corrosion resistances with an acceptable level of chemical stability and high hardness [86]. High degree of cracks happened in the coating of ZrO_2 prepared via sol–gel method is one of its main drawbacks and application of stabilizing agents can be effectively reduced this drawback. It's believed that these cracks form during the coating heat treatment due to stress formation. By in situ observing the gel film formation under heat treatment condition by Kozuka, it was revealed that the microscopic cracks formed in the heating-up stage and the starting temperature for the crack formation is highly depended on heating rate, film thickness, water-to-alkoxide ratio, and humidity. Via two methods, the film thickness can be increased without further crack development, (1) repetition of the coating procedure and (2) adding organic polymers [87]. The cracks and the non-homogeneity of the coating only containing ZrO_2 can be seen in Fig. 7. Zirconia single-component coating. These cracks formed during the heat treatment procedure and subsequently solvent evaporation, due to the high thermal expansion coefficient [88, 89]. The ZrO_2–BTA hybrid coating shows a well-distributed, nanostructured, and crack-free structure with the benzotriazole as an inhibitor agent which its reaction with the coating components can prevent creation and development of cracks thereby resulting in a homogeneous nanostructured surface.

According to the polarization curves (Fig. 8), the coating function as a physical protective barrier to block electrochemical reactions and changes in polarization plots toward a more positive corrosion potential (E_{corr}) and lower corrosion current (i_{corr}) can be seen. Electrochemical changes are observed between the uncoated samples and coated samples as well. The cathodic nature and anodic part of the polarization plot of Al 2024 are changed by the sol–gel coatings. The ZrO_2–BTA nanostructured coating shows a more positive corrosion potential and lower corrosion current density than the coatings and the corrosion resistance increases due to the anti-corrosive effects of benzotriazole. The coatings containing the appropriate concentration of benzotriazole blocks sites and increases the corrosion resistance by two steps. The first

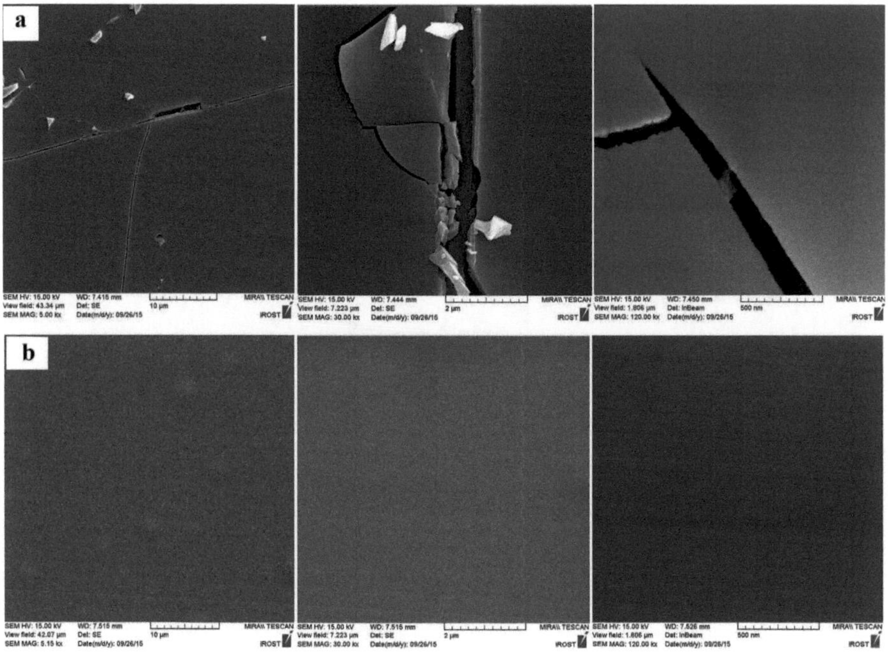

Fig. 7 FE-SEM images of the coatings deposited on Al 2024 alloy and heated at 150 °C for 2 h **a** ZrO_2 and **b** ZrO_2–benzotriazole [89]

deterring effect reaches the metal–solution interface and then absorbs the inhibitor on the metal surface. The ceramic coating based on sol–gel, works like a dielectric material and the corrosion can be only started at the defects and flaws the presence in the coating. Hence, by diffusing the water and oxygen via the cracks and reaching the substrate surface, the corrosion can be started [89].

The Nyquist and Bode plots of Al 2024, ZrO_2, and ZrO_2–BTA nanostructured coatings in the 3.5 wt.% NaCl solution after immersion for 1 and 120 h (Fig. 9) confirm that the ZrO_2 coating has acceptable corrosion resistance and BTA in the zirconia coating enhances the resistance against corrosion. The hybrid coating showed better corrosion resistance compared to the inorganic one as its surface has lower level of flaws and imperfections. The sol–gel coatings corrode only through defects which allow the electrolyte to access the metal surface [89, 90]. The healing procedure can be seen normally through the polymerization reaction of the healing agent inside of the coating. Rupturing of the microcapsules containing the healing agents, during the crack formation lead to release of encapsulated compound and these compounds will go through the polymerization process when they contact with the catalyst. During exposure to water or moisture, the healing reaction proceeds by polymerization, cross-linking, or condensation to create a film to separate the metal substrate from the corrosive environment and consequently the corrosion process was reduced. Generally, The ZrO_2–BTA nanostructured coating improves the corrosion

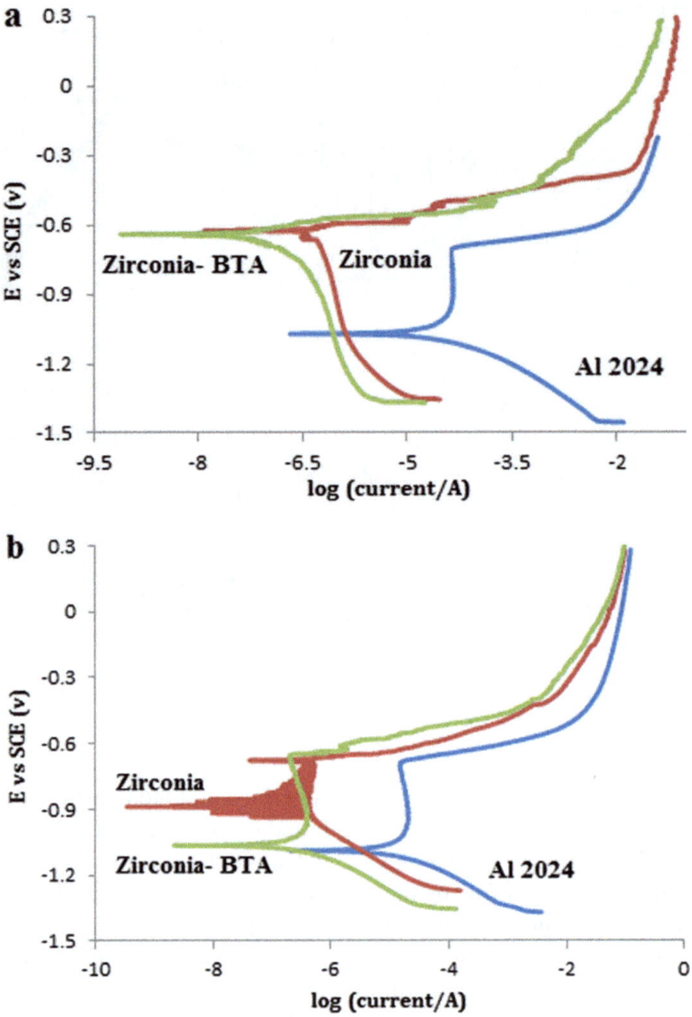

Fig. 8 Potentiodynamic polarization curves in the 3.5 wt.% NaCl solution of coatings: **a** 1 h and **b** 120 h [89]

resistance of the substrate in terms of corrosion potential, corrosion current density, passivity, and passive potential range, and the protection level of the ZrO_2–BTA nanostructured coating is better than that of the inorganic ZrO_2 coating. In general, BTA improves the corrosion behavior of the ZrO_2–BTA nanostructured coating in two ways by (1) formation of consistent and resilient coating from corrosion products and providing Warburg impedance and (2) formation of consistent and dense dual-layer to obstruct the ions and electrons movement at the interface of substrate and corrosive medium [90].

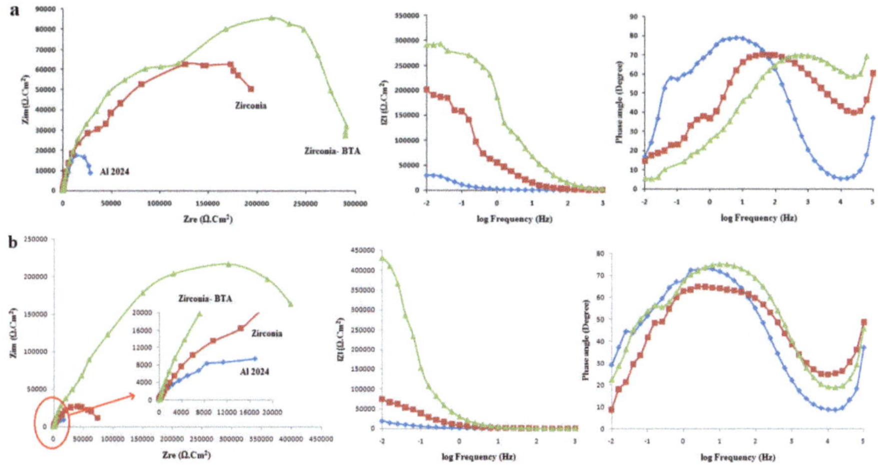

Fig. 9 Nyquist and bode-phase curves of the samples after immersion in the 3.5 wt.% NaCl solution for different times: **a** 1 h and **b**120 h [89]

5.6 Zirconia–Alumina Composite Coatings

The mismatching of physical and chemical features of ZrO_2 and the substrate, Al 2024, is the reason for poor corrosion properties of this combination as the flaws can be created on ZrO_2 coating. By adding some compounds such as magnesium, yttrium and calcium oxides as stabilizer agents, the cubic and tetragonal phases will be more stable, and consequently, the harmful volume expansion due to transforming the tetragonal phase to monoclinic one will be prevented. Here, to enhance the behavior of ZrO_2 coating, Al_2O_3 is used to make the coating more stable and enhance its function [91–93].

According to the FE-SEM images showed in Fig. 10, cracks can be seen in the ZrO_2–Al_2O_3 coating (Fig. 10a) compared to the ZrO_2–Al_2O_3–benzotriazole coating (Fig. 10b), because of huge thermal expansion coefficient of ZrO_2 and removal of organic substances upon the heat treatment procedure. By performing the drying process in isopropanol and heat treatment procedure with a slower heating rate (1°C/min), crack propagation can be controlled [93, 94]. On the other hand, by using benzotriazole, no cracks and flaws would be seen due to the better ZrO_2–Al_2O_3–benzotriazole polymerization process. As shown in Fig. 10c–d, the elemental map of O confirms uniform distributions in the coatings. There is a larger percentage of oxygen and aggregation in the coating containing benzotriazole compared to the other coatings. One of the main roles of benzotriazole in smaller flaws is preventing crack propagation and its further development and therefore, providing flawless and crack-free coating [93–95]. Its mechanism of action is releasing the inhibitor into the cracks and by its reaction to the aluminum surface, protective [benzotriazole]ads:Al layer fills out the porosities. The thickness of the ZrO_2–Al_2O_3 and ZrO_2–Al_2O_3–benzotriazole coatings are 810 and 950 nm, respectively (Fig. 10 e–f).

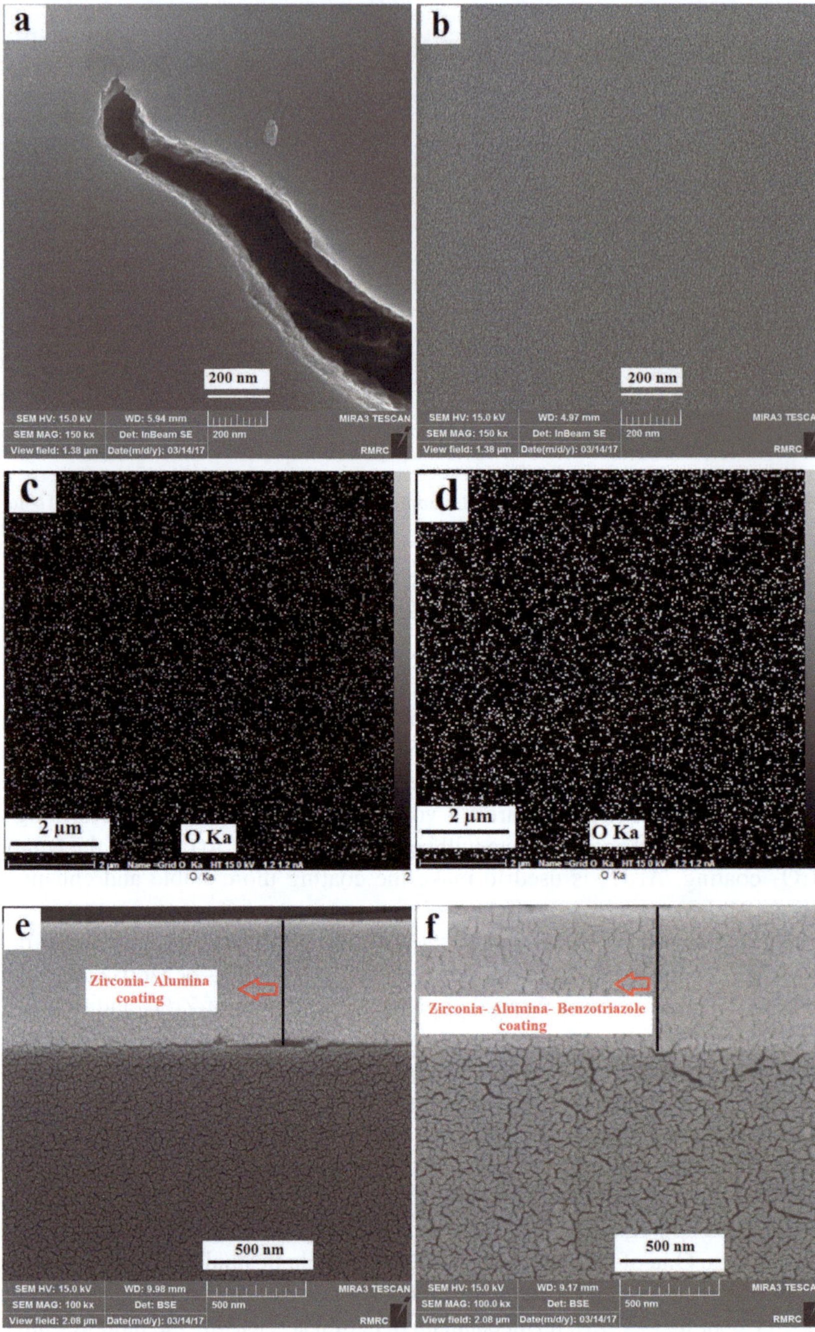

Fig. 10 FE-SEM images of **a** ZrO_2–Al_2O_3 coating, and **b** ZrO_2–Al_2O_3–benzotriazole; Elemental map of O acquired from **c** ZrO_2–Al_2O_3 coating, and **d** ZrO_2–Al_2O_3–benzotriazole coating; FE-SEM cross-sectional images of **e** ZrO_2–Al_2O_3 coating, and **f** ZrO_2–Al_2O_3–benzotriazole coating [93]

According to the Nyquist plots showed in Fig. 11, the dense ZrO_2–Al_2O_3–benzotriazole coating shows better corrosion compared to the other groups. Although, there is a race between water absorption and chloride ions diffusion, however due to benzotriazole release, different behavior can be seen at different immersion time span. In this regard, after being immersed for 1–4 h, as this race is going on side by side, there is no difference between the corrosion resistance behavior of the different coatings. However, when the immersion time reached 6 h, excessive release of benzotriazole overcomes the other factors, water absorption and aggressive ions diffusion and thus, better corrosion resistance can be seen from ZrO_2–Al_2O_3–benzotriazole coating.

The presence of some regions like Al_2Cu, Al_2CuMg, and Al_2Mg_3, as the cathodic areas make the substrate, the Al 2204, more susceptible to intergranular corrosion. As the intergranular copper rich area would be dissolved, some sediments would be seen at the boundaries of corroded grains [84, 86]. According to the Nyquist plot

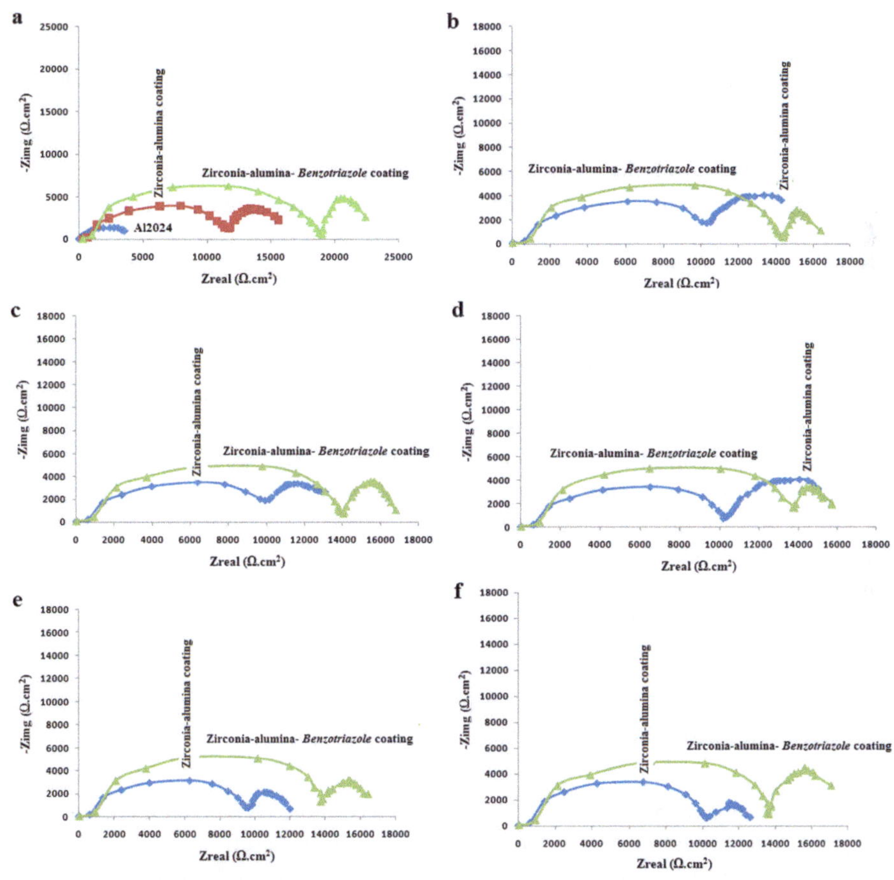

Fig. 11 Nyquist plots of the ZrO_2–Al_2O_3 and ZrO_2–Al_2O_3–benzotriazole immersed for **a** 1 h, **b** 2 h, **c** 3 h, **d** 4 h, **e** 5 h, and **f** 6 h [93]

acquired from samples immersed for 1 h, the ZrO_2–Al_2O_3 coating hampered the galvanic corrosion and therefore improved the corrosion resistance. By extending the immersion time, due to water absorption and chloride ions diffusion within the cracks and flows, the corrosion resistance would be declined. benzotriazole relieves the cathodic and anodic reactions and also by hampering OH^- formation retains the neutrality of the solution exposed to the surface for longer time. Moreover, by forming corrosion products, the surface can be protected [62, 93–96]. Two mechanisms can be considered for zirconia–alumina–benzotriazole coating: (i) chloride ions diffusion prevention and adsorbing water via a homogenous coating without any crack (immersed for 1–4 h), and (ii) release of benzotriazole from the coating. As it can be seen in Fig. 11, benzotriazole mechanism of action is based on the below two states: (1) hampering cathodic reactions as a result of corrosion products creation via reaction with oxygen and (2) improving the double layer behavior developed between the surface of coating and solution (this enhancement can be verified due to improving the capacitance of double layer (C_{dl}) and the resistance between aqueous medium and the coating, after being immersed for 4–6 h) [93].

benzotriazole protects the surface of aluminum through formation of [benzotriazole]ads–Al layer. This can be explained by formation of BTA^- due to the loss of hydrogen from nitrogen in benzotriazole and its reaction with Al ions to form [benzotriazole]ads–Al [62, 93]. Many factors have impact of inhibitor molecules absorption, such as the functional group, electron density, π orbitals, as well as charged metal surface.

According to the graph acquired from the open circuit potentials (OCP) vs. time (Fig. 12), it can be seen that the potential for ZrO_2–Al_2O_3–benzotriazole, first shifted toward the positive direction followed by its negative shift. By application of benzotriazole, it can be seen from the polarization plot that the cathodic reaction was effectively hampered, delayed oxygen or OH^- reduction and consequently the corrosion of aluminum was prevented [62, 93]. Due to better protection on the case of ZrO_2–Al_2O_3–benzotriazole coating compared to the ZrO_2–Al_2O_3, smaller potential fluctuation can be seen. The shifting of OCP to more positive values by extending the immersion period, can be a result of benzotriazole secretion into the medium. Though, the corrosion was prevented effectively due to the presence of stable and flawless ZrO_2–Al_2O_3–benzotriazole coating and consequently the variation of OCP is small and insignificant.

6 Nanomechanical Properties

In addition to improving the corrosion behavior of ceramics-base coatings containing inhibitors by self-healing mechanisms, the mechanical properties of the coatings are of great importance. The intrinsic properties of ceramics such as high hardness lead to increased hardness and reduced flexibility of ceramic coatings, and usually different components of ceramic composite coatings lead to improved mechanical properties of the coatings. It has been shown that the alumina coating enhances the mechanical

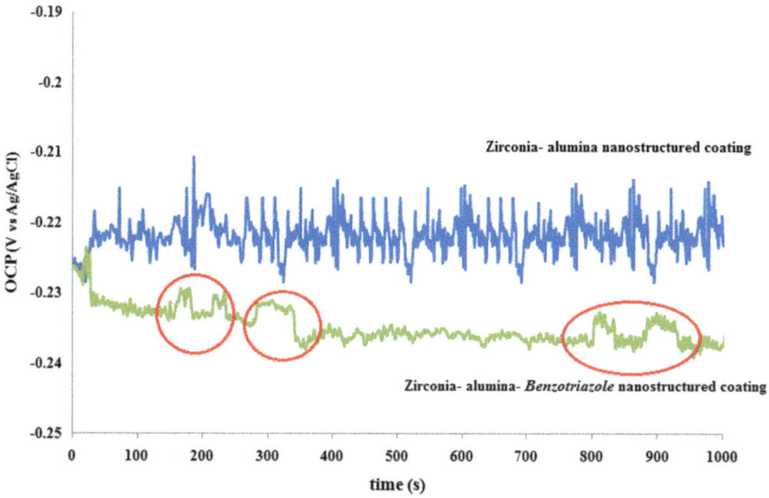

Fig. 12 Open circuit potential (OCP) versus immersion time for the ZrO_2–Al_2O_3 and ZrO_2–Al_2O_3–benzotriazole coatings [93]

properties of sol–gel coatings on stainless steel or nitrided steel in terms of hardness and wear resistance, however, the tribological behavior of the coatings would be deteriorated due to the presence of cracks and defects. Adding ZrO_2 and TiO_2 to the coatings, decreased and increased the microhardness, respectively [97, 98]. benzotriazole in the nanostructured ZrO_2–Al_2O_3–benzotriazole coating improves the corrosion resistance of Al 2024 with a protection efficiency of 76% [62]. Although, this enhancement is related to lower defects, however, a more comprehensive evaluation of the nanomechanical properties of hybrid ceramics-based coatings is necessary to understand the wear mechanism.

To evaluate the mechanical properties of the coatings, nanoindentation and nanoscratch tests are done. The curves and AFM images related to the nanoindentation tests at the applied loads of 50 and 60 μN are shown in Fig. 13 and the results, including hardness, elastic modulus and plastic deformation area are summarized in Table 3. Accordingly, results acquired for all the mentioned parameters are higher in the case of ZrO_2 compared to the ZrO_2–Al_2O_3–benzotriazole ones at the load of 50 μN. These imperfections in the ZrO_2 coating cause higher energy absorption. At the load of 60 μN, the ZrO_2–Al_2O_3–benzotriazole coating showed lower hardness while its elastic modulus and plastic deformation area are larger compared to the other groups. In general, by increasing the elastic modulus, dislocations movement will be harder and the reduction in hardness can be related to the organic compounds. The strengthening effect which has been seen in mechanical properties of ZrO_2–Al_2O_3–benzotriazole coating is related to the soli solution formation [93, 94, 96].

At the load of 60 μN, not only the ZrO_2–Al_2O_3–benzotriazole coating showed higher hardness and elastic modulus, but also, bigger plastic deformation area can

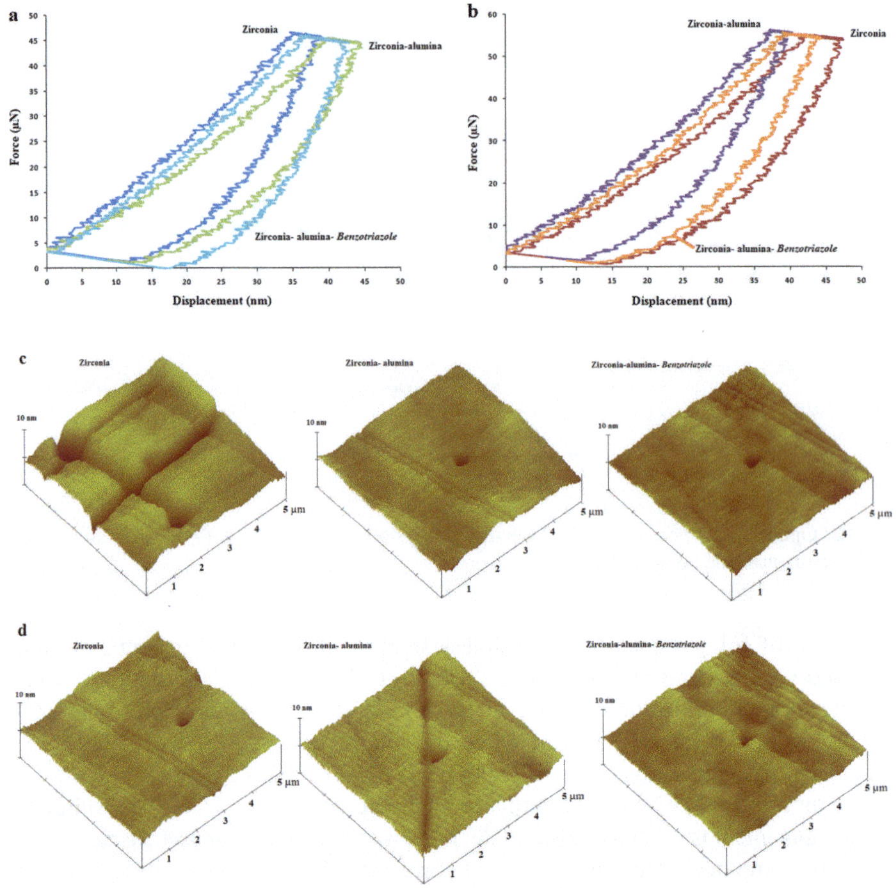

Fig. 13 Nanoindentation graphs for the coatings, exposed to two different loads: **a** 50 μN and **b** 60 μN; AFM images after nanoindentation of the coatings at different loads: **c** 50 μN and **d** 60 μN [94]

Table 3 Coatings parameters acquired from the nanoindentation test at the loads of 50 and 60 μN [94]

Depth of penetration (nm)	Hardness (H) (GPa)	Elastic modulus (E) (GPa)	Coatings	Force (μN)
39	0.93	14	ZrO_2	50
43	0.75	10.2	ZrO_2–Al_2O_3	
42	0.69	15	ZrO_2–Al_2O_3–benzotriazole	
48	0.79	12.1	ZrO_2	60
44	1.15	15	ZrO_2–Al_2O_3	
45	0.91	13.8	ZrO_2–Al_2O_3–benzotriazole	

be seen compared to the ZrO_2 coating. Less defects, cracks free, uniform and integrated coatings improve the strength of the ZrO_2–Al_2O_3 coating compared to the ZrO_2 coating. As an organic compound, benzotriazole enhanced the behavior of the coating at the load of 60 μN by making a significant contribution to the reduction of hardness, elasticity and flexibility of the coating compared to the ZrO_2–Al_2O_3 coating. In fact, by increasing the tension and the applied force, the coatings' strength and solid solution formation are reduced and enhanced, respectively, due to the presence of benzotriazole. The coating hardness and elasticity would be decreased by increasing the force from 50 to 60 μN. By applying higher load, defects play a more prominent role and make the dislocations to move easier in the ZrO_2 coating, as it can be concluded from the nanoindentation plot and maximum penetration depth. The ZrO_2–Al_2O_3–benzotriazole coating shows a dissimilar behavior compared to the ZrO_2 coating, as the hardness and elastic modulus increased by elevating the applied force from 50 to 60 μN, although those would be decreased at the deepest depth of penetration and also in the horizontal part of the nanoindentation graph. The elastic modulus of ZrO_2–Al_2O_3 coating is lower than the ZrO_2 coating at the applied load of 50 μN, however by increasing the applied force to 60 μN, the penetration depth would be increased. The higher force affects a larger area and the dislocations movements are hampered by the ZrO_2–Al_2O_3 solid solution [94, 96].

The mechanical behavior is assessed via the nanoscratch test. The image of the scratch and the roughness of ZrO_2, ZrO_2–Al_2O_3, and ZrO_2–Al_2O_3–benzotriazole at 50 and 60 μN (Fig. 14) indicate that both the ZrO_2 and ZrO_2–Al_2O_3 coatings show a more apparent plasticity encountered to indentation by increasing the load from 50 to 60 μN, as shown by surface roughness plot in the movement path of indentation. It can be due to the sluggish dislocation movement at the load of 50 μN. Under a load of 60 μN, plastic deformation and dislocations movement causes a shrinkage at the end of the indenter path. Such shrinking, and also groove edges rounding represent the high plastic deformability for the coating with benzotriazole at the load of 50 μN. Two interesting points can be seen in the ZrO_2–Al_2O_3–benzotriazole coating friction coefficient at the load of 60 μN. These repeatable fluctuations stemmed from the homogeneity of coating properties throughout the indenting path and the rise and fall of the friction coefficient can be explained by even distribution of benzotriazole in the coating.

Benzotriazole increases the flexibility of the coating and plays a dominant role in wear adhesion. By roughness increasing and the ZrO_2–Al_2O_3–benzotriazole shrinkage through the path of indentation, friction coefficient was first increased and then decreased. In general, alumina increases the ability of plastic deformation and impedes movement of dislocations, whereas the flexibility was increased due to benzotriazole, since the coating was piled up at the edge and at the end of the friction path [99–102].

Figure 15 illustrates that the average friction coefficients of the ZrO_2, ZrO_2–Al_2O_3, and ZrO_2–Al_2O_3–benzotriazole coatings are 0.38, 0.32, and 0.29 under a load of 50 μN, and 0.48, 0.51, and 0.41 under 60 μN. According to the wear curves and the coatings diagram for surface roughness at the load below 50 μN, abrasion with plowing can be seen as the dominant mechanism. The elastic and plastic deformation

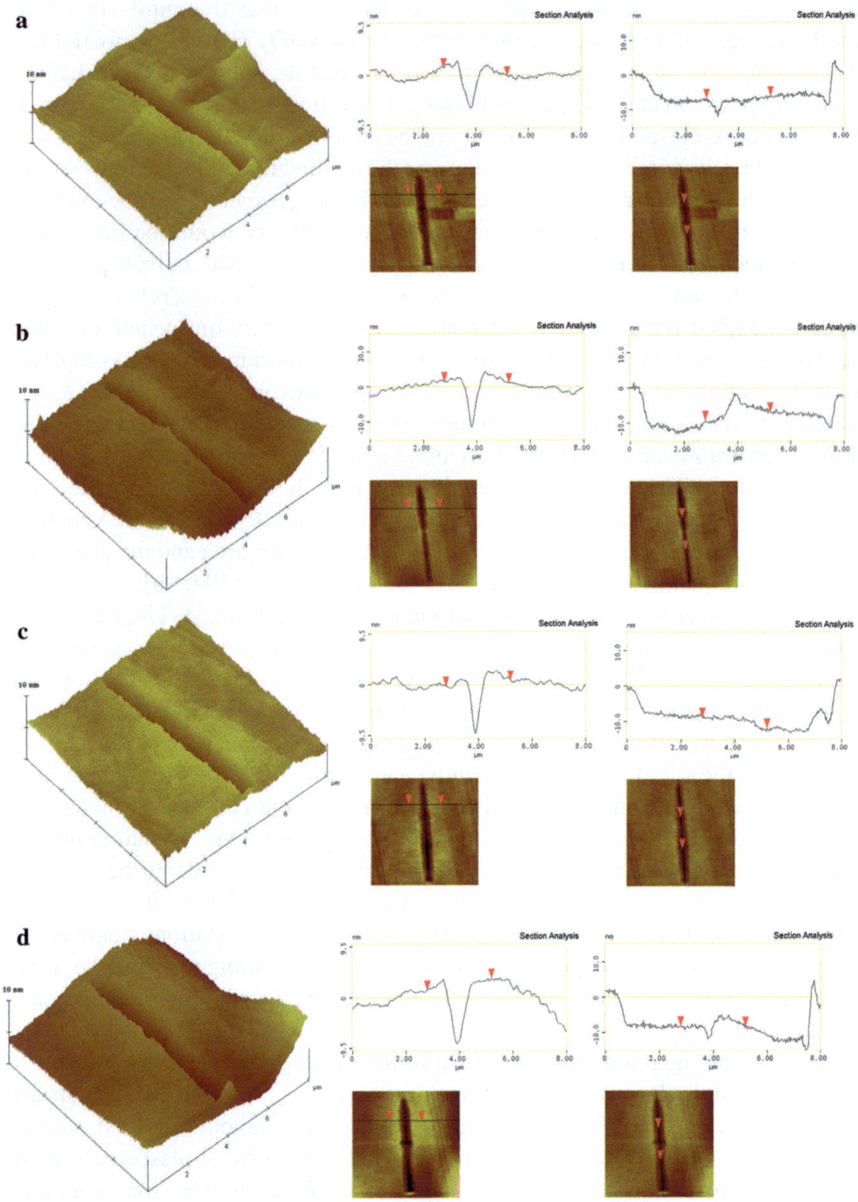

Fig. 14 The AFM image of the nanoscratched area and its relative surface roughness by applying 50 μN load on **a** ZrO_2, **b** ZrO_2–Al_2O_3, and **c** ZrO_2–Al_2O_3–benzotriazole coatings, and 60 μN load on: **d** ZrO_2, **e** ZrO_2–Al_2O_3, and **f** ZrO_2–Al_2O_3–benzotriazole coating [94]

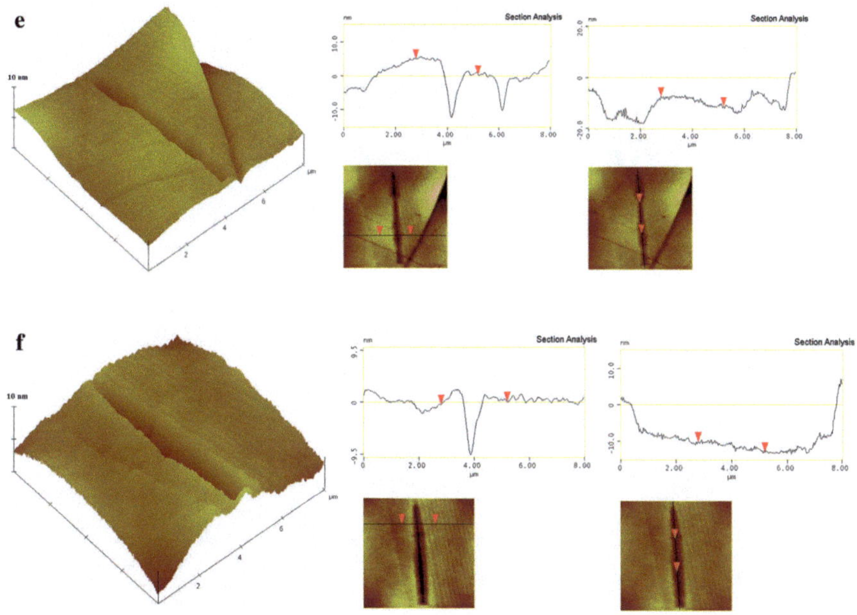

Fig. 14 (continued)

the plowing phenomenon, which can act as barriers throughout the wear test thus giving rise to larger friction coefficients for the ZrO_2–Al_2O_3–benzotriazole coating. One method to reduce the friction coefficients and improve the wear behavior is to increase the hardness of the surface and in this regard, the ZrO_2 coating has the lowest friction coefficient due to the largest hardness under 50 μN [94].

50 μN load may not apply sufficient plastic deformation and so abrasion is the dominant wear mechanism. In the beginning, small plastic deformation occurs in the indentation path due to tensions caused by the formation of the homogeneous and uniform coating. In comparison, as the adhesion can be seen in the wear area of ZrO_2–Al_2O_3–benzotriazole coating, it can be said that the dominant wear mechanism is adhesion [94]. By increasing the load to 60 μN, the dominant mechanism is changed to the abrasion with shear for the ZrO_2–Al_2O_3 and ZrO_2–Al_2O_3–benzotriazole coatings, and the proof for this fact is the noticeable peaks and valleys in the wear graphs, especially the one with benzotriazole. As the strength of ZrO_2–Al_2O_3 is higher compared to the ZrO_2–Al_2O_3–benzotriazole and the 60 μN load has an influence on dislocation movement, the former one has higher indentation resistance.

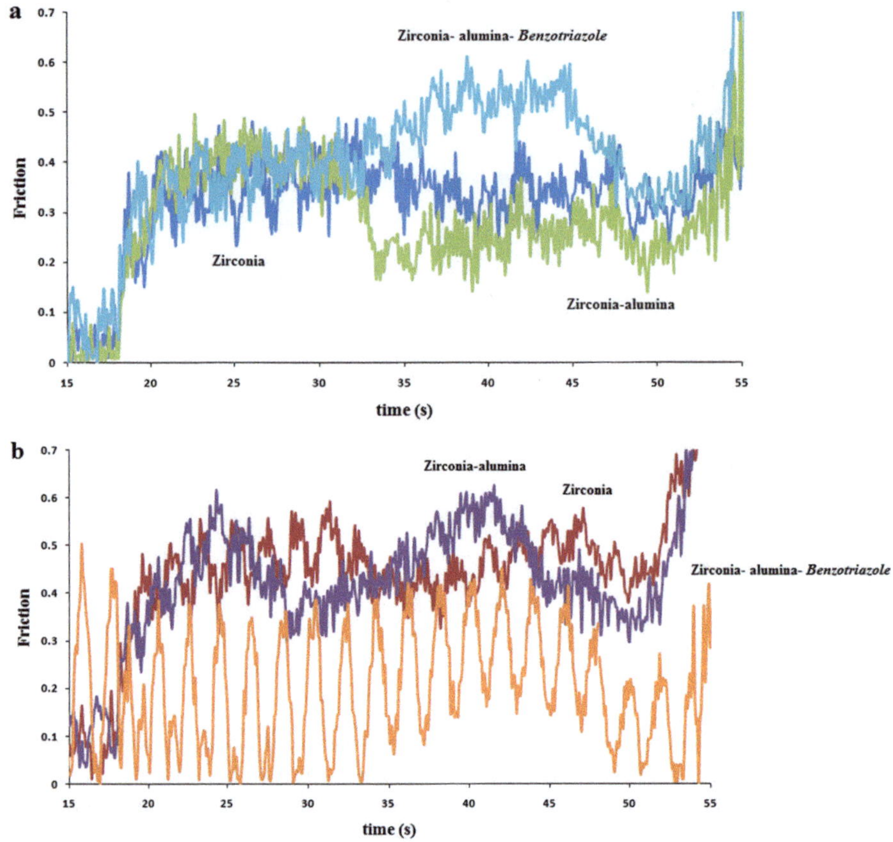

Fig. 15 Friction coefficients graphs of the ZrO_2, ZrO_2–Al_2O_3, and ZrO_2–Al_2O_3–benzotriazole coatings under **a** 50 μN, and **b** 60 μN [94]

7 Conclusion and New Perspective

Much effort has been made to develop self-healing coatings with high corrosion resistance and desirable mechanical properties in order to increase the lifetime of metallic components. In particular, ceramic coatings are widely used in the industry due to the high hardness, low friction, high corrosion resistance, and robust chemical resistance, but cracking and porosity reduce the corrosion resistance leading to possible failure in the field. In this respect, self-healing agents can be incorporated into the structure to increase the properties and longevity. Incorporation of repairing agents into ceramic coatings increases the corrosion resistance by leaving repairing agents at defects such as cavities and cracks, and the reactions with oxygen and moisture produce corrosion products that facilitate repair at defects. The proper choice of the inhibiting polymers and their products increases the flexibility and strength of the ceramic coatings along with enhanced corrosion resistance. However, it should be mentioned that these materials are still in the experimental stage and more research is required to bring the technology to industrial fruition.

At present, existing technologies for designing self-healing materials are expensive, which has hampered the widespread usage of the materials for various common applications, and with the spread of science, it is expected that new advancement will make the use of self-healing materials possible in everyday life. In future, one of the most important applications of ceramic-based self-healing materials can be their use in biomaterials along with using natural and biocompatible inhibitors so that biocompatible ceramic-based self-healing composites can improve durability of artificial bones, dentures, and so on.

References

1. I. Gonzalez-Torre, J. Norambuena-Contreras, Constr. Build. Mater. **258**, 119568 (2020)
2. S.K. Ghosh (ed.), *Self-Healing Materials: Fundamentals, Design Strategies, and Applications* ed. by S.K. Ghosh (Wiley, Weinheim, 2009). pp. 1–25
3. W. Wang, N.G. Moreau, Y. Yuan, P.R. Race, W. Pang, Comput. Mater. Sci. **168**, 180 (2019)
4. P. Fratzl, R. Weinkamer, in *Self Healing Materials: An Alternative Approach to 20 Centuries of Materials Science*, edited by S. van der Zwaag (Springer, 2007), pp. 323–335
5. F.J. Vermolen, W.G. van Rossum, E. Javierre, J.A. Adam, in *Self Healing Materials: An Alternative Approach to 20 Centuries of Materials Science*, edited by S. van der Zwaag (Springer, 2009), pp. 337–363
6. M.R. Kessler, Proc. Inst. Mech. Eng. Part G J. Aerosp. Eng. **221**, 479 (2007)
7. W. Zhang, Q. Zheng, A. Ashour, B. Han, Compos. Part B Eng. **189**, 107892 (2020)
8. K. Li, Z. Liu, C. Wang, W. Fan, F. Liu, H. Li, Y. Zhu, H. Wang, Prog. Org. Coatings **145**, 105668 (2020)
9. H. Li, Y. Feng, Y. Cui, Y. Ma, Z. Zheng, B. Qian, H. Wang, A. Semenov, D. Shchukin, Prog. Org. Coatings **145**, 105684 (2020)
10. A.G. Cordeiro Neto, A.C. Pellanda, A.R. de Carvalho Jorge, J.B. Floriano, M.A. Coelho Berton, Prog. Org. Coatings **147**, 105874 (2020)
11. S.K. Ghosh, *Functional Coatings by Polymer Microencapsulation*, ed. by S.K. Ghosh (Wiley, Weinheim, 2006), pp. 1–28
12. S. Javadian, Z. Ahmadpour, A. Yousefi, Prog. Org. Coatings **147**, 105678 (2020)
13. S. Xu, J. Li, H. Qiu, Y. Xue, J. Yang, Compos. Commun. **19**, 220 (2020)
14. S.R. White, N.R. Sottos, P.H. Geubelle, J.S. Moore, M.R. Kessler, S.R. Sriram, E.N. Brown, S. Viswanathan, Nature **409**, 794 (2001)
15. A. Danish, M.A. Mosaberpanah, M. Usama Salim, J. Mater. Res. Technol. **9**, 6883 (2020)
16. D. Mohanty, A. Sil, K. Maiti, Ceram. Int. **37**, 1985 (2011)
17. F. Tavangarian, D. Hui, G. Li, Compos. Part B Eng. **144**, 56 (2018)
18. L. Restuccia, A. Reggio, G.A. Ferro, J.M. Tulliani, Procedia Struct. Integr. **3**, 253 (2017)
19. F. Awaja, S. Zhang, M. Tripathi, A. Nikiforov, N. Pugno, Prog. Mater. Sci. **83**, 536 (2016)
20. B.S. Kim, K. Ando, M.C. Chu, S. Saito, J. Soc. Mater. Sci. Jpn. **52**, 667 (2003)
21. K. Ando, B.-S. Kim, S. Kudama, S.-P. Liu, K. Takahashi, S. Saito, Journal **52**, 1464 (2003)
22. W. Nakao, T. Osada, K. Yamane, K. Takahashi, K. Ando, J. Jpn. Inst. Met. **69**, 663 (2005)
23. CHASE and M.W., J. Phys. Chem. Reffernce Data **1529** (1998)
24. M. Nosonovsky, B. Bhushan, Mater. Sci. Eng. R. Rep. **58**, 162 (2007)
25. M. Nosonovsky, S.K. Esche, Entropy **10**, 49 (2008)
26. F. Zhang, P. Ju, M. Pan, D. Zhang, Y. Huang, G. Li, X. Li, Corros. Sci. **144**, 74 (2018)
27. G. Mu, X. Li, Q. Qu, J. Zhou, Corros. Sci. **48**, 445 (2006)

28. A.E. Hughes, in *Recent Advances in Smart Self-Healing Polymers and Composites*, ed. by G. Li, H.B.T.-R.A. in S.S.P. and C. Meng (Woodhead Publishing, 2015), pp. 211–241
29. C. Monticelli, in *Encyclopedia of Interfacial Chemistry: Surface Science and Electrochemistry*, Vol. 6, ed. by K. Wandelt (Elsevier, 2018), pp. 164–171
30. Y. Morozov, L.M. Calado, R.A. Shakoor, R. Raj, R. Kahraman, M.G. Taryba, M.F. Montemor, Corros. Sci. **159**, 108128 (2019)
31. Y. Su, S. Qiu, D. Yang, S. Liu, H. Zhao, L. Wang, Q. Xue, J. Hazard. Mater. **391**, 122215 (2020)
32. L.B. Coelho, E.B. Fava, A.M. Kooijman, Y. Gonzalez-Garcia, M.-G. Olivier, Corros. Sci. **175**, 108893 (2020)
33. D. Li, F. Wang, X. Yu, J. Wang, Q. Liu, P. Yang, Y. He, Y. Wang, M. Zhang, Prog. Org. Coat. **71**, 302–309 (2011)
34. R.B. Figueira, R. Sousa, C.J.R. Silva, in *Advances in Smart Coatings and Thin Films for Future Industrial and Biomedical Engineering Applications*, ed. by A.S.H. Makhlouf and N. Y. B. T.-A. in S. C. and T. F. for F. I. and B. E. A. Abu-Thabit (Elsevier, 2020), pp. 57–97
35. P. Wang, X. Dong, D.W. Schaefer, Corros. Sci. **52**, 943–949 (2010)
36. C. Verma, E.E. Ebenso, M.A. Quraish, J. Mol. Liq. **316**, 113874 (2020)
37. F. Ubaid, A.B. Radwan, N. Naeem, R.A. Shakoor, Z. Ahmad, M.F. Montemor, R. Kahraman, A.M. Abdullah, A. Soliman, Surf. Coatings Technol. **372**, 121 (2019)
38. S.A. Umoren, U.M. Eduok, Carbohydr. Polym. **140**, 314 (2016)
39. L.P. Kazansky, I.A. Selyaninov, Yu.I. Kuznetsov, Appl. Surf. Sci. **258**, 6807–6813 (2012)
40. Y. Chen, B. Ren, S. Gao, R. Cao, J. Colloid Interface Sci. **565**, 436–448 (2020)
41. H. Shi, F. Liu, E. Han, Mater. Chem. Phys. **124**, 291 (2010)
42. T. Hack, D. Raps, R. Supplit, U. Schubert, US 8,361,627 B2 (2013)
43. N.P. Tavandashti, S. Sanjabi, Prog. Org. Coatings **69**, 384 (2010)
44. S.B. Ulaeto, R. Rajan, J.K. Pancrecious, T.P.D. Rajan, B.C. Pai, Prog. Org. Coatings **111**, 294 (2017)
45. D.G. Shchukin, H. Möhwald, Small **3**, 926 (2007)
46. M. Toorani, M. Aliofkhazraei, A. Sabour Rouhaghdam, Surf. Coatings Technol. **352**, 561 (2018)
47. A.F. Jaramillo, L.F. Montoya, J.M. Prabhakar, J.P. Sanhueza, K. Fernández, M. Rohwerder, D. Rojas, C. Montalba, M.F. Melendrez, Prog. Org. Coatings **135**, 191 (2019)
48. A. Conde, A. Durán, J.J. de Damborenea, Prog. Org. Coatings **46**, 288 (2003)
49. X. Zhang, Y. Lv, S. Fu, Y. Wu, X. Lu, L. Yang, H. Liu, Z. Dong, Mater. Sci. Eng. C **117**, 111321 (2020)
50. M.L. Zheludkevich, R. Serra, M.F. Montemor, I.M.M. Salvado, M.G.S. Ferreira, Surf. Coatings Technol. **200**, 3084 (2006)
51. A. Ghafari, M. Yousefpour, A. Shanaghi, Appl. Surf. Sci. **465**, 427 (2019)
52. L.S. Živković, B.V. Jegdić, V. Andrić, K.Y. Rhee, J.B. Bajat, V.B. Mišković-Stanković, Prog. Org. Coatings **136**, 105219 (2019)
53. S.M. Dezfuli, M. Sabzi, Int. J. Appl. Ceram. Technol. **15**(5), 1248–1260 (2018)
54. H. Pulikkalparambil, S. Siengchin, J. Parameswaranpillai, Nano-Struct. Nano-Objects **16**, 381–395 (2018)
55. R. Corriu, T.A. Nguyên, *Molecular Chemistry of Sol-Gel Derived Nanomaterials* (Wiley Online Library, 2009)
56. A.K. Zak, W.H.A. Majid, Ceram. Int. **37**, 753 (2011)
57. S.M. Madani, M. Ehteshamzadeh, H.H. Rafsanjani, S.S. Mansoori, Mater. Corros. **61**, 318 (2010)
58. A. Stankiewicz, I. Szczygieł, B. Szczygieł, J. Mater. Sci. **48**, 8041 (2013)
59. K. Yasakau, *Active Corrosion Protection of AA2024 by Sol-Gel Coatings with Corrosion Inhibitors* (Universidade de Aveiro, 2011)
60. A. Shanaghi, M. Kadkhodaie, Corros. Eng. Sci. Technol. **52**, 332 (2017)
61. A. Shanaghi, P.K. Chu, H. Moradi, Surf. Rev. Lett. **24**, 1 (2017)
62. M.S. Sharifiyan, A. Shanaghi, H. Moradi, P.K. Chu, Surf. Coatings Technol. **321**, 36 (2017)

63. M. Farahani, H. Yousefnia, Z.S. Seyedraoufi, Y. Shajari, Ceram. Int. **45**(13), 16584–16590 (2019)
64. D. Wang, G.P. Bierwagen, Prog. Org. Coatings **64**, 327 (2009)
65. K.H. Wu, T.C. Chang, C.C. Yang, G.P. Wang, Thin Solid Films **513**, 84 (2006)
66. E.D. Mekeridis, I.A. Kartsonakis, G.C. Kordas, Prog. Org. Coatings **73**, 142 (2012)
67. H. Hassannejad, T. Shahrabi, F. Malekmohammadi, A. Shanaghi, M. Aliofkhazraei, A. Oskuie, Curr. Appl. Phys. **10**, 1022 (2010)
68. D. Zare Hossein Abadi, A. Ershad Langroudi, A. Rahimi, J. Colour Sci. Technol. **3**, 121 (2009)
69. H.-L. Wang, H.-P. Qi, X.-N. Wei, X.-Y. Liu, W.-F. Jiang, Chin. J. Catal. **37**, 2025 (2016)
70. G.C.C. Yang, C.-J. Li, Desalination **200**, 74 (2006)
71. G.X. Shen, Y.C. Chen, C.J. Lin, Thin Solid Films **489**, 130 (2005)
72. X. Nie, E.I. Meletis, J.C. Jiang, A. Leyland, A.L. Yerokhin, A. Matthews, Surf. Coatings Technol. **149**, 245 (2002)
73. C.G. Darive, A.F. Galio, in *Dev. Corros. Prot.*, ed. by M. Aliofkhazraei (INTECH, 2014), pp. 365–379
74. M.R. Mohammadi, Mater. Sci. Semicond. Process. **27**, 711 (2014)
75. X. Du, K. Men, Y. Xu, B. Li, Z. Yang, Z. Liu, L. Li, L. Li, T. Feng, W. ur Rehman, I. Ullah, S. Mao, J. Colloid Interface Sci. **443**, 170 (2015)
76. K. Kadirgama, M.M. Noor, N.M. Zuki, M.M. Rahman, M.R.M. Rejab, R. Daud, K.A. Abou-El-Hossein, Jordan J. Mech. Ind. Eng. **2**, 209 (2008)
77. S.A. Yero, M.R. Hainin, H. Yacoob, Ijrras **13**, 98 (2012)
78. G. Quintana, M.L. Garcia-Romeu, J. Ciurana, J. Intell. Manuf. **22**, 607 (2011)
79. A.K. Mainjot, A. Najjar, B.D. Jakubowicz-Kohen, M.J. Sadoun, Dent. Mater. **31**, 1142 (2015)
80. H.-C. Yu, S.B. Adler, S.A. Barnett, K. Thornton, Electrochim. Acta **354**, 136534 (2020)
81. S. Darwich, *Corrosion Protection Concepts for Aluminium and Magnesium Alloys Coated with Silica Films Prepared by Water-Based Sol-Gel Process* (Technischen Universitat Chemnitz, 2012)
82. A.R. Simbar, A. Shanaghi, H. Moradi, P.K. Chu, Mater. Chem. Phys. **240**, 122233 (2020)
83. K.F. Khaled, Electrochim. Acta **48**, 2493 (2003)
84. J. Aljourani, K. Raeissi, M.A. Golozar, Corros. Sci. **51**, 1836 (2009)
85. D. Laouali, F. Bénière, J. Mater. Environ. Sci. **3**, 34 (2012)
86. S. Saridag, O. Tak, G. Alniacik, World J. Stomatol. **2**, 40 (2013)
87. S. Sakka, in *Handbook of Advanced Ceramics. Materials, Applications, Processing, and Properties*, ed. by S. Somiya, 2nd ed. (Academic Press, 2013), pp. 1–17
88. L. Paussa, N.C. Rosero Navarro, D. Bravin, F. Andreatta, A. Lanzutti, M. Aparicio, A. Durán, L. Fedrizzi, Prog. Org. Coatings **74**, 311 (2012)
89. S.M. Dezfuli, A. Shanaghi, S. Baghshahi, Prot. Met. Phys. Chem. Surfaces **54**, 1050 (2018)
90. E. Setare, K. Raeissi, M.A. Golozar, M.H. Fathi, Corros. Sci. **51**, 1802 (2009)
91. M.A. Domínguez-Crespo, A. García-Murillo, A.M. Torres-Huerta, F.J. Carrillo-Romo, E. Onofre-Bustamante, C. Yáñez-Zamora, Electrochim. Acta **54**, 2932 (2009)
92. E. Celik, I. Keskin, I. Kayatekin, F. Ak Azem, E. Özkan, Mater. Charact. **58**, 349 (2007)
93. A. Shanaghi, A.R. Souri, P.K. Chu, J. Alloys Compd. **816**, 152662 (2020)
94. A. Shanaghi, A.R. Souri, M. Rafie, P.K. Chu, Thin Solid Films **689**, 137417 (2019)
95. S.M. Dezfuli, M. Sabzi, Adv. Mater. Technol. **7**(4), 75–92 (2019)
96. A. Shanaghi, A.R. Souri, M. Rafie, P.K. Chu, Appl. Phys. A Mater. Sci. Process. **125**, 1 (2019)
97. R. Younes, M.A. Bradai, A. Sadeddine, Y. Mouadji, A. Bilek, A. Benabbas, Trans. Nonferrous Met. Soc. China **26**, 1345 (2016)
98. B.J. McEntire, B.S. Bal, M.N. Rahaman, J. Chevalier, G. Pezzotti, J. Eur. Ceram. Soc. **35**, 4327 (2015)

99. N. Zotov, M. Bartsch, G. Eggeler, Surf. Coatings Technol. **203**, 2064 (2009)
100. M.G. Ahangari, A. Fereidoon, Mater. Chem. Phys. **151**, 112 (2015)
101. A. Shanaghi, A.R.S. Rouhaghdam, S. Ahangarani, P.K. Chu, Mater. Res. Bull. **47**, 2200 (2012)
102. J. Sun, C. Liu, R. Zhang, F. Gong, C. Wang, G. Li, Ceram. Ceram. Inter. **45**(10), 13597–13604 (2019)

Utilization of Advanced Ceramics Towards Treatment of Wastewater

Deepti, Piyal Mondal, and Mihir K. Purkait

Abstract Ceramics have played a significant role in various fields such as environmental and biomedical due to their high melting point, corrosion-resistant property, biocompatibility, cheap, and easy availability. Nowadays, researchers have developed advanced ceramics with superior properties to be utilized in optical and electrical applications along with environmental and biomedical applications. This chapter focuses on the various advanced ceramic materials, which have been utilized for wastewater treatment purposes. Ceramic adsorbents, resins, and aerosols along with ceramic membranes which have been utilized intensively for wastewater treatment purposes have also been discussed in-depth in this chapter. Moreover, the chapter includes advanced ceramics to modify membrane-based technology to handle wastewater treatment more effectively. The chapter also discusses the future scope and perspective of such advanced ceramic materials and their modifications to ensure better efficiency towards environmental remediation purposes.

Keywords Ceramic adsorbents · Ceramic resins · Membrane technology · Wastewater · Organic ceramic · Inorganic ceramic

1 Introduction

Materials have been an overwhelmingly significant determinant of individual and social competition in human history. A better understanding of the atomic and molecular structure of materials is becoming increasingly important for the development of new materials as well as the advancement of existing materials. As a result, products are being customized to suit complex applications to overcome demanding

Deepti · M. K. Purkait
Center for Environment, Indian Institute of Technology Guwahati, Guwahati, Assam 781039, India

P. Mondal (✉) · M. K. Purkait
Department of Chemical Engineering, Indian Institute of Technology Guwahati, Guwahati, Assam 781039, India
e-mail: piyal@iitg.ac.in

S. J. Ikhmayies (ed.), *Advanced Ceramics*, Advances in Material Research and Technology, https://doi.org/10.1007/978-3-031-43918-6_5

industrial and social problems in today's intensely dynamic environment [1]. The current evolution of ceramics and materials and the related material technologies is increasing quickly with new technological development. Approaches of handling, forming, and finishing are essential to maintain pace with this rapid rate of development, as new materials and technologies are established. Since humans have been capable of conceptual thinking, the current scenario has provided the most innovation in ceramics and materials science. Over this time, substantial metallurgical advances have resulted in nearly any possible mixture of metal alloys, and the strengths of such alloys are reasonably well known and utilized. However, the thrust for faster, more efficient, low-cost production techniques continues today. To keep up with technical advances, new materials capable of working at higher temperatures, higher speeds, longer life factors, and lower maintenance costs are needed as metal-based structures hit their limits. As compared to other materials, advanced ceramics is one of the finest examples to overcome such limitations [2].

Ceramics technology plays a particularly important role in this stage and thus has emerged as a pillar of technological progress in many industrial sectors. Ceramics are solid materials made up of inorganic, nonmetallic elements. All ceramic materials have one thing in common: they are exposed to high temperatures either at the time of manufacture or use (usually above 500 °C). Ceramic is generally, but not always, a metallic oxide, boride, carbide, nitride, or a combination or compound of such materials [3].

Ceramics are known for their high hardness, heat and electrical insulation properties, heat and corrosion resistance, and brittleness and fracture behaviour without deformation. Traditional and advanced ceramics are the two types of ceramics. Traditional ceramics are made using a standard process that starts with the preparation of ceramic suspensions and finishes with sintering. Glass, porcelain, clays, kaolins, feldspars, and quartz are some of the common examples for traditional ceramics. Advanced ceramics, on the other hand, are a new family of ceramics manufactured from synthetic chemicals that are highly pure. These ceramics are aimed at a variety of high-performance industrial applications [4]. Advanced ceramics are mostly made up of oxides, nitrides, and carbides, though diamond and graphite are generally included in this category. As a result, advanced ceramics are often categorized as oxide or non-oxide ceramics depending on their composition. Oxide ceramics include binary oxides, aluminates, niobates, zirconates, ferrites, and titanites. Whereas, non-oxide ceramics include borides, carbon, carbides, nitrides and ceramic composites [5]. Advanced ceramic materials are expected to become more important in several areas, including optical, electrical, medical, and environmental applications. In the fields of restorative dentistry, bone reconstruction, and artificial bones, oxide ceramics are widely used (e.g., zirconia and alumina) [6]. Several ceramics are being investigated as catalysts, adsorbents, resins, and membranes for pollution prevention, monitoring, and remediation. They are particularly appealing for pollution control applications because of their potential to be customized, low cost, thermal and chemical durability [7, 8]. This chapter offers an overview of advanced ceramic materials that are used in many types of applications. The use of advanced ceramics in environmental remediation has been highlighted. Readers will get a detailed insight on ceramic adsorbents, resins, and aerogels, as well as ceramic membranes, which have been extensively used for wastewater treatment.

2 Classifications of Ceramics Based on Application

2.1 Optical-Based Applications

Advanced ceramics are being used in a broader variety of applications than ever before. Ceramics are gaining prominence in the fields of health, technology, nuclear power, electricity generation, transportation, and storage [9].

In optical technology, ceramics are becoming increasingly relevant. The most common optical components are silica glass, silicon dioxide, silicon and oxide single crystals. Thin films and fibres are also examples of optical materials. Amplifiers, detectors, waveguide couplers, modulators, filters, and attenuator polarizers are some of the optical instruments made with ceramic oxide materials. One possibility for this is because ceramic compounds, due to their electronic nature, are insulators with a large optical distance, making them ideal for optical applications. Furthermore, the microstructure of ceramics has a significant impact on their response to light propagation. As a result, processing conditions will alter its clarity. As passive optical elements, transparent ceramics are used as lenses, optical filters, waveguides, and mirror supports. Armours, home appliances, mirrors, light-emitting diodes (LEDs), and other products made of transparent ceramics are being used to replace glasses with outstanding properties. An increasing range of active optical instruments, such as scintillators, amplifiers, power lasers, phosphors, and multifunctional materials use ceramics. Ceramic materials often used in different optical devices are depicted in Table 1 [10]. The porosity of ceramic is the most important element in its transparency. The pore surface is a transition between phases with different optical properties, reflecting and refracting light. Ceramics are opaque because of the huge number of pores present.

Pores can be inter-grain or intra-grain. Though they are submicron in size, removing intra-grain pores is a tough and time-consuming procedure than removing closed inter-grain pores [11].

Glass ceramics are a type of transparent ceramics made from crystallization of glasses in a controlled manner. Glass ceramics have a variable but a high number of crystallites contained in the amorphous matrix, ranging from 30 to 90% by volume. Glass ceramics have excellent behaviour as compared to conventional glasses or single crystals because of their properties that can be regulated by their processing and composition [12]. Two phases are included in the production of glass ceramics. The glass is produced by casting the melt in the first phase, which is then re-crystallized with the assistance of nucleation aids in a second phase [10]. Nowadays, in an industrial glass ceramic, a mixture of $Li_2O/Al_2O_3/SiO_2$ (LAS) with Na_2O, K_2O, and CaO as glass formers and ZrO_2 and TiO_2 as nucleation vehicles is used. Studies

Table 1 Ceramic materials often used in optical devices

Material	Waveguide couplers	Laser amplifiers	Detector modulators	Filter attenuators	Polarizers	Protective barriers
SiO_2	x	x		x	x	x
Glasses	x	x				
YSZ, Ta_2O_5				x		x
$LiNbO_3$	x	x	x	x	x	
$CaCO_3$	x			x		
Al_2O_3	x	x				x

reported that fabrication of phosphors for fluorescence lamps is done with ceramics doped with optically active ions. For example, halophosphate ($Ca_5(PO_4)_3$(F, Cl): Sb^{3+}, Mn^{2+}), Tb^{3+}, Ce^{3+}: $LaPO_4$, and Eu^{3+}: Y_2O_3 have been used for white light green and blue light, and red light, respectively. Similarly, Pb: $CaWO_4$, Mn: $ZrSiO_4$, and Eu: YVO_4 for blue, green, and red colour, respectively, are used for Cathodic Ray Tubes (CRT) of television and computer colour screens [13].

Zhang et al. (2021) prepared $(Tb_{1-x}Lu_x)_2O_3$ transparent ceramics that are both magnetic and optical through a solid-solution method. The prepared ceramics had high thermal conductivities and could be used for a Faraday rotator in both visible and near-infrared regions [14]. Fang et al. (2019) investigated prepared ReSC-5 optical ceramics that exhibited good thermal stability, which were used as high-temperature solar absorbers. Ceramics exhibited low radiation temperature at higher temperatures due to their conduction behaviour [15].

2.2 *Biomedical Applications*

Bioceramics are ceramic materials that are engineered to accomplish a certain physiological behaviour and are used as a building material for prosthetic devices or artificial internal organs [5]. Over the last three decades, it has been shown that certain ceramics can facilitate the regeneration of neighbouring tissue, can spontaneously bind to living tissues, and can kill cancer cells locally, allowing natural tissue regeneration after treatment. According to many studies, ceramics have been shown to play an important part in the development of good quality of life [16]. Alumina and zirconia in medical grade are outstanding examples. Bioactive glasses, including resorbable calcium phosphate ceramics, heralded the rise of a new age of biomaterials. Recent findings suggest that third-generation biomaterials, products that stimulate the genes of the tissues in contact with the materials, are now feasible [17]. Based on their chemical surface reactivity, bioceramics can be categorized

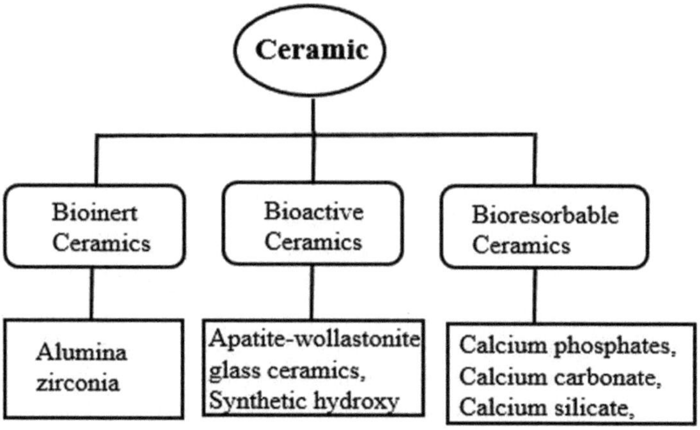

Fig. 1 Classification of ceramics based on their chemical surface reactivity. Reproduced with permission from Kapusetti et al. [18]. Copyright (2019). Springer Nature

as bioinert (e.g., alumina and zirconia), bioactive (e.g., hydroxyapatite and glass ceramics), and bioresorbable (e.g., Calcium oxide and gypsum) materials as shown in Fig. 1 [18].

The crystal structure of oxide ceramics is characterized by the tight packing of oxygen ions with metal ions found in the interstices. These ceramics are bioinert in nature. Bioactive ceramics, on the other hand, tend to form a bond with the interfacial core and to promote apatite growth. Biomaterials that show regulated chemical breakdown and resorption are bioresorbable ceramics. Compatibility of ceramics with the physiological system is crucial to their biomaterial ability. Bioceramics are compatible as they are made up of ions present in the physical system such as calcium, sodium, potassium, magnesium, and many more) as well as ions of low toxicity like aluminium and titanium. Bioceramics are used in several healthcare applications. Crowns, implants, bridges, and inlays/onlays are few examples that are used in dentistry. Appearance, colour stability, strength, thermal conductivity, and radiopacity are some of the appealing features/attributes of these materials, which are close to natural dentition properties. Bioinert ceramics include oxide ceramics such as alumina and zirconia, which are mostly used as prosthetic devices in medicine and dentistry. In the early 1990s, zirconia ceramics were first used in dentistry as endosseous implants in dental prosthetic surgery. Zirconia is biocompatible with oral cavity tissues and documented to be osteoconductive, which means it helps bone formation when it comes into contact with it. According to research, zirconia does not induce any sought of allergic reactions or taste changes. In terms of mechanical properties, zirconia is considered to have the best fracture resistance among the most commonly used ceramics, with high hardness, corrosion resistance, wear resistance, coefficient of thermal expansion, and modulus of elasticity [6].

Biphasic calcium phosphate (BCP), hydroxyapatite (HA), tricalcium phosphate (TCP), and a combination of these, have been widely studied and potential materials for bone repair among the different calcium phosphate (CaP) ceramics. CaP ceramics satisfy numerous significant criteria that are essential for the use as load-bearing clinical implants. These requirements include properties, such as toughness, hardness, and strength, to facilitate osseointegration, osteoinduction, and remodelling procedures, to have the same weight as the initial bone defect, easy accessibility without material constraints [19].

Borkowski et al. (2020) prepared fluorapatite having physical, chemical, and biological properties, which would support regeneration of bone tissue. Obtained fluorapatite was calcinated at 800 °C and characterized by slow release of fluoride at safe level for osteoblasts cell line [20]. Li et al. (2021) fabricated an enamel-like structure on the surface of zirconia-based polymer-infiltrated ceramic network (PICN) materials to improve the biological and mechanical properties. The enamel-like structure had a good antibacterial property and outstanding bio-compatibility, which can promote the proliferation and adhesion of human gingival fibroblast cells (HGFs) [21].

2.3 *Environmental Remediation Applications*

Ceramics have also been accepted as environmentally sustainable materials, enhancing their potential even further. To lower their greenhouse gas footprints, efforts are also being made to synthesize ceramics with minimal energy inputs. The enhanced synthesis of ceramics has seen significant advancements. A new application of ceramics and advanced ceramic materials has emerged as an intriguing focus on the reduction, regulation, and remediation of pollution in the air, water, and land media [22]. Ceramics are categorized as traditional or advanced, advanced ceramics are now commonly used in the materials science community. Such ceramic materials necessitate a degree of manufacturing technology and engineering well above those used in the manufacture of traditional ceramics. Many ceramics are being investigated as catalysts and photocatalysts, adsorbents, membranes, resins, and aerogels for pollution prevention, control, and remediation applications, and they can play important roles in this regard. Because of their low cost, thermal and chemical reliability, and potential to be tuned and tailored, they are particularly appealing for pollution control applications [2]. Ceramic materials are the ideal materials for many critical reactions in environmental and energy applications due to their lightweight and versatility in chemical composition, which allows them to be easily modified [23]. Ceramics, in the form of catalytic filters, membranes, and adsorbents, have become a significant moment in improving air quality by removing toxic emissions from the atmosphere. Li et al. (2015) used the incipient-wetness impregnation process to prepare a Fe–Mn–Ce catalyst and tested it for mercury removal in the simulated coal-fired flue gas. According to their findings, formulated ceramic catalysts had greater conductivity, thermal stability as well as higher corrosion resistivity and mechanical efficiency [24]. Many studies reported in the area of ceramic adsorbents and membranes, similar to ceramic catalysts. Jeong et al. (2017) fabricated a ceramic membrane from pyrophyllite and alumina coated with an alumina powder suspension to obtain a narrow pore size distribution. The membranes that were developed were used to treat low-strength domestic wastewater [25]. A detailed study of ceramic materials used for environmental remediation is discussed in the subsequent sections.

3 Recent Advances in Ceramics for Wastewater-Based Applications

3.1 Ceramic Adsorbents

Ceramic materials either conventional or advanced are of great importance in many fields as discussed in the previous sections. Even though each of the specified fields of application makes use of one or more distinct properties of ceramic materials, the phenomenon of adsorption is a common feature of all of them. Adsorption is a process, when a liquid or gas accumulates on the surface of a solid or liquid which is termed as adsorbent, forming a film (adsorbate). Adsorption is usually divided into two types based on the type of attraction between adsorbent and adsorbate; physical and chemical adsorption [26]. Surface chemical activity, thermal stability, exterior condition, and porosity determine the adsorption properties of materials [27]. There are many ceramic-based adsorbents available for water and wastewater treatment. Amongst them, clay is one of the oldest ceramic materials used as an adsorbent for decades. Ceramic material for adsorption includes non-metallic and inorganic materials such as carbides, oxides, and nitrides. For instance, metal oxides like aluminium oxide, iron oxides, magnesium oxides are all considered ceramic materials for adsorption.

Chen et al. (2012) investigated the removal of arsenate from aqueous solution by a ceramic adsorbent (mixture of Fe_2O_3, Akadama mud, and wheat starch). They found the adsorbent very advantageous with the following properties; (a) enough mechanical stability even after a prolonged time of adsorption, (b) low cost preparation process, (c) no secondary contamination, and (d) better removal performance over a wider pH range [28]. Similarly, Ma et al. (2016) developed a porous ceramic adsorbent of zirconium-pillared montmorillonite for the removal of chromium (Cr^{3+}) from wastewater. Different techniques like X-ray diffraction (XRD), Brunauer–Emmett–Teller (BET), and Field Emission Scanning Electron Microscope (FESEM) were used to study their chemical and physical characteristics. They found that zirconium-pillared montmorillonite as an adsorbent was advantageous with respect to its mechanical and thermal stability, working efficiency with a wide range of pH, and cost-effective [29]. Zhao et al. (2013) used kanuma clay and akadama clay for the adsorption of ammonium from ammonium-contaminated wastewater. They have also evaluated the effects of used adsorbent as a fertilizer. The study also revealed that the developed adsorbent was promising in terms of low cost, better soil amendment, high adsorption capacity [30].

3.2 Ceramic Aerogels

Ceramic aerogels are extremely porous materials with structures (three-dimensional) assembled by nanostructures that belong to cellular solids family [31]. Samuel Stephens Kistler coined the word "aerogel" in 1932 to describe gels in which the liquid phase is replaced by a gas phase without the gel network collapsing [32]. Properties such as high porosity, low density, large open pores, high surface area, are gaining attention in wastewater treatment. Ceramic aerogels have the potential to be used in high-temperature and oxygen-rich conditions due to their higher chemical stability. Traditional ceramic aerogels, which are made up of oxide nanoparticles, are brittle and suffer from high-temperature volume shrinkage. Brittleness is caused primarily by the fragile nature of ceramics and inefficient neck-like links between nanoparticles. The volume shrinkage caused by the high-temperature sintering action of the oxide nanoparticles causes the high-temperature instability. To enhance high-temperature stability, attempts have been made to fabricate carbides, nitrides, or ultrahigh-temperature ceramic aerogels. The brittleness of these materials, however, remains a limitation in practical applications. As a result, ceramic aerogels that are both mechanically robust and temperature stable are highly desirable [31, 33]. There are various kinds of ceramic aerogels available, namely, carbon-based aerogels (carbon nanofiber aerogels, carbon nanotube aerogel, graphene aerogel) [34], cellulose-based aerogels [35], polymer-coated ceramic aerogels [36]. Different types of preparation methods lead to different microstructures of ceramic aerogels. Some of the reported preparation methods include solution blow spinning, sol–gel, chemical vapour deposition, freeze-drying, and sacrificial template method [37]. Figure 2 represents the steps involved in the synthesis of aerogels by sol–gel technology [32].

Aerogels are made from a variety of materials, which include metal oxides, metals, carbon, and polymer-based materials, and are widely used in environmental remediation, especially for wastewater treatment [38]. Pham et al. (2021) derived aerogels from waste paper that are used to remove organic pollutants such as phenol and 2-chlorophenol from wastewater. The carbon-based aerogels had the highest surface area and could withstand higher temperatures and they are very cost-effective [39]. Zhan et al. (2019) prepared a hybrid graphene/polydopamine modified multiwalled carbon nanotube with robust and ultra-lightweight aerogels through a green route. The synthesized aerogel was utilized as an adsorbent for the removal of copper and

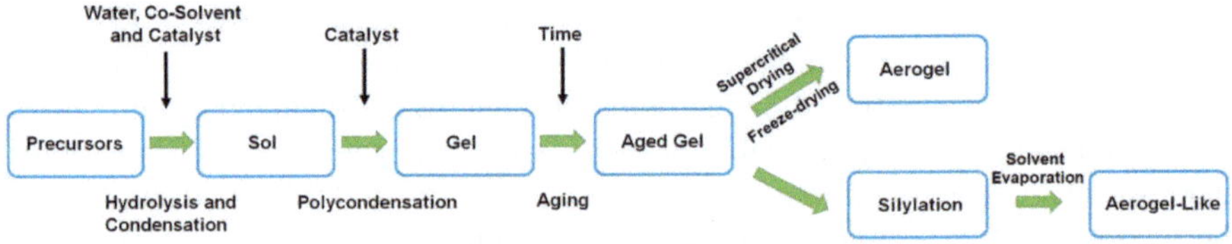

Fig. 2 Steps involved in the synthesis of aerogels through sol–gel technology. Reproduced with permission from Almeida et al. [32]. Copyright (2020), Elsevier

lead. The hybrid aerogels showed outstanding structural stability, high specific area, availability of active adsorption sites, and high adsorption capacity [40].

3.3 Ceramic Membranes

A membrane is a thin barrier, which allows one or more constituents to selectively pass from one medium to the other in the presence of an appropriate driving force [41]. Ceramic membranes were first developed in the early 1960s for beer filtration, extract recovery, and gas separation [42]. Separation is aided by the active layer, and the membrane is supported mechanically by the substrate. Flat-sheet and tubular types are the two main configurations used in the membrane process. Both types are employed in many wastewater treatment-based applications, such as organic pollutant removal, colour removal, heavy metal remediation, and desalination [43]. The schematic of ceramic membranes along with their properties is depicted in Fig. 3 [44]. The performance and function of the membrane depend upon the materials utilized for the fabrication of the membrane. Ceramic materials for membranes in water and wastewater treatment are chosen based on their chemical and thermal stabilities, hydrophilicity, porosity, pore size, mechanical strength, cost-effectiveness, and other water treatment-related characteristics [45]. Many studies have been reported on a variety of raw materials namely alumina, titania, silica, zirconium dioxide, and zeolite for the fabrication of membrane. Also, there are many clay-type materials used for the fabrication such as bentonite, kaolin, ball clay. Recently, numerous studies have been carried out to investigate low-cost ceramic products such as coal ash, fly ash, and steel slag from industrial wastes [46].

A. Barbosa et al. (2018) fabricated zeolite membrane on alumina disc for the separation of oil/water emulsion. The performance of the fabricated membrane exhibited outstanding results due to the formation of the zeolite layer on the surface of alumina ceramic supports. Because of the properties of zeolite, the efficiency of membranes improved [47].

M. Changmai et al. (2019) fabricated a fly ash-based ceramic membrane. The polymer was coated on the ceramic membrane using the dip coating method. Membrane showed excellent properties like chemical and thermal stability, good rejection of proteins [48].

Membrane characterization is a critical step in membrane development and one of the most important factors in better understanding the structure, functional performance, stability and overall performance. Field emission scanning electron microscopy (FESEM), scanning electron microscopy (SEM), atomic force microscopy (AFM), scanning tunnelling microscopy (STM), and transmission electron microscopy (TEM) are some of the general characterization techniques used to determine the morphological details of membranes, such as surface porosity, cross-sectional area, pore size and shape.

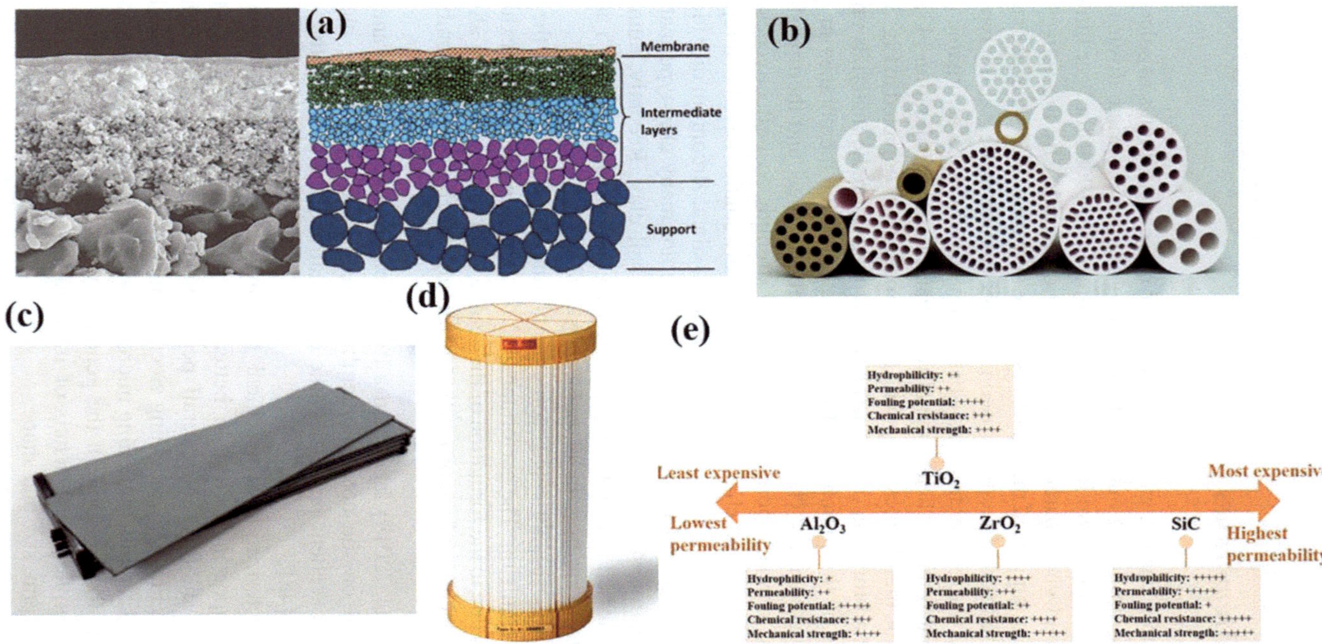

Fig. 3 **a** The schematic representation of ceramic membrane cross-section, **b** commercially available single/multichannel tabular ceramic membrane photographs, **c** flat-sheet ceramic membrane, **d** hollow fibre ceramic membrane. **e** Properties of ceramic membrane fabricated using different types of ceramic material. Reproduced with permission from Asif and Zhang [44]. Copyright (2021), Elsevier

3.4 Ceramic Membranes

A membrane is a thin barrier, which allows one or more constituents to selectively pass from one medium to the other in the presence of an appropriate driving force [41]. Ceramic membranes were first developed in the early 1960s for beer filtration, extract recovery, and gas separation [42]. Separation is aided by the active layer, and the membrane is supported mechanically by the substrate. Flat-sheet and tubular types are the two main configurations used in the membrane process. Both types are employed in many wastewater treatment-based applications, such as organic pollutant removal, colour removal, heavy metal remediation, and desalination [43]. The schematic of ceramic membranes along with their properties is depicted in Fig. 3 [44]. The performance and function of the membrane depend upon the materials utilized for the fabrication of the membrane. Ceramic materials for membranes in water and wastewater treatment are chosen based on their chemical and thermal stabilities, hydrophilicity, porosity, pore size, mechanical strength, cost-effectiveness, and other water treatment-related characteristics [45]. Many studies have been reported on a variety of raw materials, namely, alumina, titania, silica, zirconium dioxide, and zeolite for the fabrication of membrane. Also, there are many clay-type materials used for the fabrication such as bentonite, kaolin, ball clay. Recently, numerous studies have been carried out to investigate low-cost ceramic products such as coal ash, fly ash, and steel slag from industrial wastes [46].

A. Barbosa et al. (2018) fabricated zeolite membrane on alumina disc for the separation of oil/water emulsion. The performance of the fabricated membrane exhibited outstanding results due to the formation of the zeolite layer on the surface of alumina ceramic supports. Because of the properties of zeolite, the efficiency of membranes improved [47].

Changmai et al. (2019) fabricated a fly ash-based ceramic membrane. The polymer was coated on the ceramic membrane using the dip coating method. Membrane showed excellent properties like chemical and thermal stability, good rejection of proteins [48].

Membrane characterization is a critical step in membrane development and one of the most important factors in better understanding the structure, functional performance, stability and overall performance. Field emission scanning electron microscopy (FESEM), scanning electron microscopy (SEM), atomic force microscopy (AFM), scanning tunnelling microscopy (STM), and transmission electron microscopy (TEM) are some of the general characterization techniques used to determine the morphological details of membranes, such as surface porosity, cross-sectional area, pore size and shape.

3.5 Ceramic Membranes

A membrane is a thin barrier, which allows one or more constituents to selectively pass from one medium to the other in the presence of an appropriate driving force [41]. Ceramic membranes were first developed in the early 1960s for beer filtration, extract recovery, and gas separation [42]. Separation is aided by the active layer, and the membrane is supported mechanically by the substrate. Flat-sheet and tubular types are the two main configurations used in the membrane process. Both types are employed in many wastewater treatment-based applications, such as organic pollutant removal, colour removal, heavy metal remediation, and desalination [43]. The schematic of ceramic membranes along with their properties is depicted in Fig. 3 [44]. The performance and function of the membrane depend upon the materials utilized for the fabrication of the membrane. Ceramic materials for membranes in water and wastewater treatment are chosen based on their chemical and thermal stabilities, hydrophilicity, porosity, pore size, mechanical strength, cost-effectiveness, and other water treatment-related characteristics [45]. Many studies have been reported on a variety of raw materials, namely, alumina, titania, silica, zirconium dioxide, and zeolite for the fabrication of membrane. Also, there are many clay-type materials used for the fabrication such as bentonite, kaolin, ball clay. Recently, numerous studies have been carried out to investigate low-cost ceramic products such as coal ash, fly ash, and steel slag from industrial wastes [46].

Barbosa et al. (2018) fabricated zeolite membrane on alumina disc for the separation of oil/water emulsion. The performance of the fabricated membrane exhibited outstanding results due to the formation of the zeolite layer formed on the surface of alumina ceramic supports. Because of the properties of zeolite, the efficiency of membranes improved [47].

Changmai et al. (2019) fabricated a fly ash-based ceramic membrane. The polymer was coated on the ceramic membrane using the dip coating method. Membrane showed excellent properties like chemical and thermal stability, good rejection of proteins [48].

Membrane characterization is a critical step in membrane development and one of the most important factors in better understanding the structure, functional performance, stability and overall performance. Field emission scanning electron microscopy (FESEM), scanning electron microscopy (SEM), atomic force microscopy (AFM), scanning tunnelling microscopy (STM), and transmission electron microscopy (TEM) are some of the general characterization techniques used to determine the morphological details of membranes, such as surface porosity, cross-sectional area, pore size and shape.

4 Advanced Ceramics in Membrane Technology

Membranes with excellent physical and chemical properties, high tunability, and reusability have gained a thriving interest. Comprehensively, advanced materials for the ceramic membrane are usually inorganic. Inorganic materials include oxides of alumina, titania, silica or mixtures of these components, metal oxide framework, zeolites, and few industrial wastes. Few different forms of inorganic ceramic membranes such as low-cost membranes, metal oxide-based membranes, composite membranes like metal oxide framework membranes are discussed in the subsequent sections.

4.1 Metal Oxide-Based Membranes

Ceramic membrane based on metal oxide has a multi-layered asymmetric structure consisting of a denser support layer with large pores. This support provides mechanical support for the membrane structures, an intermediate layer to decrease pore size to mesoporous sizes, and a layer with small and selective pores for separation. Ceramic membranes containing a diverse range of insoluble oxides exhibited intriguing separation and processing properties [49].

Commonly used ceramic material for fabricating membranes is Alumina (Al_2O_3). α, γ, η, and θ alumina are the thermodynamically stable phases of alumina. Among these phases, γ- and α alumina-based membranes are commonly used for water, and wastewater treatment. Alumina functioned as the active layer, substrate, an intermediate layer in a ceramic membrane due to its properties such as thermal and chemical stability and high strength. Sol–gel, dip coating and sintering are the commonly used fabrication methods [47, 50]. Similarly, Titania is another ceramic material widely used for the fabrication of ceramic membranes due to its intrinsic properties such as high chemical resistance, and mineralization of organic compounds under UV irradiation [51].

Several advantages are incorporated into a single membrane, such as high permeability and antifouling and antibacterial capacities. Other ceramic membranes, for example alumina membrane, do not have a significant anti-bacterial advantage as alumina has no photocatalytic property. Titania has high chemical stability compared to other ceramic materials and is therefore capable of treating wastewater related to corrosive media [52].

Zirconia (ZrO_2) is another ceramic material used for water and wastewater treatment. The property of zirconia is that it can change from monoclinic crystal to tetragonal and cubic phases from room temperature to higher temperature, which helps to improve the toughness of the zirconia that is stabilized [51]. Furthermore, it has the greatest hydrophilicity and thermal resistance, making it suitable for wastewater treatment. The treatment of wastewater such as oily and saline wastewater is one of the most important applications of zirconia membrane [53].

Nataraj et al. (2011) developed α-alumina tubular ceramic membrane modules coated with cellulose acetate. Distillery wastes, paper, and pulp wastes, and the sugar industry wastes were treated using tubular membrane. They found that coating an alumina ceramic membrane with cellulose acetate made the composite membranes less prone to membrane fouling and offered more potential for wastewater treatment [54]. Similarly, Seres et al. (2016) used alumina ceramic membrane for the treatment of vegetable oil refinery wastewater. The membrane had a pore size of 200 nm. The membrane served its best in terms of higher flux and high rejection [55]. Zhu et al. (2011) in his investigation prepared a TiO_2 membrane by applying TiO_2 nanoparticles to tubular alumina support using a dip-coating method. The fabricated membrane was used with ozonation to treat the wastewater from a wastewater treatment plant to reduce colour and trace organics [56]. Da et al. (2016) fabricated a pure tetragonal ZrO_2 nano-filtration membrane with high permeable and separation performance by sol–gel method for the treatment of high saline wastewater [57].

4.2 Metal–Organic Framework (MOF) Membranes

Metal–organic frameworks (MOFs) are a type of crystalline porous materials with intermittent network structure made up of inorganic metal-containing nodes linked by bridged organic ligands created by self-assembly [58]. They have grown in popularity over the last two decades as a result of their versatility and wide range of potential applications. MOFs have high tuneable porosities and large available surface areas, as well as the ability to effortlessly merge species without modifying the framework topology. MOFs are growing rapidly in gas storage, drug delivery, and catalysis due to their remarkable properties. Furthermore, MOFs are commonly used in the field of membrane separation. MOFs have been extensively used in membrane applications such as forward osmosis, reverse osmosis, and nanofiltration [59]. Now, the most common substrates used to prepare MOF membranes are inorganic compounds and polymers. As a result, MOF membranes are classified into two types: pure MOF membranes consisting of MOF layers and a porous inorganic substrate, and mixed matrix membranes (MMMs) created by mixing MOF particles with an organic matrix. Many techniques for preparing MOF membranes have been developed including, layer-by-layer growth, in-situ growth, secondary growth (seed method), solvothermal, electrodeposition, vacuum filtration, and electrochemical synthesis. Many studies on the fabrication of membranes containing MOFs for the treatment of different types of wastewater have been reported [60].

Gong et al. (2020) fabricated a nanofiltration membrane incorporating titanium-based MOFs into polyethyleneimine (PEI) and trimesic acid (TMA) cross-linking system. MOF-based membrane was used to remove some heavy metals. The membrane showed excellent permeability and high rejection [61]. Samari et al. (2020) synthesized polyethersulfone (PES) ultrafiltration (UF) membrane, using a melamine-modified zirconium-based metal–organic framework (MOF). The membrane exhibited high thermal stability and permeability. They observed an excellent performance of the membrane by introducing the MOF to the matrix. The synthesized membrane was used to treat oily wastewater and found it effective in terms of permeability and high oil rejection [62].

4.3 Low-Cost Membranes

Ceramic membranes, as discussed in previous sections, provide special advantages in membrane processes due to their properties, such as mechanical strength, thermal stability, and chemical resistance and decreased fouling propensity [63]. Many applications do not require the advantages offered by high-purity refractory oxides, such as the treatment of domestic or industrial wastewater, where the required quality of the treated effluent can be achieved with membranes synthesized with more conventional materials, namely, low-cost ceramic membranes. This would result in a considerable reduction in raw materials and processing costs, lowering the cost of the ceramic membranes produced. Low-cost ceramic membranes are made up of lower-cost raw materials that are typically found in conventional ceramic products. The sintering temperatures used, on the other hand, are lower than those used in industrial ceramic membranes, lowering processing costs [64].

Likewise, fly ash, charred wastes, coal ash, steel slag grouped under industrial wastes and agricultural wastes like rice husk and sugarcane bagasse have been studied and reported as raw materials that are economically convenient for the fabrication of ceramic membranes [65]. The compositions of these materials are advantageous for the fabrication of membrane. These materials encourage further research towards cost-effective membranes of great demand.

Recently, Deepti et al. (2020) have reported work on the preparation of microfiltration ceramic membranes from Linz Donawitz (LD) slag, waste from the steel industry. They found that the fabricated membrane had excellent flexural strength, chemical stability, high flux, and capacity to treat the steel industry wastewater [8]. Manni et al. (2020) prepared a ceramic microfiltration membrane from natural magnesite to treat textile wastewater. Magnesite-based membrane showed good thermal stability along with higher flux. Filtration results showed that the membrane efficiently removed all turbidity from textile wastewater [66]. Hubadillah et al. (2018) developed low-cost ceramic membranes from silica-based rice husk which is an agricultural waste. Rice husk was modified into rice husk by the burning process at 600 and 1000 °C. They reported that silica content and sintering temperature had a substantial effect on membrane surface roughness, mechanical strength, structure, porosity, and flux [65].

Industrial applicability of membrane technology is determined by its cost. It is reported that polymeric membranes available for industrial use cost around \$50–200/m^2. Ceramic membranes, on the other hand, are said to be 8 to 10 times more expensive than polymeric membranes (\$50–200/m^2). Ceramic symmetric membrane tubes fabricated using α-alumina range between \$500 and 1000/m^2 [67]. However, certain ceramic membranes made from low-cost raw materials will be less expensive than polymeric membranes used in industrial applications. Considering the raw materials cost, the cost of the fabricated membranes using LD slag was estimated in the range of 32.55–55.7 $/m^2$ [8]. Similarly, Nandi et al. (2007) reported that the cost of fabricated membrane with kaolin as the main raw material is estimated to be \$130/m^2. It is necessary to retain that membrane prices differ by region, as well as by membrane properties and surface area. By utilizing low-cost raw materials, the cost of a ceramic membrane may be greatly decreased.

5 Future Research Perspective

Ceramics are now well-accepted materials as a result of ongoing changes and advances. Ceramic materials are used as effective and efficient materials in a variety of applications. Many of the remarkable applications, however, need more advances and improvements in the use of advanced ceramics, since they cannot be achieved easily and reliably with available ceramics. As a result, ceramic materials, especially advanced ceramic materials, must be developed for use in modern applications such as health care, optical, electrical, and environmental.

Even though their uses in water and wastewater treatment are expanding, the higher capital expense, low mechanical strength and poor dimensional stability under high temperatures (above 600 °C) is regarded as the key impediments to their wider application. However, mixing with other ceramic thermally resistant phases is a viable approach to improve their stability. The cost of the ceramics may be drastically decreased using low-cost natural materials such as kaolin, pyrophyllite, and dolomite, as well as certain industrial wastes listed in previous sections. However, the stability and handling of low-cost ceramics during operation is difficult. As a consequence, research into their fabrication methods and module design are important for commercial applications. When designing structures, components, and products, the properties of advanced ceramic materials must be taken into account. The design and material selection must be cost-efficient, technically accurate, and, preferably, an upgrade on current technologies. Prior performance knowledge is advantageous; but, in many emerging implementations, prior knowledge will not be usable, close observation and documentation of experimental model or plant trial performance characteristics is needed. In this respect, new methods such as Finite Element Analysis have proved useful.

6 Summary

In this chapter, an overview of the advanced ceramic materials along with their applications in different fields particularly in environmental remediation is presented. Because of their environmental compatibility, ceramics have been accepted as environmentally friendly materials, which contribute to their potential when compared to other materials. Advanced ceramics prove to be attractive alternatives for use in medical applications such as dentistry, bone reconstruction, and so on. The use of advanced ceramics in optics and environmental remediation demonstrated outstanding results. Advanced ceramic materials are now commonly used in several applications. The advantages of advanced ceramic materials are clearly shown by increased efficiency, service life, operating cost savings, and maintenance cost savings. Cost economics is just twice the cost of existing component prices, giving advanced ceramics materials a substantial benefit.

They play an ever more significant role in a variety of fields in the future. Availability of advanced ceramic materials with customized properties, as well as the development of new ceramics, provides essential resources for the design of alternative production processes suitable for sustainable growth. With the development of material technology, it is possible to produce high-performance, inexpensive and ceramic materials in the future. With this regard, this chapter contributes to an extensive discussion for understanding of various advanced ceramics with their remarkable properties along with their diverse applications. This chapter will also be helpful to the readers for gaining expertise on application of advanced ceramic materials on the reduction, regulation, and remediation of wastewater with some notable achievements.

References

1. X. Zhu, W. Yang, Mixed conducting ceramic membranes: Fundamentals, Materials and Applications (Green Chemistry and Sustainable Technology), 1st edn. (Springer, 2017), pp. 1–10
2. L. Gurnani, A. Mukhopadhyay, in Handbook of Advanced Ceramics and Composites, ed. by Y. Mahajan, J. Roy (Springer, Cham,2019), pp. 1–46
3. A.R. Boccaccini, in *Biomaterials, Artificial Organs and Tissue Engineering.* A volume in Woodhead Publishing Series in Biomaterials, ed. by L.L. Hench, J.R. Jones (Woodhead Publishing Limited, 2005), pp. 26–36
4. D. Salamon, in *Advanced Ceramics for Dentistry.* ed. by J. Shen, T. Kosmac, (Elsevier Inc. 2014), pp. 104–122
5. S.F. Hulbert, L.L. Hench, D. Forbers, L.S. Bowman, History of bioceramics. Ceram. Int. **8**, 131–140 (1982). https://doi.org/10.1016/0272-8842(82)90003-7
6. T. Kokubo, S. Yamaguchi, Development of novel glass-ceramics for dental applications by the Höland group. J. Non Cryst. Solids **558**, 120656 (2021). https://doi.org/10.1016/j.jnoncrysol.2021.120656
7. M. Hasanpour, M. Hatami, Application of three dimensional porous aerogels as adsorbent for removal of heavy metal ions from water/wastewater: a review study. Adv. Colloid Interface Sci. **284**, 102247 (2020). https://doi.org/10.1016/j.cis.2020.102247

8. Deepti, A. Sinha, P. Biswas, S. Sarkar, U. Bora, M.K. Purkait, Utilization of LD slag from steel industry for the preparation of MF membrane. J. Environ. Manage. **259**, 110060 (2020). https://doi.org/10.1016/j.jenvman.2019.110060
9. S. Somiya, *Handbook of Advanced Ceramics: Materials, Applications, Processing, and Properties*, 2nd edn. (Academic Press, 2013), pp. 883–910
10. V.M. Orera, R.I. Merino, Ceramics with photonic and optical applications. Bol. La Soc. Esp. Ceram. y Vidr. **54**, 1–10 (2015). https://doi.org/10.1016/j.bsecv.2015.02.002
11. Z. Xiao, S. Yu, Y. Li, S. Ruan, L.B. Kong, Q. Huang, Z. Huang, K. Zhou, H. Su, Z. Yao, W. Que, Y. Liu, T. Zhang, J. Wang, P. Liu, D. Shen, M. Allix, J. Zhang, D. Tang, Materials development and potential applications of transparent ceramics: a review. Mater. Sci. Eng. R Rep. **139**, 100518 (2020). https://doi.org/10.1016/j.mser.2019.100518
12. J. Zhong, D. Chen, Y. Peng, Y. Lu, X. Chen, X. Li, Z. Ji, A review on nanostructured glass ceramics for promising application in optical thermometry. J. Alloys Compd. **763**, 34–48 (2018). https://doi.org/10.1016/j.jallcom.2018.05.348
13. I. Alekseeva, O. Dymshits, M. Tsenter, A. Zhilin, V. Golubkov, I. Denisov, N. Skoptsov, A. Malyarevich, K. Yumashev, Optical applications of glass-ceramics. J. Non Cryst. Solids **356**, 3042–3058 (2010). https://doi.org/10.1016/j.jnoncrysol.2010.05.103
14. J. Zhang, H. Chen, J. Wang, D. Wang, D. Han, J. Zhang, S. Wang, Preparation of $(Tb1\text{-}xLux)_2O_0$ transparent ceramics by solid solution for magneto-optical application. J. Eur. Ceram. Soc. **41**, 2818–2825 (2021). https://doi.org/10.1016/j.jeurceramsoc.2020.12.034
15. Z. Fang, L. Chen, Y. Ni, C. Lu, Z. Xu, Robust optical properties of $Re_{0.5}Sr_{0.5}CoO_{3-\delta}$ (Re = Nd, Eu, Gd) ceramics for high temperature solar absorber applications. Appl. Surf. Sci. **469**, 76–81 (2019). https://doi.org/10.1016/j.apsusc.2018.11.009
16. L. George, R.P. Rusin, G.S. Fischman, V. Janas, Bioceramics, in *Ceramic Transactions, Bioceramics: Materials and Applications III*, vol. 110 (American Ceramic Society, Westerville, 1999), pp. 113–27
17. S. Balasubramanian, B. Gurumurthy, A. Balasubramanian, Biomedical applications of ceramic nanomaterials: a review. International Journal of Pharmaceutical Sciences and Research (IJPSR,) **8,** 4950–4959 (2017). https://doi.org/10.13040/ijpsr.0975-8232.8(12).4950-59
18. V.K. Pandey, S. Paul, in *Biomedical Engineering and Its Applications in Healthcare* (Springer, 2019), pp. 89–104.
19. P. Dee, H.Y. You, S.H. Teoh, H. Le Ferrand, Bioinspired approaches to toughen calcium phosphate-based ceramics for bone repair. J. Mech. Behav. Biomed. Mater. **112**, 104078 (2020). https://doi.org/10.1016/j.jmbbm.2020.104078
20. L. Borkowski, A. Przekora, A. Belcarz, K. Palka, G. Jozefaciuk, T. Lübek, M. Jojczuk, A. Nogalski, G. Ginalska, Fluorapatite ceramics for bone tissue regeneration: synthesis, characterization and assessment of biomedical potential. Mater. Sci. Eng. C **116**, 111211 (2020). https://doi.org/10.1016/j.msec.2020.111211
21. K. Li, H. Kou, J. Rao, C. Liu, C. Ning, Fabrication of enamel-like structure on polymer-infiltrated zirconia ceramics. Dent. Mater. **7**, 245–255 (2021). https://doi.org/10.1016/j.dental.2021.01.002
22. H. Adamu, S. Shaaibu, Y.O.S adiq, ASHAKWU Journal of Ceramics 17, 1 (2020)
23. R.B. Heimann, Classic and advanced ceramics: from fundamentals to applications (Wiley-VCH Verlag GmbH & Co. KGaA, 2010), pp. 175-252.
24. J. Li, J. Chen, Y. Yu, C. He, Fe-Mn-Ce/ceramic powder composite catalyst for highly volatile elemental mercury removal in simulated coal-fired flue gas. J. Ind. Eng. Chem. **25**, 352–358 (2015). https://doi.org/10.1016/j.jiec.2014.11.015
25. Y. Jeong, S. Lee, S. Hong, C. Park, Preparation, characterization and application of low-cost pyrophyllite-alumina composite ceramic membranes for treating low-strength domestic wastewater. J. Memb. Sci. **536**, 108–115 (2017). https://doi.org/10.1016/j.memsci.2017.04.068
26. N.B. Singh, G. Nagpal, S. Agrawal, Rachna, Water purification by using adsorbents: a review. Environ. Technol. Innov. **11**, 187–240 (2018). https://doi.org/10.1016/j.eti.2018.05.006
27. N. Ya'acob, M. Abdullah, M. Ismail, in *Trends in Telecommunications Technologies*. ed. by C. J. Bouras (IntechOpen, 1989), pp. 137–144. https://doi.org/10.5772/8474.

28. R. Chen, Z. Zhang, Z. Lei, N. Sugiura, Preparation of iron-impregnated tablet ceramic adsorbent for arsenate removal from aqueous solutions. DES. **286**, 56–62 (2012). https://doi.org/10.1016/j.desal.2011.11.004
29. H. Ma, Y. Hei, L. Hua, Y. Guo, Y. Yang, C. Yu, Fabrication of zirconium-pillared montmorillonite porous ceramic as adsorbents for Cr 3 þ removal and recycling. Ceram. Int. **42**, 14903–14909 (2016). https://doi.org/10.1016/j.ceramint.2016.06.130
30. Y. Zhao, Y. Yang, S. Yang, Q. Wang, C. Feng, Z. Zhang, Journal of Colloid and Interface Science Adsorption of high ammonium nitrogen from wastewater using a novel ceramic adsorbent and the evaluation of the ammonium-adsorbed-ceramic as fertilizer. J. Colloid Interface Sci. **393**, 264–270 (2013). https://doi.org/10.1016/j.jcis.2012.10.028
31. L. Su, M. Niu, D. Lu, Z. Cai, M. Li, H. Wang, A review on the emerging resilient and multifunctional ceramic aerogels. J. Mater. Sci. Technol. **75**, 1–13 (2021). https://doi.org/10.1016/j.jmst.2020.10.018
32. C.M.R. Almeida, M.E. Ghica, L. Durães, An overview on alumina-silica-based aerogels. Adv. Colloid Interface Sci. **282** (2020). https://doi.org/10.1016/j.cis.2020.102189
33. N. Hüsing, U. Schubert, Aerogels—airy materials: chemistry, structure, and properties. Angew. Chemie Int. Ed. **37**, 22–45 (1998). https://doi.org/10.1002/(sici)1521-3773(19980202)37:1/2%3c22::aid-anie22%3e3.0.co;2-i
34. Y. Liu, J. Liu, P. Song, Recent advances in polysaccharide-based carbon aerogels for environmental remediation and sustainable energy. Sustain. Mater. Technol. **27**, e00240 (2021). https://doi.org/10.1016/j.susmat.2020.e00240
35. L. Melone, L. Altomare, I. Alfieri, A. Lorenzi, L. De Nardo, C. Punta, Ceramic aerogels from TEMPO-oxidized cellulose nanofibre templates: synthesis, characterization, and photocatalytic properties. J. Photochem. Photobiol. A Chem. **261**, 53–60 (2013). https://doi.org/10.1016/j.jphotochem.2013.04.004
36. F. Ram, K. Suresh, A. Torris, G. Kumaraswamy, K. Shanmuganathan, Highly compressible ceramic/polymer aerogel-based piezoelectric nanogenerators with enhanced mechanical energy harvesting property. Ceram. Int. **47**, 15750-15758 (2021). https://doi.org/10.1016/j.ceramint.2021.02.147
37. H. Baqer, R. Irmawati, I. Ismail, N. Azah, Utilization of waste engine oil for carbon nanotube aerogel production using fl oating catalyst chemical vapor deposition. J. Clean. Prod. **261**, 121188 (2020). https://doi.org/10.1016/j.jclepro.2020.121188
38. D.K. Sam, E.K. Sam, A. Durairaj, X. Lv, Z. Zhou, J. Liu, Synthesis of biomass-based carbon aerogels in energy and sustainability. Carbohydr. Res. **491**, 107986 (2020). https://doi.org/10.1016/j.carres.2020.107986
39. T.H. Pham, S.H. Jung, Y.J. Kim, T.Y. Kim, Adsorptive removal and recovery of organic pollutants from wastewater using waste paper-derived carbon-based aerogel. Chemosphere **268**, 129319 (2021). https://doi.org/10.1016/j.chemosphere.2020.129319
40. W. Zhan, L. Gao, X. Fu, S.H. Siyal, G. Sui, X. Yang, Green synthesis of amino-functionalized carbon nanotube-graphene hybrid aerogels for high performance heavy metal ions removal. Appl. Surf. Sci. **467–468**, 1122–1133 (2019). https://doi.org/10.1016/j.apsusc.2018.10.248
41. M.K. Purkait, R. Singh, *Membrane technology in separation science* (CRC Press 2018), pp. 1-43.
42. S.K. Hubadillah, M.H.D. Othman, T. Matsuura, A.F. Ismail, M.A. Rahman, Z. Harun, J. Jaafar, M. Nomura, Fabrications and applications of low cost ceramic membrane from kaolin: a comprehensive review. Ceram. Int. **44**, 4538–4560 (2018). https://doi.org/10.1016/j.ceramint.2017.12.215
43. B. Das, B. Chakrabarty, P. Barkakati, Preparation and characterization of novel ceramic membranes for micro-filtration applications. Ceram. Int. **42**, 14326–14333 (2016). https://doi.org/10.1016/j.ceramint.2016.06.125
44. M.B. Asif, Z. Zhang, Ceramic membrane technology for water and wastewater treatment: a critical review of performance, full-scale applications, membrane fouling and prospects. Chem. Eng. J. **418**, 129481 (2021). https://doi.org/10.1016/j.cej.2021.129481

45. M. Ben Ali, N. Hamdi, M.A. Rodriguez, K. Mahmoudi, E. Srasra, Preparation and characterization of new ceramic membranes for ultrafiltration. Ceram. Int. **44**, 2328–2335 (2018). https://doi.org/10.1016/j.ceramint.2017.10.199
46. S. Jana, A. Saikia, M.K. Purkait, K. Mohanty, Chitosan based ceramic ultrafiltration membrane: preparation, characterization and application to remove Hg(II) and As(III) using polymer enhanced ultrafiltration. Chem. Eng. J. **170**, 209–219 (2011). https://doi.org/10.1016/j.cej.2011.03.056
47. A. dos S. Barbosa, A. dos S. Barbosa, T.L.A. Barbosa, M.G.F. Rodrigues, Synthesis of zeolite membrane (NaY/alumina): effect of precursor of ceramic support and its application in the process of oil–water separation. Sep. Purif. Technol. **200**, 141–154 (2018). https://doi.org/10.1016/j.seppur.2018.02.001
48. M. Changmai, M.K. Purkait, Detailed study of temperature-responsive composite membranes prepared by dip coating poly (2-ethyl-2-oxazoline) onto a ceramic membrane. Ceram. Int. **44**, 959–968 (2018). https://doi.org/10.1016/j.ceramint.2017.10.029
49. P.S. Goh, A.F. Ismail, A review on inorganic membranes for desalination and wastewater treatment. Desalination **434**, 60–80 (2018). https://doi.org/10.1016/j.desal.2017.07.023
50. A.K. Fard, G. McKay, A. Buekenhoudt, H. Al Sulaiti, F. Motmans, M. Khraisheh, M. Atieh, Inorganic membranes: preparation and application for water treatment and desalination. Materials (Basel) **11** (2018). https://doi.org/10.3390/ma11010074
51. Z. He, Z. Lyu, Q. Gu, L. Zhang, J. Wang, Ceramic-based membranes for water and wastewater treatment. Colloids Surf. A Physicochem. Eng. Asp. **578**, 123513 (2019). https://doi.org/10.1016/j.colsurfa.2019.05.074
52. S. Ghannadi, H. Abdizadeh, A. Babaei, Fabrication of porous titania sheet via tape casting: microstructure and water permeability study. Ceram. Int. **46**, 8689–8694 (2020). https://doi.org/10.1016/j.ceramint.2019.12.103
53. B.A. da Silva, V. de Souza Godim de Oliveira, M. Di Luccio, D. Hotza, K. Rezwan, M. Wilhelm, Characterization of functionalized zirconia membranes manufactured by aqueous tape casting. Ceram. Int. **46**, 16096–16103 (2020). https://doi.org/10.1016/j.ceramint.2020.03.162
54. S.K. Nataraj, S. Roy, M.B. Patil, M.N. Nadagouda, W.E. Rudzinski, T.M. Aminabhavi, Cellulose acetate-coated α-alumina ceramic composite tubular membranes for wastewater treatment. Desalination **281**, 348–353 (2011). https://doi.org/10.1016/j.desal.2011.08.016
55. Z. Šereš, N. Maravić, A. Takači, I. Nikolić, D. Šoronja-Simović, A. Jokić, C. Hodur, Treatment of vegetable oil refinery wastewater using alumina ceramic membrane: optimization using response surface methodology. J. Clean. Prod. **112**, 3132–3137 (2016). https://doi.org/10.1016/j.jclepro.2015.10.070
56. B. Zhu, Y. Hu, S. Kennedy, N. Milne, G. Morris, W. Jin, S. Gray, M. Duke, Dual function filtration and catalytic breakdown of organic pollutants in wastewater using ozonation with titania and alumina membranes. J. Memb. Sci. **378**, 61–72 (2011). https://doi.org/10.1016/j.memsci.2010.11.045
57. X. Da, X. Chen, B. Sun, J. Wen, M. Qiu, Y. Fan, Preparation of zirconia nanofiltration membranes through an aqueous sol-gel process modified by glycerol for the treatment of wastewater with high salinity. J. Memb. Sci. **504**, 29–39 (2016). https://doi.org/10.1016/j.memsci.2015.12.068
58. Y. Deng, Y. Wu, G. Chen, X. Zheng, M. Dai, C. Peng, Metal-organic framework membranes: recent development in the synthesis strategies and their application in oil-water separation. Chem. Eng. J. **405**, 127004 (2021). https://doi.org/10.1016/j.cej.2020.127004
59. B.M. Jun, Y.A.J. Al-Hamadani, A. Son, C.M. Park, M. Jang, A. Jang, N.C. Kim, Y. Yoon, Applications of metal-organic framework based membranes in water purification: a review. Sep. Purif. Technol. **247**, 116947 (2020). https://doi.org/10.1016/j.seppur.2020.116947
60. Z. Jia, S. Hao, J. Wen, S. Li, W. Peng, R. Huang, X. Xu, Electrochemical fabrication of metal–organic frameworks membranes and films: a review. Microporous Mesoporous Mater. **305**, 110322 (2020). https://doi.org/10.1016/j.micromeso.2020.110322
61. X.Y. Gong, Z.H. Huang, H. Zhang, W.L. Liu, X.H. Ma, Z.L. Xu, C.Y. Tang, Novel high-flux positively charged composite membrane incorporating titanium-based MOFs for heavy metal removal. Chem. Eng. J. **398**, 125706 (2020). https://doi.org/10.1016/j.cej.2020.125706

62. M. Samari, S. Zinadini, A.A. Zinatizadeh, M. Jafarzadeh, F. Gholami, Designing of a novel polyethersulfone (PES) ultrafiltration (UF) membrane with thermal stability and high fouling resistance using melamine-modified zirconium-based metal-organic framework (UiO-66-NH_2/MOF). Sep. Purif. Technol. **251**, 117010 (2020). https://doi.org/10.1016/j.seppur.2020.117010
63. M. Changmai, M. Pasawan, M.K. Purkait, A hybrid method for the removal of fluoride from drinking water: parametric study and cost estimation. Sep. Purif. Technol. **206**, 140–148 (2018). https://doi.org/10.1016/j.seppur.2018.05.061
64. S. Mestre, A. Gozalbo, M.M. Lorente-Ayza, E. Sánchez, Low-cost ceramic membranes: a research opportunity for industrial application. J. Eur. Ceram. Soc. **39**, 3392–3407 (2019). https://doi.org/10.1016/j.jeurceramsoc.2019.03.054
65. S.K. Hubadillah, M.H.D. Othman, A.F. Ismail, M.A. Rahman, J. Jaafar, Y. Iwamoto, S. Honda, M.I.H.M. Dzahir, M.Z.M. Yusop, Fabrication of low cost, green silica based ceramic hollow fibre membrane prepared from waste rice husk for water filtration application. Ceram. Int. **44**, 10498–10509 (2018). https://doi.org/10.1016/j.ceramint.2018.03.067
66. A. Manni, B. Achiou, A. Karim, A. Harrati, C. Sadik, M. Ouammou, S. Alami Younssi, A. El Bouari, New low-cost ceramic microfiltration membrane made from natural magnesite for industrial wastewater treatment. J. Environ. Chem. Eng. **8**, 103906 (2020). https://doi.org/10.1016/j.jece.2020.103906
67. B.K. Nandi, R. Uppaluri, M.K. Purkait, Preparation and characterization of low cost ceramic membranes for micro-filtration applications. Appl. Clay Sci. **42**, 102–110 (2008). https://doi.org/10.1016/j.clay.2007.12.001

Nanoscale Rare-Earth-Based Mixed-Metal Oxides for Solar Photocatalytic Applications

Sahar Zinatloo-Ajabshir and Seyyed Javad Heydari-Baygi

Abstract The research and development have increased regarding rare-earth-based mixed-metal oxide nanostructures as a notable category of materials with unique characteristics, and as one of the most substantial fields of research in the last decades. Many studies have demonstrated their outstanding usage in different fields, specifically solar photocatalytic applications. Recent advances in the production techniques and photocatalytic usages of $Ln_2B_2O_7$ (B=Zr, Ce, and Sn) nanostructures have been summarized in this chapter. Diverse methods have been presented for the fabrication of lanthanide zirconate, lanthanide stannate, and lanthanides for ceramics production. Briefly, the privileges and disadvantages of every fabrication method have been introduced in this chapter. Ultimately, solar photocatalytic applications of rare-earth-based mixed-metal oxide nanostructures are reviewed as well.

Keywords Rare-earth-based mixed-metal oxides · Solar photocatalytic applications · Nanostructures · Ceramics

1 Introduction

The tunable production of nanostructures with favorable features has been one of the most important challenges in nanoscience. Several approaches (e.g. chemical and physical methods) have been presented to adjust the features of nanostructures such as their morphology, dimensions, crystal structure, and purity during production [1, 2].

$A_2B_2O_7$ is a class of rare-earth-based nanostructures with pyrochlore and also fluorite structures [2, 3]. The A site indicates metal elements such as yttrium, lanthanum, praseodymium, neodymium, samarium, europium, gadolinium, terbium,

S. Zinatloo-Ajabshir (✉)
Department of Chemical Engineering, University of Bonab, P.O. Box. 5551395133, Bonab, Iran
e-mail: s.zinatloo@ubonab.ac.ir

S. J. Heydari-Baygi
Chemical Engineering Department, Tarbiat Modares University, Tehran, Iran

S. J. Ikhmayies (ed.), *Advanced Ceramics*, Advances in Material Research and Technology, https://doi.org/10.1007/978-3-031-43918-6_6

dysprosium, holmium, erbium, ytterbium, and lutetium. Further, the B site represents metal elements such as tin, titanium, zirconium, hafnium, ruthenium, and cerium [1, 4].

In this chapter of the book, we will review a specific class of rare-earth-based mixed-metal oxides, namely, $Ln_2B_2O_7$ nanostructures (B=Zr, Sn, and Ce) [5, 6]. This chapter addresses each production technique with its privileges and disadvantages, and properties of as-gained nanostructures. The approaches are as follows: hydrothermal [7], solid-state reaction [8], sol–gel [9], co-precipitation [8], combustion [7], etc. Finally, the solar photocatalytic usages of $Ln_2B_2O_7$ nanostructures such as photodegradation of contaminants will be discussed [10]. Other usages as follows: electrochemical water splitting [11], oxidative coupling of methane (OCM) [12], catalytic conversion of carbon dioxide [13], and other related purposes [14].

2 Synthetic Methods for the Production of Rare-Earth-Based Mixed-Metal Oxides

There are several available approaches for preparing rare-earth-based nanostructures involving of sol–gel, solid-state reaction, hydrothermal, solvothermal, pechini, co-precipitation, etc. [15–17]. The employed methods considerably affect the particle morphology, dimensions, and the photocatalytic properties of the produced nanostructures.

2.1 *Solid-State Reaction*

In the solid-state reaction as a conventional way for the fabrication of $Ln_2Zr_2O_7$, rare-earth oxide and also zirconium dioxide are physically combined and the obtained materials are calcined at extremely thermal conditions for a long time [18]. Qiang et al. produced $Dy_2Zr_2O_7$ mixed-metal oxide by dysprosium oxide and also zirconia as precursor substances at 1600 °C for 10 h [19]. The outcomes demonstrated that the fabricated $Dy_2Zr_2O_7$ mixed-metal oxide had a fluorite structure. Likewise, Tong et al. provided $La_2Zr_2O_7$ and $Nd_2Zr_2O_7$ using grinding reactants such as potassium carbonate and zirconyl chloride octahydrate ($ZrOCl_2{\cdot}8H_2O$) together with a mortar. Subsequently, the as-obtained powders were calcined at 900 °C within 1 h [20]. The synthesized lanthanum zirconate sample had a square-like shape. Different wet chemistry methods have been improved for preparing rare-earth-based mixed-metal oxides with favorable properties. In the wet chemical ways, precursor materials are combined at microscopic dimensions, thus leading to the formation of ultimate nanostructures with high homogeneity. Hagiwara produced $Eu_2Zr_2O_7$ and $La_2Zr_2O_7$ nanocrystals using solid-state reaction method using zirconium dioxide, europium, and lanthanum oxide as precursors [21].

In another research, Hongsong et al. prepared $(Sm_{1-x}Gd_x)_2Ce_2O_7$ ceramics via a solid-state reaction approach. They evaluated the structure improvement and thermo-physical features of $(Sm_{1-x}Gd_x)_2Ce_2O_7$ nanostructure to achieve the fluorite-structure nanoparticles, and selected cerium(IV) oxide, gadolinium oxide, and samarium(III) oxide as the starting materials. The rare-earth-based nanopowders were heated at 1000 °C for 5 h. Then, the dried materials were changed to granulated particles and compacted under pressure of 50 MPa. Finally, the created structures were sintered in the present air at 1600 °C within 10 h [22].

2.2 *Hydrothermal*

In the hydrothermal technique, the hydroxides are precipitated from precursor solutions, and the powders are made by the nucleation process and particle growth according to a certain temperature and pressure. It is a comfortable, facile, and usefully controlled method [23–27]. In this way, water is the solvent that is an inexpensive, renewable, and abundant structure [28]. Gao et al. synthesized $Gd_2Zr_2O_7{:}Tb^{3+}$ phosphors via a hydrothermal route using $Gd(NO_3)_3$, N_3O_9Tb, $ZrOCl_2$, and NH_3 solutions as precursors at 200 °C for 20 h [29]. The produced nanocrystal with a fluorite structure had a radius of about 15 nm and a length of 150–300 nm. Moreover, Wang et al. synthesized $Sm_2Zr_2O_7$ pyrochlore nanostructure by a hydrothermal method using Sm_2O_3, $ZrOCl_2{\cdot}8H_2O$, nitric acid, and aqueous NH_3 as starting materials for 1 day [30]. They studied the effect of process temperature on the particle morphology and also purity of $Sm_2Zr_2O_7$. Based on X-ray diffraction (XRD) patterns, and transmission electron microscope (TEM) images, $Sm_2Zr_2O_7$ nanostructures could be formed at 190 °C. Wang et al. synthesized $La_2Ce_2O_7$ ceramic using a hydrothermal approach with nitric acid, lanthanum oxide, polyethylene glycol as stabilizing agent, and cerium(III) nitrate hexahydrate at 180 °C within 24 h [31]. The obtained samples had cubic shape and fluorite structure. They found that the type of stabilizing agent could affect the surface attributes and crystallite dimension of $La_2Ce_2O_7$ [31].

2.3 *Combustion*

Combustion reactions containing metal precursors and a fuel material can lead to the formation of rare-earth-based nanocrystals [32]. Zhang et al. produced $La_2Zr_2O_7$ nanostructures by a combustion method. To this end, they prepared a mixture including zirconium acetate, nitric acid, lanthana, and urea as fuel, and then aqueous NH_3 was added to this solution to get gel structure. The as-obtained gel was heated at 600 °C for 15 min. $La_2Zr_2O_7$ sample has a pyrochlore structure with nanoparticle diameter in the range of 45–70 nm. Du et al. synthesized $Y_2Zr_2O_7{:}Dy^{3+}$, $Y_2Zr_2O_7{:}Dy^{3+}$ and Li^+ phosphors with a sol–gel combustion method utilizing

NH_4NO_3 as the fuel at 800 °C [33]. Likewise, Tong et al. produced $Nd_2(Zr_{1-x} Sn_x)_2O_7$ nanomaterials with a salt-assisted combustion technique by KCl and $C_2H_5NO_2$ as the fuel at 700 °C [34, 35]. In another study, Yi et al. fabricated $LaYZr_2O_7$ pyrochlore nanostructure by a combustion route by applying glycine as the fuel at 1200 °C [36]. Lei studied the shape and size of nanoparticles through XRD, Fourier-transform infrared spectroscopy (FT-IR), thermogravimetry–differential thermal analysis (TG–DTA), scanning electron microscopy (SEM), and TEM and observed the phase structure and the sintering densification of nanocrystals. He produced $Gd_2Zr_2O_7$ nanostructures with fluorite structure by two combustion routes. The diameters of the synthesized nanoparticles by these methods were 10 and 50 nm [37]. Hong et al. further prepared a set of ceramic nanostructures related to $Nd_2Ce_2O_7$ through a citrate–nitrate gel combustion synthesis [38]. To start this process, precursors such as neodymium(III) nitrate hydrate, cerium(III) nitrate hexahydrate, magnesium nitrate hexahydrate, cobalt(II) nitrate hexahydrate, strontium (II) nitrate hexahydrate, and barium nitrate hexahydrate were applied for producing mixed-metal oxides. Additionally, citric acid as the chelator was added to the complex solution with a molar ratio of 1.5:1 to all ions. Then, the ultimate solution for obtaining the viscous gel was fired at 343 K. The gel was then heated again on the furnace at 550 °C within 60 min. Finally, rare-earth-based mixed-metal oxides were synthesized by the calcination process at 800 °C for 3 h [38].

In a study by Sun et al., lanthanum cerate nanostructures were prepared using a combustion route. Lanthanum oxide and cerium(III) nitrate hexahydrate with 99% purity were provided as the initial materials. First, soluble lanthanum oxide was added into HNO_3 solution under heating and mixing processes, followed by cerium(III) nitrate hexahydrate. Next, $C_6H_8O_7$ was added with molar ratios of 1.5:1 (citric acid: metal ions), and the pH value was adjusted near 7 with ammonium hydroxide. The solution was placed under a temperature of 70 °C. Ultimately, the as-gained nanopowders were calcined at 800 °C for 3 h to get $La_2Ce_2O_7$ nanoparticles with high purity [39]. Also, Salehi et al. produced $Dy_2Ce_2O_7$ nanostructures employing a combustion by applying NaCl as a dispersant, dysprosium nitrate, and cerium (IV) ammonium nitrate as starting materials, as well as the succinic acid and various amino acids as the capping agent [40].

They concluded that the kind of amino acid could apply a substantial effect on the size and particle morphology of the $Dy_2Ce_2O_7$ nanostructure. The schematic of the fabrication technique of $Dy_2Ce_2O_7$ nanostructures is shown in Fig. 1. The final rare-earth-based cerate ceramic was characterized by TEM images, and the TEM image of $Dy_2Ce_2O_7$ is illustrated in Fig. 2.

Zinatloo et al. synthesized $Dy_2Ce_2O_7$ samples by employing various values of the extract of Comosus pineapple. Then, the gel was formed after the vaporization of the mixture, and finally, the dried structure was prepared at 450 °C for 3.5 h. Tong et al. produced $Er_2Sn_2O_7$ nanocrystals [41] by a combustion method using erbium oxide and glycine as the fuel and $SnCl_4{\cdot}5H_2O$ as a source of tin.

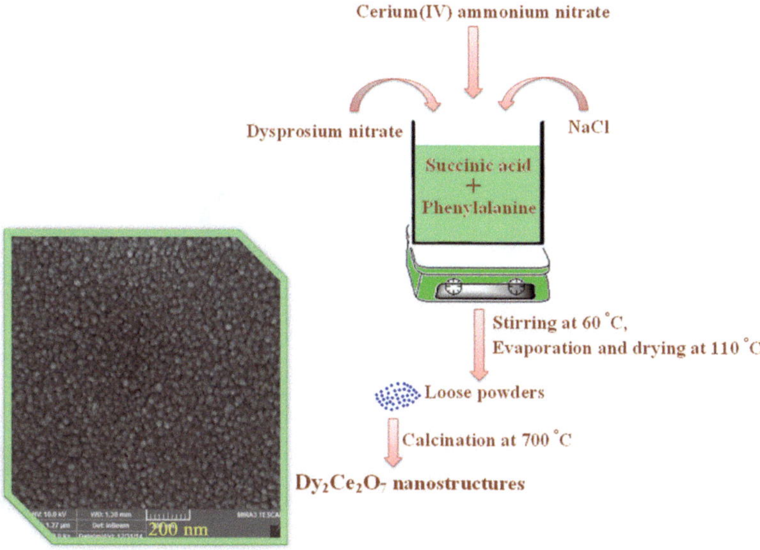

Fig. 1 The schematic of the fabrication technique of $Dy_2Ce_2O_7$ nanostructure [40]

2.4 *Co-precipitation*

In the co-precipitation method, a base material is added to the precursor solution, and then metal hydroxides are precipitated simultaneously. Next, the obtained hydroxide is calcined to form a rare-earth zirconium ceramic. Liu et al. synthesized $Sm_xZr_{1-x}O_{2-x/2}$ ($0.1 < x < 0.5$) nanostructure using the co-precipitation method and the thermal process at 800 °C for 300 min [42]. In another research, $(Sm_xGd_{1-x})_2Zr_2O_7$ ($0 < x < 1.0$) nanostructures were prepared using the co-precipitation approach and the calcination process at 1073 K for 360 min [43], and the nanopowders had a pyrochlore phase and structure. Moreover, Hu et al. fabricated $(Gd_{1-x}Eu_x)_2Zr_2O_7$ ($0 < x < 0.100$) nanomaterials by the co-precipitation technique and the calcination process at 1073 K with a fluorite structure [44].

2.5 *Sol–Gel*

In the sol–gel method, wet gels are obtained with the hydrolysis-condensation processes of metal alkoxides as the precursors. Then, the obtained gels are transformed to rare-earth-based oxides through the thermal stage [45–47]. The sol–gel method is economic, facile, and eco-friendly compared to other approaches [31, 48, 49]. Sohn et al. synthesized $Ln_2Zr_2O_7$ (lanthanide elements, e.g. samarium, europium, gadolinium, and terbium) nanopowders through the sol–gel method by zirconium nitrate and tetraisopropoxide as the precursors and isopropyl alcohol as

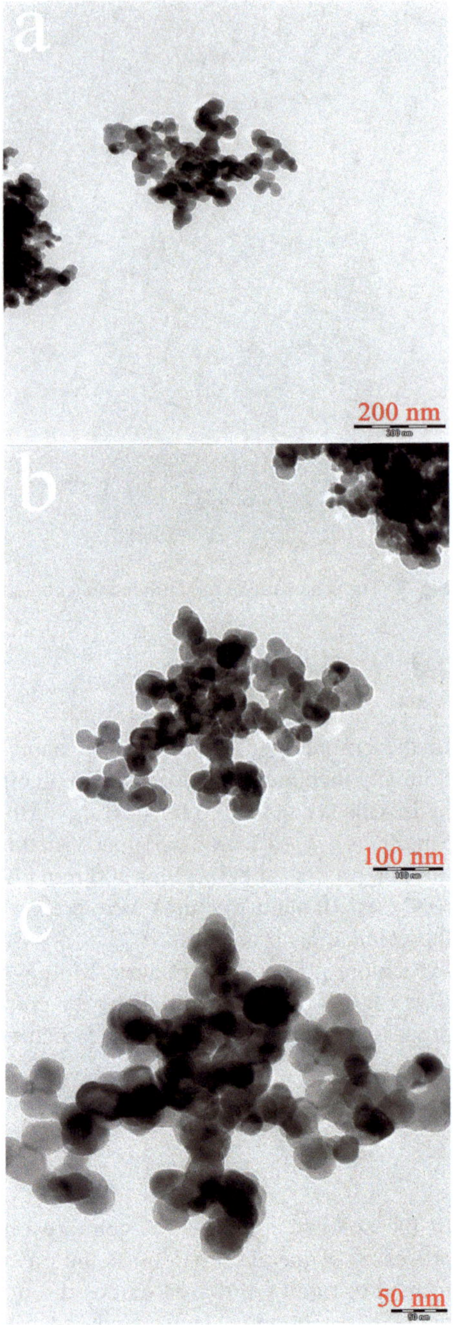

Fig. 2 TEM images of the nanocrystalline $Dy_2Ce_2O_7$ [40]

the solvent at 1473 K. They fabricated the nanostructures with a pyrochlore structure [49]. In another research, Saitzek et al. created $Ln_2Zr_2O_7$ pyrochlore on strontium titanate layers with a sol–gel route [50]. Furthermore, Jovaní et al. showed that $Y_2Zr_2O_7$ nanostructures with fluorite structure were synthesized by the sol–gel method at 1573 K for 12 h [51]. Tong produced lanthanide zirconate (Ln=La, Nd, Sm, Dy, and Er) mixed-metal oxide with a cubic shape using a sol–gel approach [52, 53]. Salts containing $NO^{-}{}_3$ ions were applied as starting materials and stearic acid was used as a solvent. The molar ratio of zirconia and lanthanide oxide was chosen equally [54].

In another study, Zinatloo et al. [55] employed modified sol–gel approach to synthesize $Dy_2Sn_2O_7$–SnO_2 nanocomposite by increasing $C_9H_6O_6$ (trimesic acid) in stannic chloride pentahydrate and dysprosium(III) nitrate pentahydrate solutions at 623 K within 120 min. They studied different effects of stabilization substances and calcination process temperature on the dimensions and size of dysprosium stannate-stannic oxide nanocomposites. The schematic diagram of the production of $Dy_2Sn_2O_7$–SnO_2 nanostructures is shown in Fig. 3.

Salehi et al. also produced $Dy_2Ce_2O_7$ nanostructure using cerium (IV) ammonium nitrate and dysprosium nitrate [31] and reported the effect of kind of the connecting agent and the chelating material on the particle morphology and size of dysprosium cerate. The schematic diagram of the production of $Dy_2Ce_2O_7$ nanostructure is depicted in Fig. 4.

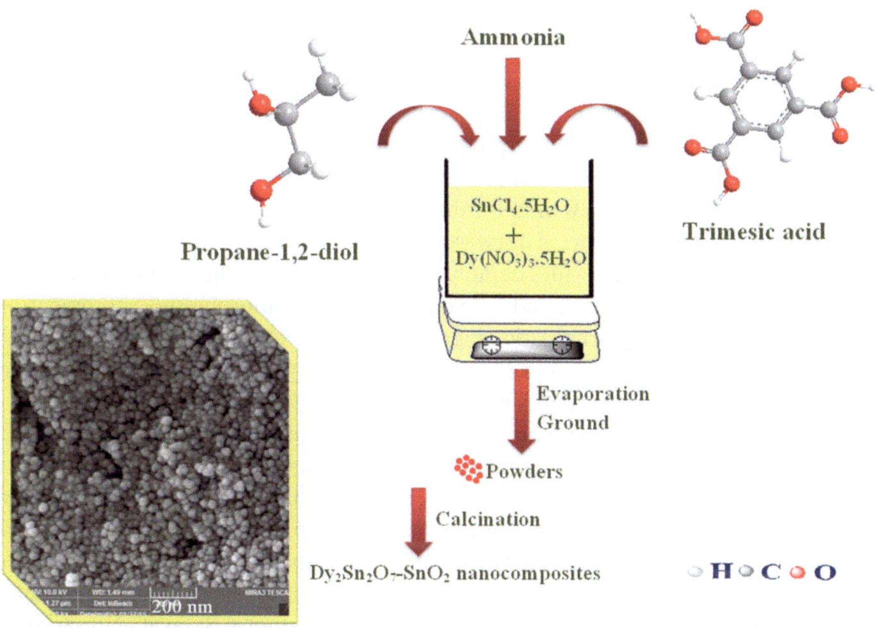

Fig. 3 Schematic diagram for the production of $Dy_2Sn_2O_7$–SnO_2 nanostructure [55]

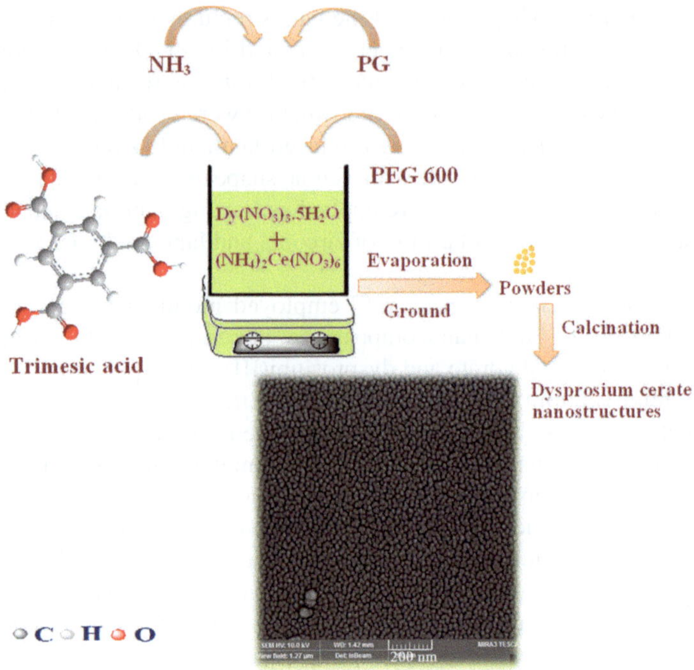

Fig. 4 Schematic diagram of the production of $Dy_2Ce_2O_7$ nanostructure [31]

In another study, Zinatloo et al. prepared $Nd_2Sn_2O_7$ nanostructure through an easy, novel, and low-temperature method by pomegranate juice [56]. Initially, the appropriate content of the juice of pomegranate as the new fuel was mixed into the solution containing $SnCl_4{\cdot}5H_2O$ and $Nd(NO_3)_3$ in water, and the obtained mixture was mixed at 45 °C within 20 min. Further, the viscous gel was dried via the evaporation of the ultimate mixture. Afterward, the calcination process was carried out at 500 °C within 4.5 h. The results demonstrated the effect of the important variable, production temperature, on the adjustment of particle's morphology, composition, and purity of $Nd_2Sn_2O_7$. Schematic diagrams for the creation of $Nd_2Sn_2O_7$ and field emission scanning electron microscope (FESEM) images of oxide nanostructures prepared at different temperatures are displayed in Figs. 5 and 6 [56].

Salavati-Niasari et al. also fabricated neodymium zirconate ($Nd_2Zr_2O_7$) ceramic by a modified sol–gel method by applying succinic acid ($C_4H_6O_4$) as stabilization material as well as propane-1,2-diol and ($C_8H_{23}N_5$) as a pH controller at temperature 700 °C within 240 min [57]. They investigated the effect of the kind of the pH regulator and propane-1,2-diol concentrations on the size and dimensions of neodymium zirconate and observed that these factors could be significantly effective in determining the morphology of $Nd_2Zr_2O_7$ ceramics. Furthermore, Salavati-Niasari et al. synthesized lanthanide-based oxide $Pr_2Zr_2O_7$ using a facial approach with $C_3H_8O_2$

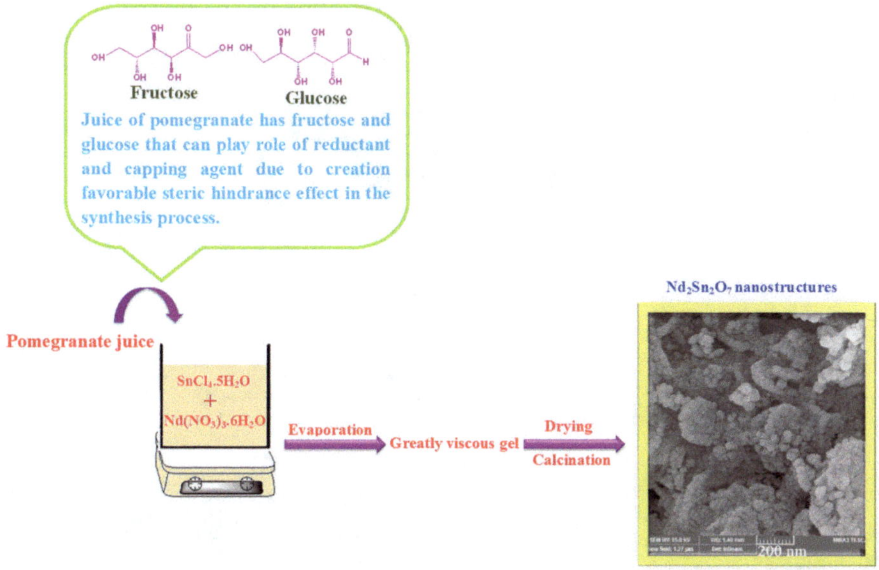

Fig. 5 Schematic diagrams for the production of $Nd_2Sn_2O_7$ [56]

as the connecting agent and $C_9H_6O_6$ as the stabilization structure at a low-thermal treatment [58].

Salavati-Niasari et al. produced $Nd_2Zr_2O_7$ nanoparticles via a modified Pechini route using succinic acid ($C_4H_6O_4$) as the capping agent and propane-1,2-diol as the connecting agent at 973 K for 4 h [57]. They studied the effect of the kind of acidity regulator and propane-1,2-diol dosage on the particle's morphology of neodymium zirconate and found that the investigated parameters have a remarkable effect on adjusting the dimensions and shape of neodymium zirconate. Figure 7 displays the schematic diagram of the fabrication of neodymium zirconate ceramic.

In addition, Salavati-Niasari et al. provided praseodymium zirconate nanostructures using a facile method with benzene tricarboxylic acid as the stabilization agent in the presence of $C_3H_8O_2$ as the cross-linking agent at low temperature [58]. Their results indicated that the particle's morphology and crystallinity of praseodymium zirconate nanostructure were controlled by altering the kind of the stabilization agent. The FESEM images of the prepared $Pr_2Zr_2O_7$ nanostructures utilizing various carboxylic acids are depicted in Fig. 8. Salavati-Niasari et al. also synthesized $Nd_2Zr_2O_7$–Nd_2O_3 nanocomposite with an easy modified Pechini approach using salicylic acid as a novel chelating agent [17].

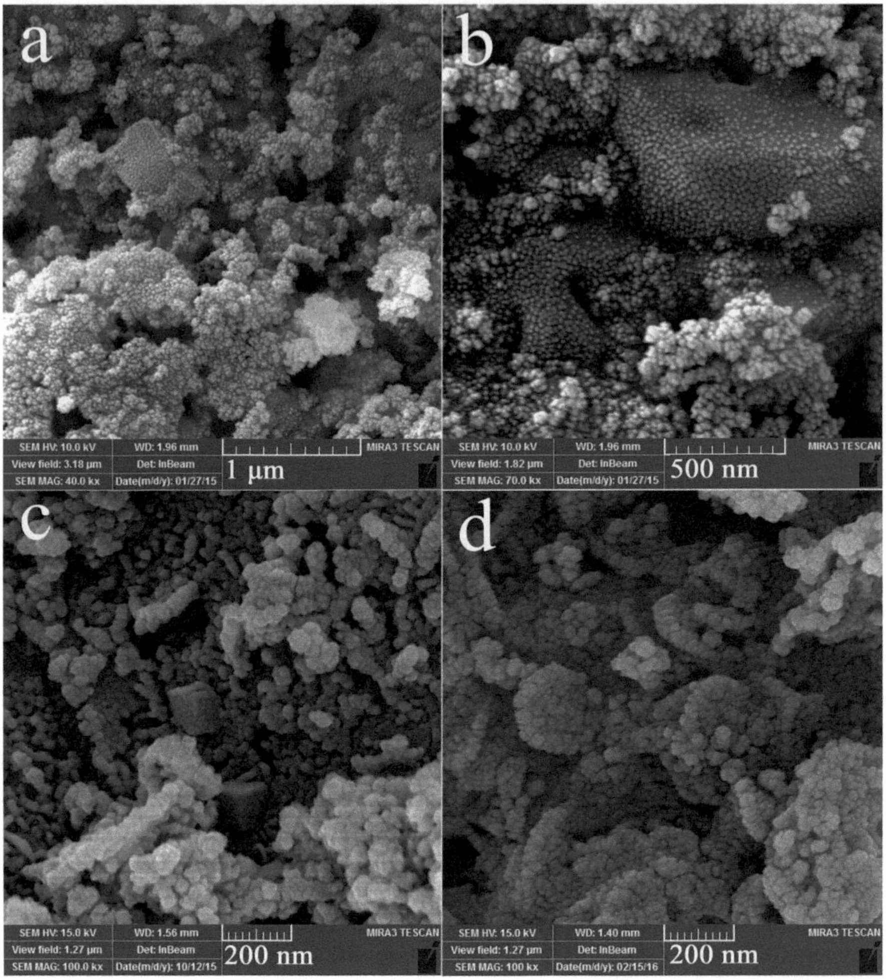

Fig. 6 FESEM images of samples 1, 2, and 3 produced at 700 (**a** and **b**), 600 (**c**), and 500 °C (**d**), respectively [56]

2.6 Other Preparation Approaches

2.6.1 Molten Salt Synthesis

In the molten salt route, the solution of raw materials can be simplified with dissolvable agents in the molten salt solution, and thus the reaction is conducted at a relatively low temperature [59, 60]. Huang et al. also fabricated lanthanum zirconate pyrochlore via the molten salt method utilizing zirconium dioxide and lanthanum oxide as the starting materials in the presence of sodium chloride (NaCl), potassium chloride (KCl), and sodium fluoride (NaF) mixture as the reaction solution at 1100 °C within

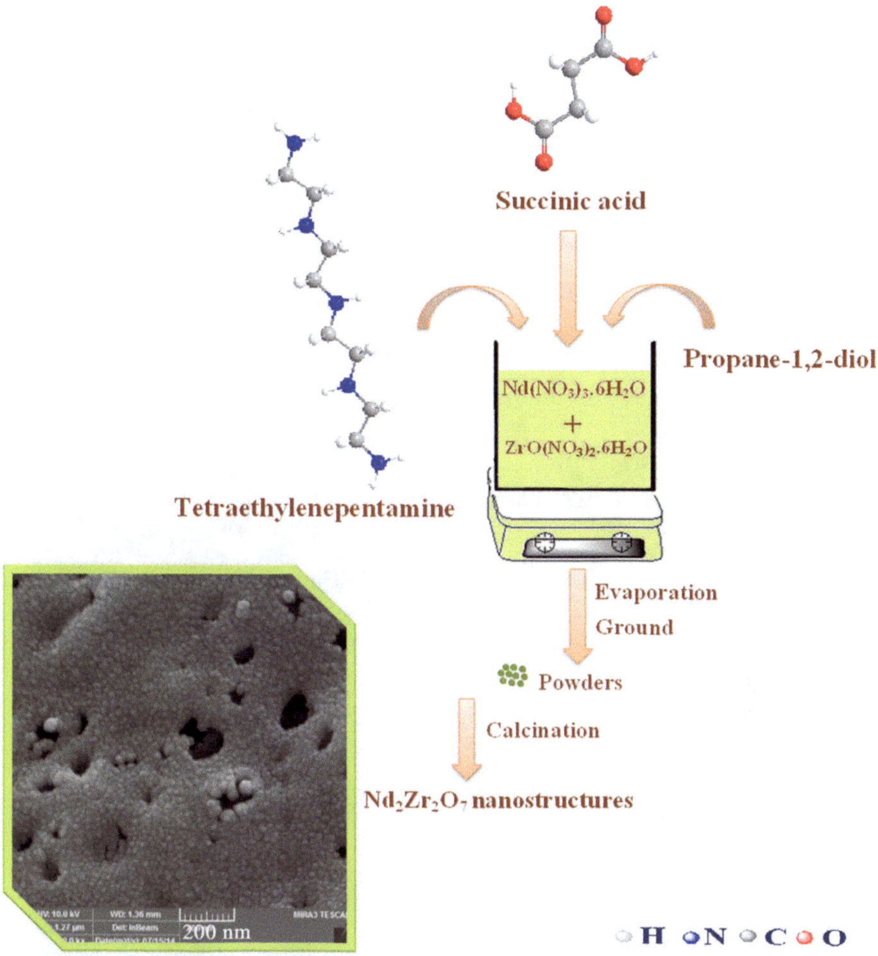

Fig. 7 Schematic diagram of fabrication of neodymium zirconate nanostructure [57]

180 min [61]. They investigated the effect of parameters such as process temperature and the salt to raw material ratio on the production of $Ln_2Zr_2O_7$.

2.6.2 Co-ions Complexation Method

Xu et al. produced lanthanum zirconate pyrochlore via the co-ions complexation technique at 1300 °C. The method was applied based on three steps. Firstly, solidification through the construction of the complexed compound including zirconium acetate and lanthanum acetate, then lyophilization, and finally, the thermal treatment process known as calcination [62].

Fig. 8 FESEM images of the samples synthesized in the presence of the salicylic acid (**a** and **b**), maleic acid (**c** and **d**), succinic acid (**e** and **f**) [58]

2.6.3 Cathode Plasma Electrolysis

Liu et al. prepared $Ln_2Zr_2O_7$ ceramic with both pyrochlore and fluorite structures utilizing plasma electrochemical decomposition in the electrolyte of $Zr(NO_3)_4$ and $La(NO_3)_3$ in the absence of thermal treatment [63].

2.6.4 Precursor Route

Payne produced $Nd_2Zr_2O_7$ nanostructure utilizing a precursor route. To achieve the end, an aqueous solution was provided through combining neodymium nitrate and zirconium (IV) oxynitrate. Then, ammonium hydroxide was added to fabricate the gel structure. Additionally, the obtained gel compound was subjected to mixing, washing, drying, and then annealing operations. In this route, structure conversion from fluorite to the pyrochlore was observed at high temperatures after the annealing process [64].

2.6.5 Chemical Solution Deposition

$Ln_2Ce_2O_7$ nanostructures with different structures were produced via chemical solution deposition. The nanopowders were obtained via pouring 15 mL of the starting material solution into an aluminum oxide crucible close to a hot furnace. The wet gel structure was formed at 1000 °C for 4 h [65].

2.7 Advantages and Disadvantages of Fabrication Methods

The advantages and disadvantages of every fabrication method were briefly reviewed in this chapter. Several synthesized samples have been produced through different routes using the Zinatloo research group. There are different chemical methods to prepare the rare-earth-based mixed-metal oxides with remarkable characteristics. In the wet chemistry method, initial structures are combined, and therefore, the target nanostructures have desirable structural homogeneity. However, the solid-state reaction method is one of the above-mentioned methods that can lead to unsuitable features of synthesized nanoparticles, including undesirable control of size and particle morphology, and nonhomogeneity of the composition [20]. In the hydrothermal approach, formation of homogeneous nucleation reactions is a notable benefit that can lead to the production of nanostructures with favorable dimensions without needing the calcination process. Moreover, the formation rate of rare-earth-based mixed-metal oxides is slow thus the amount of thermal energy at low temperature is essential after the hydrothermal stage for running the solid-state reaction [66]. The combustion method can result in nanostructure formation with excellent purity, appropriate morphology, and uniform composition. However, the control of combustion is difficult. Additionally, the combustion reaction can cause the aggregation of the obtained particles at high temperatures [35]. The co-precipitation method can lead to the production of excellent purity combinations with a uniform structure due to low temperature although the obtained nanomaterials have large dimensions. Accordingly, a grinding stage is necessary for creating a powder with appropriate morphology [18, 67]. The sol–gel technique is a suitable method for the structural and compositional controlling of the mixed-metal oxides that were produced via a calcination process with a fine size [68]. Since the kind of preparation approach can be very effective in regulating the features of photocatalytic nanopowders and thus its performance. Hence, the selection of an appropriate technique to produce nanopowders with the fine particle size, high uniformity, and good purity can be very helpful. In addition, energy consumption and economic prospects are notable for selecting the fabrication strategy. Thus, for large-scale applications, the usage of a simple, reproducible, and environmentally friendly approach to fabricate photocatalytic nanopowders with superior efficiency can be necessary. In the next section, we summarize the applications of rare-earth-based mixed-metal oxide.

3 Application of Rare-Earth-Based Mixed-Metal Oxide Nanostructures

Mixed-metal nanostructures have commonly been admitted as a considerable category of chemical catalysis employed for the decomposition of organic contaminants and wastewater treatment [25, 69, 70]. The number of photocatalysts having efficient and outstanding capability is still restricted, thus the design and improvement of photocatalysts can play an essential role in the degradation of organic pollutants, wastewater treatment, and water purification [68, 71]. In recent studies, rare-earth-based mixed-metal oxide nanostructures have been offered as an appropriate, impressive, and engaged alternative of heterogeneous photocatalyst owing to their great ability for photodegradation and high stability [58, 72–74]. It has been illustrated that the photocatalytic role of $Ln_2B_2O_7$ (B=Zr, Ce, Sn, and the like) structures largely relies on their morphology, purity, shape, and amount of the surface area.

3.1 Rare-Earth-Based Mixed-Metal Oxide Nanostructures as Photocatalyst

Studies have been conducted on the elimination of ability dysprosium cerate ($Dy_2Ce_2O_7$) produced with the extract of comosus pineapple for eriochrome black T degradation under visible radiation. There is no decomposition without applying visible light or samples of dysprosium cerate. They observed that 91.7% photocatalytic efficiency occurs within 1 h [41].

Tong et al. applied $Ln_2Zr_2O_7$ (Ln=La and Nd) ceramics as photocatalysts with a concentration of 1 g/L for the degradation of the methyl orange with a concentration of 20 mg/L under ultraviolet (UV) radiation by a mercury lamp [20]. The results demonstrated that the solar photocatalytic performance of the methyl orange contaminant activated with $La_2Zr_2O_7$ nanocrystals for 1 h was greater than that of neodymium zirconate pyrochlore. The lanthanum zirconate metal oxide structures could exert a strong catalytic role due to their better purity and wide catalytic surface.

In another research, the photocatalytic proficiency of $Ln_2Zr_2O_7$ (Ln=La, Nd, Sm, Dy, and Er) nanoparticles was also evaluated by the photodecomposition of the methyl orange organic pollutant under ultraviolet light for 2 h [54]. The results demonstrated that the photocatalytic efficiency of $Ln_2Zr_2O_7$ nanopowders is strongly associated with the kind of lanthanide elements, increasing in the order of lanthanum > samarium > erbium > neodymium > dysprosium. The photodegradable performance of $Y_2Zr_2O_7$ ceramic structure, produced at various temperatures by Tong et al., was compared based on methyl orange destruction under UV light in 120 min [53]. The outcomes indicated that the catalytic yield decreased by the improvement of calcination temperature owing to the diminution in the surface area. In another investigation, Bai et al. evaluated the photocatalytic output of $(La_xFe_{1-x})_2Zr_2O_7$ ceramic by the photocatalytic degradation of methyl orange contaminant under UV light for 2 h [75]. Based on the results, the activity of rare-earth zirconate ceramic structures could be enhanced by Fe doping. In addition, $(La_{0.9}Fe_{0.1})_2Zr_2O_7$ operated as the most proper photocatalyst due to the desired value of the doped iron. Similarly, Wang et al. utilized the samarium zirconate ($Sm_2Zr_2O_7$) nanocrystal with a concentration of 0.05 g/L for the photodecomposition of the congo red contaminant under Xenon lamp radiation [30]. The results indicated that the synthesized nanostructures could degrade more than three-quarters of congo red contaminant at 30 min and prove the better photocatalytic function rather than bulk-samarium zirconate due to the improper morphology and large surface. Likewise, Liu et al. used $La_2Zr_2O_7$ microspheres with a concentration of 1 g/L for the photodegradation of methyl orange contaminant with the loading of 20 mg/L under visible radiation. The photodegradation efficiency of methyl orange was activated with $La_2Zr_2O_7$ ceramic structure near 50% within 1 h. Salavati-Niasari et al. have recently presented praseodymium zirconate ($Pr_2Zr_2O_7$) as a novel photocatalytic nanostructure with similar or greater performance compared to those currently known nanostructures. Salavati-Niasari et al. also investigated the influence of some factors such as the morphology of praseodymium zirconate nanostructures, type of pollutant, and type of radiation (UV or visible light) on the photoactivity yield [58]. The outcomes revealed that the formed praseodymium zirconate ceramics with fine dimensions had great ability to be used as a new, efficient, and desirable kind of photocatalyst under visible and UV lights for the removal of organic contaminants. The solar photoactivity outcomes of praseodymium zirconate ceramics are displayed in Fig. 9.

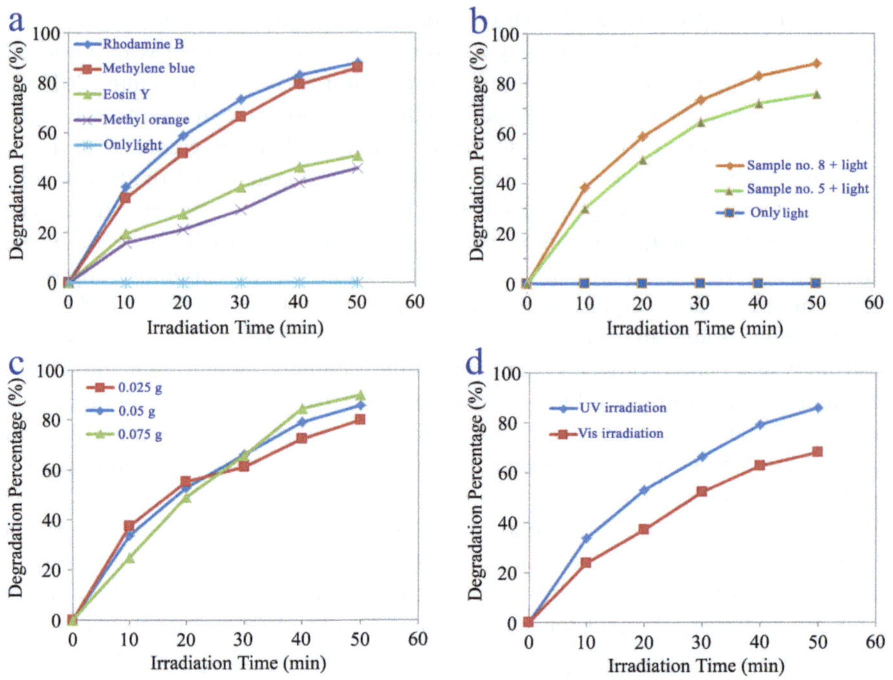

Fig. 9 **a** The photocatalytic efficiency of $Pr_2Zr_2O_7$ nanostructures (sample No. 8) on decomposition of various contaminants (cationic and anionic types), **b** The photocatalytic performance of the prepared $Pr_2Zr_2O_7$ nanostructures (sample Nos. 5 and 8), **c** The influence of the dosage of the nanostructured $Pr_2Zr_2O_7$ (sample No. 8) on the methylene blue decomposition percentage and **d** The photocatalytic performance of the nanostructured $Pr_2Zr_2O_7$ (sample No. 8) under UV and visible lights [58]

The prepared $Nd_2Zr_2O_7$ nanostructures by Salavati-Niasari et al. were applied for the photocatalytic decomposition of erythrosine and eriochrome black T contaminants under UV radiation [57]. Eriochrome black T and erythrosine photodecomposition were measured to be 84% and 88% for 50 min, respectively. The results showed that nanostructured neodymium zirconate produced by succinic acid as a new stabilization structure and propane-1,2-diol with fine crystalline dimensions could be employed as a favorable photocatalyst type under UV radiation. Figure 10 illustrates the photocatalytic erythrosine degradation of $Nd_2Zr_2O_7$ nanoparticles and the recovery examinations of the synthesized $Nd_2Zr_2O_7$ nanostructures for the decomposition of erythrosine under UV light.

$Nd_2Zr_2O_7$–ZrO_2 nanocomposites were prepared using Salavati-Niasari et al. for the photodecomposition of methylene blue and 2-naphthol as water pollutants under ultraviolet light [76]. Neodymium zirconate-zirconium dioxide mixed-metal oxide nanostructure with a lower energy gap indicated desired photocatalytic efficiency rather than the nanostructured ZrO_2. The recombination rate of charges could be reduced due to the content of lower energy gap. The reactions performed to degrade

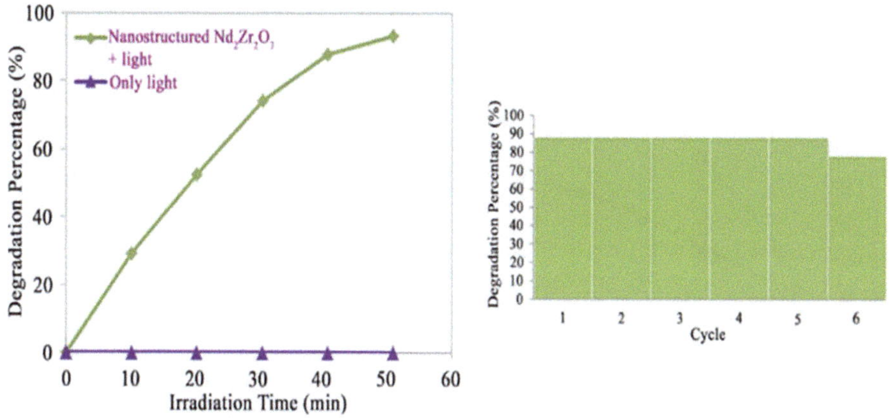

Fig. 10 (left) Photocatalytic erythrosine degradation of the created $Nd_2Zr_2O_7$ nanostructures and (right) recovery tests of the produced $Nd_2Zr_2O_7$ nanostructures for decomposition of erythrosine under UV light [57]

methylene blue and 2-naphthol as water pollutants can be considered as follows:

$$\text{Nanostructured photocatalyst} + h\nu \rightarrow \text{Nanostructured photocatalyst}^* + e^- + h^+ \quad (1)$$

$$h^+ + H_2O \rightarrow OH^{\cdot} \quad (2)$$

$$e^- + O_2 \rightarrow O_2^{-\cdot} \quad (3)$$

$$OH^{\cdot} + O^{2-\cdot} + \text{Methylene blue or 2-naphthol} \rightarrow \text{Degradation products} \quad (4)$$

Photocatalytic methylene blue and 2-naphthol destruction of the formed neodymium zirconate-zirconium dioxide nanocomposites and nanostructured zirconium dioxide are depicted in Fig. 11.

In Table 1, the solar photocatalytic applications for several lanthanide-based nanostructures especially $Ln_2B_2O_7$ category are summarized.

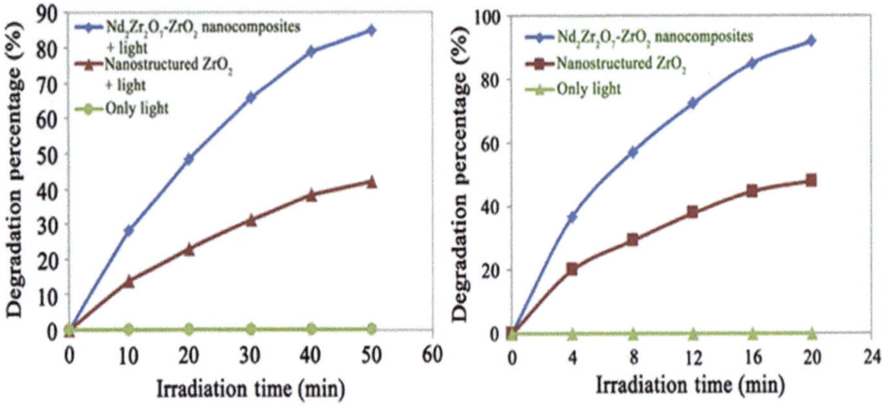

Fig. 11 Photocatalytic (left) methylene blue and (right) 2-naphthol decomposition of the formed $Nd_2Zr_2O_7$–ZrO_2 nanocomposites and nanostructured zirconium dioxide [76]

Table 1 Rare-earth-based mixed-metal oxide nanostructures for photodegradation of different pollutants

Nanostructure	Fabricating method	Light source (W)	Pollutant (Degradation (%), time duration (min))	Reference
$Nd_2Zr_2O_7$–ZrO_2	Modified Pechini	Mercury lamps (400)	Methylene blue (85, 50)	[76]
$Dy_2Ce_2O_7$	Green method	Osram lamps (125)	Methylene blue (92.8, 70)	[77]
$Pr_2Ce_2O_7$	Pechini	Mercury lamps (400)	Methyl orange (92.1, 50)	[78]
$Nd_2Sn_2O_7$–SnO_2	Combustion	Osram lamps (125)	RhB (88.7, 70)	[79]
$Nd_2Sn_2O_7$	Green method	Osram lamp (125)	Methyl violet (96.7, 70)	[56]
$Dy_2Ce_2O_7$	Green method	Osram lamps (125)	Eriochrome black T. (91.7, 60)	[41]

4 Conclusion

This chapter includes the last advances relevant to fabrication methods and solar photocatalytic applications of rare-earth-based mixed-metal oxides. Many techniques have been presented for the production of $Ln_2B_2O_7$ and $A_2B_2O_7$ nanostructures owing to the technical importance of this category of nanomaterials. Various morphologies of rare-earth-based nanostructures have been obtained by different methods. These nanostructures have been reported as a considerable, useful, and functional type of heterogeneous photocatalyst in recent years. According to literature, the photocatalytic performance of $Ln_2B_2O_7$ nanostructures strongly relies on their particle morphology, crystallinity, and surface area. Extremely limited numbers of routes have been presented for the creation of rare-earth zirconate structures with more favorable properties for photocatalytic applications. Furthermore, $Ln_2B_2O_7$ nanostructures have been used for the elimination and decomposition of limited number of contaminants. Thus, easy and economic methods should be further evaluated to produce rare-earth-based structures with more favorable properties as photocatalysts for the degradation and removal of different pollutants.

References

1. Y. Zhou, Phys. Sci. Rev. **5**, 35 (2019)
2. J. Sankar, S. Kumar, Curr. J Appl. Sci. Technol. **21**, 601 (2021)
3. D.R. Rittman, K.M. Turner, S. Park, A.F. Fuentes, J. Yan, R.C. Ewing, W.L. Mao, J. Appl. Phys. **121**, 045902 (2017)
4. J. Xu, R. Xi, X. Xu, Y. Zhang, X. Feng, X. Fang, X. Wang, J. Rare Earths **38**, 840 (2020)
5. P.M. Kowalski, Scr. Mater. **189**, 7 (2020)
6. Z. Li, J. Yang, Y. Xing, C. Wan, S. Watanabe, W. Pan, J. Am. Ceram. Soc. **104**, 27 (2021)
7. Y.-F. Huo, N. Qin, C.-Z. Liao, H.-F. Feng, Y.-Y. Gu, H. Cheng, J. Cent, South Univ. **26**, 1416 (2019)
8. S. Zinatloo-Ajabshir, M. Salavati-Niasari, A. Sobhani, Z. Zinatloo-Ajabshir, J. Alloys Compd. **767**, 1164 (2018)
9. S. Sardar, Manufacturing of pyrochlore ($A_2B_2O_7$) nanoparticles using sol gel method, in (University of Leeds, 2020)
10. A. Sánchez-Coronilla, J. Navas, E.I. Martín, T. Aguilar, J.J. Gallardo, D. de los Santos, R. Alcántara, C. Fernández-Lorenzo, Insights into the photovoltaic and photocatalytic activity of Cu-, Al-, and Tm-doped TiO_2, in *Emerging Photovoltaic Materials: Silicon and Beyond* (2018). https://doi.org/10.1002/9781119407690.ch6
11. M.-S. Balogun, Y. Huang, W. Qiu, H. Yang, H. Ji, Y. Tong, Mater. Today **20**, 425 (2017)
12. H. Yun, T. Paik, Nanomaterials **9**, 1243 (2019)
13. S. Remiro-Buenamañana, L. Navarrete, J. García-Fayos, S. Escorihuela, S. Escolástico, J.M. Serra, in *Engineering Solutions for CO_2 Conversion* (Wiley-VCH Verlag GmbH & Co. KGaA, 2021). https://doi.org/10.1002/9783527346523.ch3
14. A. Abuchenari, H. Ghazanfari, M. Siavashi, M. Sabetzadeh, S. Talebi, Z.K. Chemeh, A. Jamavari, J. Compos. Compd. **2**, 147 (2020)
15. S. Zinatloo-Ajabshir, M. Salavati-Niasari, M. Hamadanian, RSC Adv. **5**, 33792 (2015)
16. S. Zinatloo-Ajabshir, M. Baladi, O. Amiri, M. Salavati-Niasari, Sep. Purif. Technol. **248**, 117062 (2020)
17. S. Zinatloo-Ajabshir, M. Salavati-Niasari, Z. Zinatloo-Ajabshir, Mater. Lett. **180**, 27 (2016)
18. W. Duarte, A. Meguekam, M. Colas, M. Vardelle, S. Rossignol, J. Mater. Sci. **50**, 463 (2015)
19. X. Qiang, P. Wei, W. Jingdong, Q. Longhao, M. Hezhuo, K. Mori, T. Torigoe, Mater. Lett. **59**, 2804 (2005)
20. Y. Tong, J. Zhu, L. Lu, X. Wang, X. Yang, J. Alloys Compd. **465**, 280 (2008)
21. T. Hagiwara, K. Nomura, H. Kageyama, J. Ceram. Soc. Jpn. **125**, 65 (2017)
22. Z. Hongsong, L. Xiaochun, L. Gang, L. Zhenjun, Ceram. Int. **40**, 4567 (2014)
23. S. Zinatloo-Ajabshir, S. Mortazavi-Derazkola, M. Salavati-Niasari, J. Mater. Sci. Mater. Electron. **28**, 17849 (2017)
24. F. Beshkar, S. Zinatloo-Ajabshir, S. Bagheri, M. Salavati-Niasari, PLoS ONE **12**, e0158549 (2017)
25. S. Mortazavi-Derazkola, S. Zinatloo-Ajabshir, M. Salavati-Niasari, Adv. Powder Technol. **28**, 747 (2017)
26. F. Razi, S. Zinatloo-Ajabshir, M. Salavati-Niasari, J. Mol. Liq. **225**, 645 (2017)
27. A. Sobhani, M. Salavati-Niasari, J. Nanostruct. **7**, 141 (2017)
28. H. Emadi, M. Salavati-Niasari, A. Sobhani, Adv. Colloid Interface Sci. **246**, 52 (2017)
29. L. Gao, H. Zhu, L. Wang, G. Ou, Mater. Lett. **65**, 1360 (2011)
30. Q. Wang, Q. Li, R. Liu, M. Zheng, J. Wen, G. Zhao, J. Proteom. **136**, 48 (2016)
31. Z. Salehi, S. Zinatloo-Ajabshir, M. Salavati-Niasari, J. Rare Earths **35**, 805 (2017)

32. S.T. Aruna, C. Sanjeeviraja, N. Balaji, N.T. Manikandanath, Surf. Coat. Technol. **219**, 131 (2013)
33. Q. Du, G. Zhou, J. Zhou, X. Jia, H. Zhou, J. Alloys Compd. **552**, 152 (2013)
34. Y. Tong, Y. Wang, Mater Charact **60**, 1382 (2009)
35. Y. Tong, S. Zhao, W. Feng, L. Ma, J. Alloys Compd. **550**, 268 (2013)
36. H. Yi, Z. Wang, G. Zhou, J. Zhang, S. Wang, Ceram. Int. **42**, 2070 (2016)
37. L. Ma, W. Ma, X. Sun, J. Liu, L. Ji, H. Song, J. Rare Earths **33**, 195 (2015)
38. Z. Zhong, Y. Jiang, Z. Lian, X. Song, K. Peng, Ceram. Int. **46**, 12675 (2020)
39. W. Sun, S. Fang, L. Yan, W. Liu, Fuel Cells **12**, 457 (2012)
40. Z. Salehi, S. Zinatloo-Ajabshir, M. Salavati-Niasari, J. Mol. Liq. **222**, 218 (2016)
41. S. Zinatloo-Ajabshir, Z. Salehi, M. Salavati-Niasari, J. Alloys Compd. **763**, 314 (2018)
42. Z.-G. Liu, J.-H. Ouyang, B.-H. Wang, Y. Zhou, J. Li, Ceram. Int. **35**, 791 (2009)
43. Z.-G. Liu, J.-H. Ouyang, Y. Zhou, J. Alloys Compd. **472**, 319 (2009)
44. K.-J. Hu, Z.-G. Liu, J.-Y. Wang, T. Wang, J.-H. Ouyang, Mater. Lett. **89**, 276 (2012)
45. L. Kong, I. Karatchevtseva, R.D. Aughterson, J. Davis, Y. Zhang, G.R. Lumpkin, G. Triani, Ceram. Int. **41**, 7618 (2015)
46. J.Y. Li, H. Dai, Q. Li, X.H. Zhong, X.F. Ma, J. Meng, X.Q. Cao, Mater. Sci. Eng., B **133**, 209 (2006)
47. A. Joulia, M. Vardelle, S. Rossignol, J. Eur. Ceram. Soc. **33**, 2633 (2013)
48. F. Ansari, A. Sobhani, M. Salavati-Niasari, Compos. B Eng. **85**, 170 (2016)
49. S. Zinatloo-Ajabshir, Z. Salehi, O. Amiri, M. Salavati-Niasari, J. Alloys Compd. **791**, 792 (2019)
50. S. Saitzek, Z. Shao, A. Bayart, A. Ferri, M. Huvé, P. Roussel, R. Desfeux, J. Mater. Chem. C **2**, 4037 (2014)
51. M. Jovaní, A. Sanz, H. Beltrán-Mir, E. Cordoncillo, Dyes Pigm. **133**, 33 (2016)
52. Y. Tong, Y. Wang, Z. Yu, X. Wang, X. Yang, L. Lu, Mater. Lett. **62**, 889 (2008)
53. Y. Tong, P. Xue, F. Jian, L. Lu, X. Wang, X. Yang, Mater. Sci. Eng., B **150**, 194 (2008)
54. Y. Tong, L. Lu, X. Yang, X. Wang, Solid State Sci. **10**, 1379 (2008)
55. S. Zinatloo-Ajabshir, M.S. Morassaei, M. Salavati-Niasari, J. Colloid Interface Sci. **497**, 298 (2017)
56. S. Zinatloo-Ajabshir, M.S. Morassaei, M. Salavati-Niasari, J. Clean. Prod. **198**, 11 (2018)
57. S. Zinatloo-Ajabshir, M. Salavati-Niasari, Sep. Purif. Technol. **179**, 77 (2017)
58. S. Zinatloo-Ajabshir, M. Salavati-Niasari, Z. Zinatloo-Ajabshir, Sep. Purif. Technol. **177**, 110 (2017)
59. X. Wang, Y. Zhu, W. Zhang, J. Non-Cryst, Solids **356**, 1049 (2010)
60. M. Pokhrel, M.G. Brik, Y. Mao, J. Am. Ceram. Soc. **98**, 3192 (2015)
61. Z. Huang, F. Li, C. Jiao, J. Liu, J. Huang, L. Lu, H. Zhang, S. Zhang, Ceram. Int. **42**, 6221 (2016)
62. C. Xu, H. Jin, Q. Zhang, C. Huang, D. Zou, F. He, S. Hou, J. Eur. Ceram. Soc. **37**, 2871 (2017)
63. C. Liu, J. Zhang, S. Deng, P. Wang, Y. He, J. Colloid Interface Sci. **474**, 146 (2016)
64. J.L. Payne, M.G. Tucker, I.R. Evans, J. Solid State Chem. **205**, 29 (2013)
65. V. Narayanan, P. Lommens, K. De Buysser, D.E. Vanpoucke, R. Huehne, L. Molina, G. Van Tendeloo, P. Van Der Voort, I. Van Driessche, J. Mater. Chem. **22**, 8476 (2012)
66. W. Shi, S. Song, H. Zhang, Chem. Soc. Rev. **42**, 5714 (2013)
67. L. Kong, I. Karatchevtseva, D.J. Gregg, M.G. Blackford, R. Holmes, G. Triani, J. Am. Ceram. Soc. **96**, 935 (2013)
68. M. Mahdiani, A. Sobhani, M. Salavati-Niasari, Sep. Purif. Technol. **185**, 140 (2017)
69. W. Du, G. Zhao, H. Chang, F. Shi, Z. Zhu, X. Qian, Mater Charact **83**, 178 (2013)
70. W. Du, H. Zhu, X. Zhang, H. Lu, J. Wei, J. Du, Kuei Suan Jen Hsueh Pao/. J. Chin. Ceram. **41**, 1134 (2013)
71. S. Zinatloo-Ajabshir, M. Salavati-Niasari, J. Mol. Liq. **243**, 219 (2017)
72. Y. Tong, S. Zhao, F. Li, C. Li, Characterization and photocatalytic activity of $La_{1.6}Ln_{0.4}Zr_2O_7$ (Ln= La, Nd, Dy, Er) nanocrystals by stearic acid method. Adv. Mater. Res. 123–125, 631–634 (2010). https://doi.org/10.4028/www.scientific.net/amr.123-125.631

73. F.N. Sayed, V. Grover, K. Bhattacharyya, D. Jain, A. Arya, C.G.S. Pillai, A.K. Tyagi, Inorg. Chem. **50**, 2354 (2011)
74. Y.P. Tong, Y.Q. Zhao, H.Y. Huo, H. Yang, Synthesis, structure and catalytic activity of TiO_2/$La_2Zr_2O_7$ nanocomposites. Adv. Mater. Res. 239–242, 2878–2881 (2011). https://doi.org/10.4028/www.scientific.net/amr.239-242.2878
75. Y. Bai, L. Lu, J. Bao, J. Inorg. Organomet. Polym. Mater. **21**, 590 (2011)
76. S. Zinatloo-Ajabshir, Z. Zinatloo-Ajabshir, M. Salavati-Niasari, S. Bagheri, S.B.A. Hamid, J. Energy Chem. **26**, 315 (2017)
77. S. Zinatloo-Ajabshir, Z. Salehi, M. Salavati-Niasari, Ceram. Int. **44**, 3873 (2018)
78. S. Zinatloo-Ajabshir, Z. Salehi, M. Salavati-Niasari, RSC Adv. **6**, 107785 (2016)
79. S. Zinatloo-Ajabshir, M.S. Morassaei, M. Salavati-Niasari, J. Environ. Manag. **233**, 107 (2019)

Electrophoretic Deposition of Hydroxyapatite Incorporated Composite Coatings on Metallic Substrates: A Review of the Fundamentals

Sandeep Singh, Gurpreet Singh, and Niraj Bala

Abstract The electrophoretic deposition (EPD) of hydroxyapatite (HA) ceramic material is an effective strategy for a wide range of biomedical applications. This review chapter encompasses an overview of the fundamental and technical aspects of EPD approach, factors influencing the deposition process, suspension preparation, control of suspension mechanism, and deposition of composite coatings obtained by EPD. This review comprehensively analyzes the kinetics involved in EPD of HA-reinforced coatings, and the different factors such as applied voltage and deposition time, which can influence the surface morphology, corrosion behavior and in vitro bioactivity assessment. The parameters are described based on the up-to-date detailed overview of the recent research development in the area of EPD coated HA composite coatings on various metallic substrates.

Keywords Electrophoretic deposition · Hydroxyapatite coating · Metallic substrates

1 Introduction

EPD is a promising technique for colloidal coating processes. Figure 1 represents a schematic mechanism of an EPD unit. Under the effect of electric field, the charged particles dispersed in a fluid travel toward the electrode and deposit [1, 2]. Based on surface charge, two forms of EPD can be described. If the particles are positively charged, they will be attracted to the cathode, and the mechanism is known as cathodic EPD. On the other hand, if the particles are negatively charged, they will be attracted

S. Singh (✉) · G. Singh
Department of Mechanical Engineering, Punjabi University, Patiala, Punjab, India
e-mail: sandeep_me@pbi.ac.in

N. Bala
Department of Entrepreneurship Development and Industrial Coordination, National Institute of Technical Teachers Training and Research, Chandigarh 160019, India

S. J. Ikhmayies (ed.), *Advanced Ceramics*, Advances in Material Research and Technology, https://doi.org/10.1007/978-3-031-43918-6_7

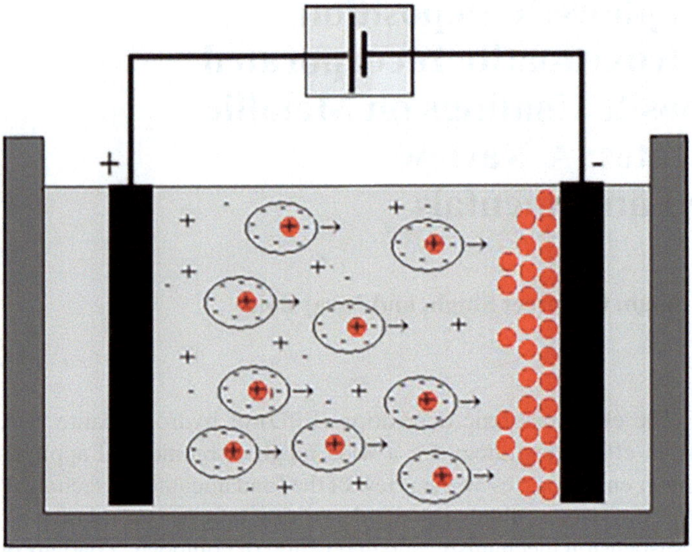

Fig. 1 Schematic diagram of EPD of charged particles with electrodes. Reprinted from Ma et al. [4], with permission from Elsevier. Copyright (2018)

to the anode and this mechanism is called anodic EPD. A current will flow between the electrodes under the application of a potential (Fig. 1). The current magnitude that flows is affected by the number of parameters, such as the solution conductivity, electrochemical kinetics and the cell constant. The suspension electric field strength (E) is analogous to the current density (J) and the suspension conductivity (k), and these three parameters are represented by Ohm's law ($J = kE$) [3].

Basic EPD is developed from organic suspensions which have many advantages like suspension chemical stability, low conductivity, and avoidance of the electrochemical reactions. This contributes to the formation of better quality coatings. However, the use of organic solvents is combined with several problems such as volatility, cost, flammability, and toxicity. Additionally, an organic solvent has less dissociation energy; this produces a limited particle charge density. Therefore, high electric field strengths are needed to transfer the particles toward the electrode [5].

The main application of EPD in the field of biomaterials has been the production of HA coatings on metallic substrates, where scientists intend to improve the bioactivity of surfaces and thereby facilitate the implants integration through surrounding tissues. As the main component of bone, HA demonstrates prominent osteoconductivity and biocompatibility [6], and has been extensively studied as effective a coating for bio-implants. It is easy to regulate the deposited HA thickness and stoichiometry over a broad range through EPD by adjusting processing parameters such as voltage, time and concentration. However, HA coatings are typically subjected to cracking during sintering process [7]. It was documented that implants failure occurs due to the increased degradation rate of HA coating, which decreases the implant's long-term stability in the body environment [8]. Palanivelu et al. [9] reported that the HA-reinforced coatings have a better ability to increase the implant's life relative to the pure HA coating. In order to refine the HA coating properties, various incorporations had been analyzed in combination with polymer and ceramic phases like zinc oxide [10], silica [11], iron oxide [12], titanium oxide [13], polylactide [14], chitosan [15], and polycaprolactide [16]. In this chapter, we have considered the benefits of EPD processing, the specific and adaptive biological functions of HA, and the HA composite coatings incorporated on different substrates. Specific attention is paid to discuss the effects of particular EPD parameters such as voltage, and deposition time on surface morphology, corrosion resistance, and in vitro bioactivity of HA composite coatings.

2 Effects of EPD Parameters on Coatings

The EPD process works on the principle of electrophoresis mechanism in which the movement of charged particles starts between the electrodes due to the effect of the applied electric field. The EPD method involves two types of group parameters: (a) suspension parameters (such as particle size, suspension conductivity, viscosity, and zeta potential), and (b) process parameters (such as voltage, time). The effects of these parameters are discussed in detail in the following sub-sections.

2.1 Suspension Parameters

Numerous parameters should be considered to understand the suspension properties such as powder surface properties, additive concentration (mainly dispersant), the influence of the type of additives, and physio-chemical nature of both liquid medium and suspended particles.

2.1.1 Particle Size

To mention the relevant particle sizes for EPD, there was no general thumb rule. The primary issue with greater sizes is that the particles contribute to sediment due to the gravitational forces resulting in thick deposits in vertical cells at the base, and thin films on top. Similarly, very fine particles tend to aggregate and thus produce non-uniform deposits [17].

2.1.2 Suspension Conductivity

In EPD experiments, the conductivity of suspension plays a vital role. It has been reported that the motion of particles was low if the conductivity is too high [18]. On the other hand, the stability of particles was lost if the suspension was too resistive, because of the electronically charged particles. The conductivity increases by enhancing the applied current in the suspension [18]. It is necessary that the conductivity lies within the suitable range for the successful assessment of the EPD process.

2.1.3 Suspension Viscosity

To examine the dispersion rate, the viscosity property cannot be considered due to the negligible use of solid particles for the EPD process [19]. However, low conductivity, high dielectric constant, and low viscosity are some of the desired properties for suspension stability.

2.1.4 Zeta Potential

In the EPD process, the zeta potential plays a pivotal role in acquiring a uniform charge of suspended particles. The different role of zeta potential in the fabrication of coating is as follow: (a) Determining the superficial behavior of powder when suspended in aqueous media. (b) Determining the movement and direction of particles in suspension. (c) Resolving the interaction between particles by stabilization of the suspension [20].

2.1.5 Suspension Stability

Electrophoresis is the occurrence of particles motion in a colloidal solution under the effect of an electric field. There are generally two types of colloidal particles in suspension. Firstly, the particles having sizes of 1 μm or less in diameter, where these particles sustain in suspension for a longer period due to Brownian effect only [21]. Secondly, particles having sizes more than 1 μm, where the hydrodynamic agitation is required. The two major factors which characterize the suspension stability are settling rate and aptness to undergo flocculation [22]. Mori et al. [23] reported that the suspension was not stable until it shows no tendency to; flocculate, make adhering deposit, and settle down slowly at the bottom of the container.

Fig. 2 Weight of deposited coating (hydroxyapatite on Ti-6Al-4V substrate) versus the applied voltage for different periods: **a** 30 s, and **b** 120 s. Reprinted from Besra and Liu [29], with permission from Elsevier. Copyright (2007)

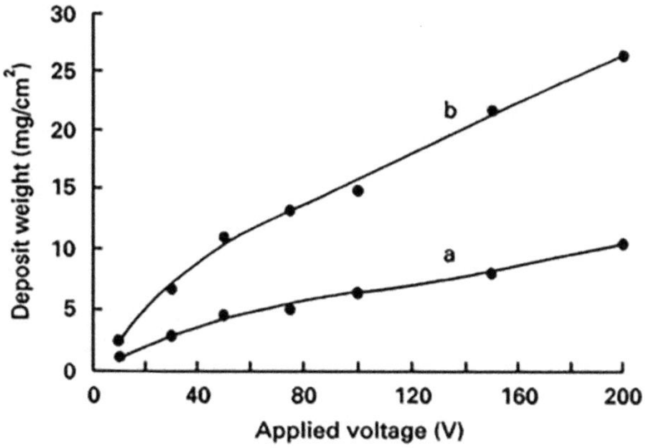

2.2 *Process Parameters*

2.2.1 Effects of Applied Voltage

In the EPD process, it is generally considered that the increase in applied voltage enhances the amount of deposition. However, higher voltage leads to film roughness [24], wrinkling, bubble formation, heterogeneity in the surface of the coating, and generates porosity [25]. Figure 2 depicts the deposition of HA coating on Ti-6Al-4V substrate. It was observed that the weight deposited increases by enhancing the applied voltage, but it can also affect the quality of the coating. Basu et al. [26] reported that the film deposited was more uniform, at a potential in the range (25–60) V/cm, whereas the crack occurs in the coating if the applied voltage was more than 60 V/cm. The non-uniformity of film originates at high voltage due to the anisotropy of the electric field on the substrate. Due to this reason, the deposition of particles takes place at the edges [27] of electrodes and generates aggregates in suspension [28].

It should also be noticed that too low voltage will not improve the film quality. On the other hand, more increase in applied voltage leads to the formation of the porous structure. Tabesh et al. [30] analyzed the porosity increase for HA-reinforced composite coatings by increasing the voltage from 5 to 40 V/cm, and concluded that the applied voltage plays an important role in the porosity of deposited composite coatings.

2.2.2 Effect of Deposition Time

The microstructure of electrophoretically deposited HA composite coatings is also affected by another important parameter, i.e., deposition time [31–33]. Karimi et al. [34] reported that the surface roughness of HA films enhanced with longer deposition time due to the agglomeration of particles in suspension. Similarly, in the earlier

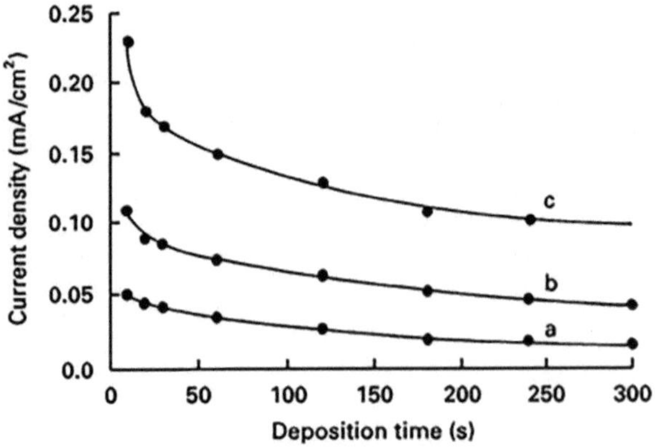

Fig. 3 Current density against deposition time of HA coating at different voltages **a** 10 V; **b** 20 V; **c** 30 V. Reprinted from Besra and Liu [29], with permission from Elsevier. Copyright (2007)

studies [35, 36] it was also observed that the non-uniform thick layer and porosity occurred on the surface of the substrate with longer deposition time.

From Fig. 3 it was revealed that at a constant voltage EPD, the effect of electric field on electrophoresis decreases with an increase in the deposition time due to the generation of particles insulating layer on the surface of electrodes. But during the starting phase of the process, the relationship between mass deposition and time was linear.

2.2.3 Effects of Gap Between Electrodes

The gap between electrodes plays an important role in the electrophoretic deposition process. If the gap is too large there were no depositions due to the resistance offered by an electrolyte. On the other hand, if the gap was too small, there were inconveniences to a current flow in the electrolyte. Based on Table 1, many of the researcher's maintain a gap within the range of 10–20 mm between the main and counter electrode [37]. The use of counter electrode in the EPD process provides potential to the working electrode within the suspension medium. To keep the counter electrode dissolving in suspension medium it is usually made of inert material such as graphite on noble metal [38].

2.2.4 Effects of Post-EPD Treatment

For post-EPD treatment of HA-reinforced coatings, different results have been reported by various researchers. The earlier studies on HA composite coatings have shown that the heat treatment leads to surface smoothing, compaction, and reduction of the interlayer spacing between the coating and substrate by removing the trapped water molecules. In addition, the formation of crevices [39], flaky surfaces, and protruding composite coating edges are also outcomes of post-EPD treatment.

Table 1 Chronological summary of recent studies (2007–2020) of HA-based composite coatings fabricated by EPD and brief description of their surface properties

Sr. no	Year	HA-reinforced matrix	Substrate	Voltage (V) and time (min)	SEM(1) morphology	Corrosion behavior	In vitro bioactivity assessment	References
1	2007	HA-BG(2)	Ti-6Al-4V	30–75, 2–10	The coating performed at 30 V and 5 min was found to be optimum and homogeneous	–	–	[44]
2	2008	HA-BG	Ti-6Al-4V	30–90, 2–5	Spherical particles of HA and BG consist of great number of agglomerated nanosized needle-like sub-particles	Current density of coated sample was reduced to 0.22 A/cm^2 to from 0.32 A/cm^2 for the uncoated sample	–	[45]
3	2009	HA-BG-CS-Alginate(3)	Pure Ti	10–30, 10	Uniform and crack free	–	–	[46]
4	2009	HA	Ti-6Al-4V	20, 30	Less uniform with micropores	Coated sample has 40% more corrosion rate than Ti-6Al-4V	Apatite formation after 4 weeks in Hank's solution	[47]
5	2010	HA-CNTs(4)	Ti-6Al-4V	2 mA, 8 s	No microcracks but rougher surface was observed at higher content of CNTs	Current densityof coated sample (HA-10% of CNTs) declined to 0.15 μA/cm^2 compared to 8 μA/cm^2 of uncoated sample	After 3 weeks of immersion in SBF(5) apatite was formed on the entire surface of coated samples	[48]
6	2010	HA-CNTs 2010	Ti	10–40, 0.5–8	Thick coating free of cracks obtained at 20 V	Coated samples at 20 V have lower current density	CNTs surface can act as nucleation sites resulting in the formation of HA layer	[49]
7	2011	HA-TiO_2 2011	Ti	165, 5	Increasing HA concentration leads to less micropores structure	Least current density (1.786×10^{-7} A/cm^2) at HA concentration (5 g/L)	Spheroid shape apatite crystallized form on HA surface after 6 days immersion in SBF	[50]

(continued)

Table 1 (continued)

Sr. no	Year	HA-reinforced matrix	Substrate	Voltage (V) and time (min)	SEM(1) morphology	Corrosion behavior	In vitro bioactivity assessment	References
8	2011	HA 2011	Ti	20–140, 1–7	At low voltage (20 and 30 V) no cracks and surface was found to be uniform	–	–	[51]
9	2012	HA-TiO_2 2012	Ti-13Nb-13Zr	30, 5	Densely packed with less porosity	Declined by one order of magnitude	–	[52]
10	2012	HA-MgO 2012	AZ31 Mg	140–550, 5	Composite coating has less porosity at low voltage than at higher voltage	Current density decreased one fold of order (2.25×10^{-3} to 7.76×10^{-4})	Bioactivity improved at pH value 4.5	[53]
11	2012	HA 2012	NiTi alloy	40–80, 2	60 V results in dense, uniform pores formation at 80 V	–	Surface covered with apatite crystals after 1 week of immersion in SBF	[54]
12	2013	n-HA	AZ91 Mg alloy	30, 5	Formation of dense and uniform HA layer	C.R. of coated sample was 48.2% more than the uncoated sample	Bone-like apatite crystals formed after immersion in SBF	[55]
13	2013	n-HA	AZ91 Mg alloy	50, 15	Almost crack free and homogeneous coatings	Current density reduced to 2.21 $\mu A/cm^2$ from 22.14 $\mu A/cm^2$	Apatite formation due to the creation of nucleation sites	[56]
14	2013	HA	AZ31 Mg alloy	10–30, 20	Fairly uniform coating at 10 V	corrosion of coated sample decreased to 9×10^{-6} A/cm^2 from 1.2×10^{-4} A/cm^2	–	[57]
15	2013	HA-MgO 2013	Pure Ti	0.85 mA, 20–30	Smooth and dense	C.R. of coated sample 14% higher than the Ti sample	Spherical apatite formation after 5 days immersion in SBF	[58]

(continued)

Table 1 (continued)

Sr. no	Year	HA-reinforced matrix	Substrate	Voltage (V) and time (min)	SEM[1] morphology	Corrosion behavior	In vitro bioactivity assessment	References
16	2013	HA coat on Ti-CaP composite coating 2013	Ti	10, 6	CaP particles are distributed uniformly in the Ti matrix	C.R. of coated sample was 56% higher than the Ti sample	–	[59]
17	2014	HA-GO [6] 2014	Pure Ti	30, 1–5	Minor cracks in pure HA but filled by addition of GO	Current density declined by two folds of magnitude than pure Ti	Presence of GO fillers promotes the bioactivity and cell proliferation	[60]
18	2014	HA 2014	Pure Ti	10–40, 2	Porous HA coating with no observable cracks	–	Bioactivity enhanced by the combined effect of crystallinity and Ca^{2+} content of the surface	[61]
19	2014	HA-PEG[7] 2014	316L SS	30, 10	Uniformly distributed and homogeneous	Effectively improve the passive ability of 316L SS	Potentially enhance bioactivity and induce implant osteointegration with living cells of human	[62]
20	2015	HA-Gr[8] 2015	Ti	60, 2	Fewer microcracks on composite coating compared to pure HA coating	HA-Gr coating has superior C.R. than its pure HA counterpart	enhanced precipitation of a newly formed apatite layer with more concentration of Gr in HA coating	[63]
21	2015	HA-BG-CS 2015	Pure Ti	20, 10	Smooth distribution of HA, BG particles along with CS molecules	Current density of coated sample reduced to 0.073 from 10.5 ($\mu m/cm^2$)	Higher bioactivity at pH value 3.3 as compared to pH value 4.5 and 5.0	[64]
22	2015	FHA[9]	AZ91 Mg alloy	100, 3	FHA coating is rough and porous with net-like surface structure	Current density of coated sample reduced to 12.5 from 63,1000 (nA/cm^2)	Spherical apatite formation after 2 weeks after immersion in SBF	[65]

(continued)

Table 1 (continued)

Sr. no	Year	HA-reinforced matrix	Substrate	Voltage (V) and time (min)	SEM[1] morphology	Corrosion behavior	In vitro bioactivity assessment	References
23	2015	HA-BG 2015	Pure Ti	30, 2	Rough microstructure contains micropores	Current density decreased by one fold of magnitude (from 2.76 to 0.27 μA/cm^2)	Required time for apatite formation of HA-BG composite coating less than the pure HA coating	[66]
24	2015	HA 2015	Ti-6Al-4V	5–20, 5–20	Homogeneous structure at 10 V compared to 15 and 20 V which contributes to heterogeneous and porous structure	–	–	[67]
25	2015	HA-Mg 2015	Ti-6Al-4V	10, 5	Micropores observed on needle and plate like morphology	C.R. of coated sample with least content of Mg was 70% more than Ti-6Al-4V sample	After 7 days immersion in SBF, samples were covered with newly grown apatite layer	[68]
26	2016	HA-BG-CS-HAL[10] 2016	Pure Ti	30, 10	Spherical morphology with the presence of cracks and porosity	Corrosion current density diminished from (91 to 0.56 μA/cm^2)	Bioactivity improves due to unique properties of HAl like augmenting drug delivery	[69]
27	2016	CS 2016	Ti-13Nb-13Zr	5–30, 4	Homogeneous coating at 10 V	Corrosion passivation of coated sample was more than uncoated sample	–	[70]
28	2016	HA 2016	Mg-3Zn alloy	20, 10	Fewer cracks observed for sample annealed at 300 °C than 400 °C	C.R. improved up to an impressive 25 times than Mg alloy	Fast apatite nucleation of HA coating due to similarity in crystal structure than Mg	[71]
29	2016	HA 2016	WE43Mg	30, 10	Formation of thick and porous coating	Current density declined to (0.090 mA/cm^2)	–	[72]

(continued)

Table 1 (continued)

Sr. no	Year	HA-reinforced matrix	Substrate	Voltage (V) and time (min)	SEM(1) morphology	Corrosion behavior	In vitro bioactivity assessment	References
30	2017	HA-MgP-ZnP 2017	AZ31 Mg	20, 10	Compact and dense uniform appearance	Current density decreased by 6 order of magnitude than AZ31 Mg sample	After 21 days of immersion least rate of corrosion due to formation of fully apatite layer	[73]
31	2017	HA-CS 2017	Ti-13Nb-13Zr	8–20, 1–6	Homogeneous and continuous coating at 10 V whereas pores were observed at 20 V	Current density reduced from 20 $\mu A/cm^2$ for the uncoated sample to 6 $\mu A/cm^2$ for the coated sample	Creation of cauliflower-like hydroxyapatite deposits on the surface of coated samples	[74]
32	2017	HA-Zn-Silk Fibron 2017	Pure Ti	30, 10	Uniform, homogeneous surface and free from microcracks	Better corrosion protection with coating than uncoated sample	After 3 days of immersion new apatite layer with rod shaped crystals covered all surface of samples	[75]
33		HA-Cu	Ti-13Nb-13Zr	30, 1–2	Inhomogeneous and several HA agglomerates at 1 min time whereas more homogenous at 2 min time	–	Increase of Cu nanoparticles decreased the bioactivity	[76]
34	2018	TiO_2-MAO-Zn-HA(11) 2018	Mg alloy	15, 60	Large number of pores on TiO_2 doped MAO sample and few agglomerates of fine particles on ZH(12) coating	Current density of ZH coating was declined to $1.2 \pm 0.3\ \mu A/cm^2$ compared to Mg alloy	Agglomeration of the apatite with spherical morphology has found	[77]
35	2018	HA-CS-COL(13)-h-BN(14) 2018	Ti-6Al-4V	15, 5	Creation of pores due to collagen orientation	Corrosion current density of coated samples reduced by two order of magnitude than uncoated sample	Synergetic effect of collagen leads to the formation of nucleation sites	[78]

(continued)

Table 1 (continued)

Sr. no	Year	HA-reinforced matrix	Substrate	Voltage (V) and time (min)	SEM[1] morphology	Corrosion behavior	In vitro bioactivity assessment	References
36	2019	HA-CS-GO 2019	Ti	30, 3	Spherical agglomerates of HA particles form clusters	current density of 21 nA/cm^2 significantly lower than that of the Ti about 200 nA/cm^2	Oxygenic functional group of GO nanosheets helps to improve the bioactivity	[34]
37	2020	HA	ZK60 Mg alloy	100, 10	Spherical closely linked agglomerates of HA particles	HA coating on ZK60 reduced the current density by one order of magnitude and resulted in more than five times lower uniform corrosion rate	–	[79]
38	2020	HA-BG-CS-HNT[15] 2020	Ti	10–50, 5–25	Formation of cracks and porosity at higher voltage and deposition time	Least current density of coated sample at 30 V which was reduced to 0.578 $\mu A/cm^2$ from 91.2 $\mu A/cm^2$ of Ti sample	Silanol functional group (Si–O-Si) of BG can induce the spherical apatite formation	[80]
39	2020	HA-Zein 2020	316L SS	6–15, 3–9	Homogeneous porous film was observed at 15 V and 3 min	Current density of 316L SS was 0.32 $\mu A/cm^2$ and declined to 0.046 $\mu A/cm^2$	Formation of HA crystals after 3 days of immersion in SBF	[81]
40	2020	HA	ZMX410 and ZM21 Mg alloys	37, 5	Observed homogeneity and uniformity of the HA coating on both alloys	Corrosive protection efficiency by HA coating to ZM21 and ZMX410 was 40.6% and 4.42% respectively	Coated samples were hydrophilic and bioactive in nature and generate apatite layer after immersion in SBF solution	[82]

[1] SEM: Scanning electron microscope; [2] BG: Bioglass; [3] CS; Chitosan; [4] CNT: Carbon nanotube; [5] SBF: Simulated body fluid; [6] GO: Graphene oxide; [7] PEG: poly(ethylene glycol); [8] Gr: Graphene; [9] FHA: Fluoridated hydroxyapatite; [10] HAL: Halloysite; [11] MAO: Micro-arc oxidation; [12] ZH: Zinc-doped hydroxyapatite; [13] COL: Collagen; [14] h-BN: Hexagonal boron nitride; HNT: [15] Halloysite nanotube

However, at high temperature (above 1200 °C), the dimensions of the substrate change, and the result is shrinkage of the coating [40]. Therefore cracks occur in the coating. During the cooling process, additional cracking occurs due to the difference between thermal expansion coefficients of the sintered coating and the substrates. Other problems that occur at high temperatures are oxidation and phase transformations between the coatings and substrate [41]. These problems can be eliminated by selecting a suitable range of temperatures (550–800) °C for post-EPD treatments [42]. Another alternative to avoid these troubles is by the development of inorganic-polymer composite coatings. The use of polymer involves processing at low temperature and avoiding the inconvenience of heat treatment process [43].

3 EPD on Metallic Substrates with Hydroxyapatite Reinforced Composite Coatings

Table 1 lists a chronological summary of recent studies (2007–2020) of HA-based composite coatings fabricated by EPD and brief description of their surface properties such as surface morphology, corrosion behavior [in terms of corrosion current density (I_{corr}) and corrosion potential (E_{corr})] and bioactivity assessment. From the analysis of the literature review it was observed that different parameters such as voltage, deposition time can influence the surface properties of HA-reinforced composite coatings which are discussed below:

The discussion of Table 1 on coating structure, corrosion behavior and in vitro bioactivity results obtained for HA-based coatings developed by EPD.

Stojanovic et al. [44] developed the HA coating on Ti-6Al-4V alloy using EPD. HA particles suspension was formulated by the stirring of 0.5 g of HA in 100 ml ethanol. The effect of EPD parameters, i.e., voltage (30, 50 and 75 V) and time (2, 5 and 10 min) on coating morphology and deposited weight were determined. It was observed that raising the applied voltage leads to more deposition yields. Figure 4 shows the relationship of the deposited weight against time and voltage.

The coating was performed at 30 V and 5 min and was found to be optimal compared to 50 and 75 V. It was found that although the coating thickness was enhanced at higher voltage, cracks formed in the coating. Figure 5 depicts the surface morphology of coatings developed by EPD at 30 and 50 V. Therefore, it was concluded that by regulating the voltage and time, the weight and thickness of coating can be controlled.

Balamurugan et al. [45] formulated a BG-HA composite coating on Ti-6Al-4V by EPD. The prepared coatings have been characterized for their structural and electrochemical properties. The coatings produced at 20 V contributed to irregular coatings, while the coatings processed for 70 V and 5 min were thicker and extremely porous. The possible mechanism is a spontaneous deposition of particles that are loosely packed and isolated by a more volume fraction of porosity, resulting in

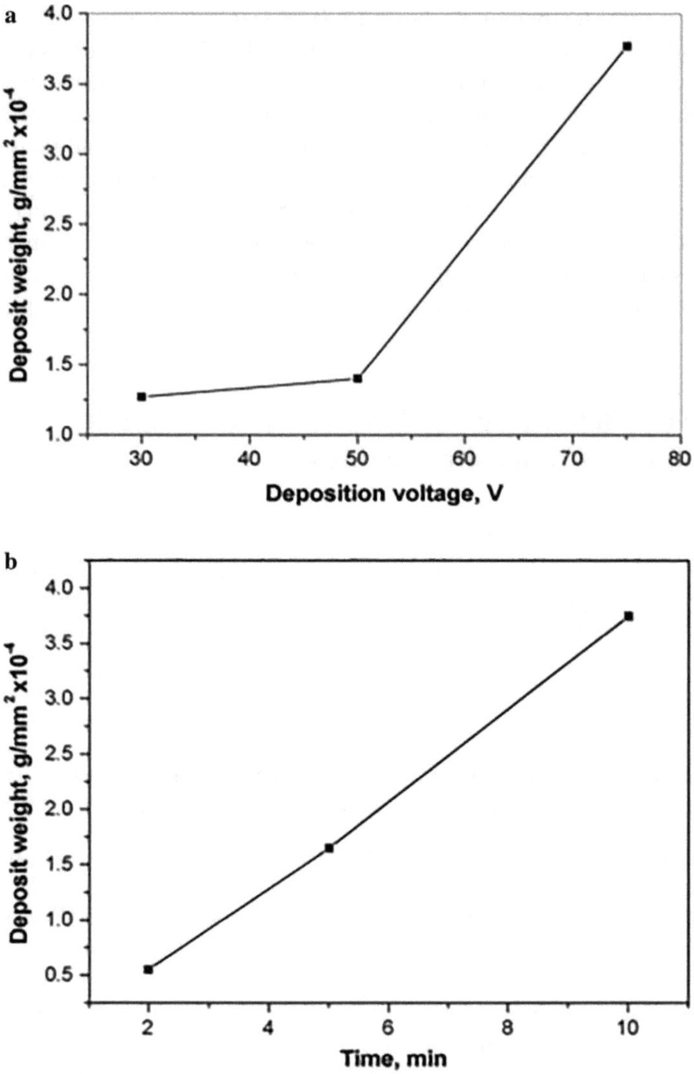

Fig. 4 **a** Dependence of deposited coating weight on deposition voltage at constant time of 10 min. **b** Dependence of deposited coating weight on time at a voltage of 75 V and **c** Dependence of deposited coating weight on time at constant voltage of 30 V. Reprinted after Stojanovic et al. [44], with permission from Elsevier. Copyright (2007)

the formation of inter-particle cracks and eventual detachment of particles from the substrate. The E_{corr} and I_{corr} for Ti-6Al-4V were −0.97 V and + 0.32 A/cm^2, respectively. For HA, bioglass (BG), and BG-HA-coated samples under 70 V for 5 min, the E_{corr} values were observed to be (0.325 ± 0.5) V, (0.380 ± 0.5) V, and (0.475 ± 0.5) V, while the values of I_{corr} were 0.60, 0.33, and 0.22 A/cm^2,

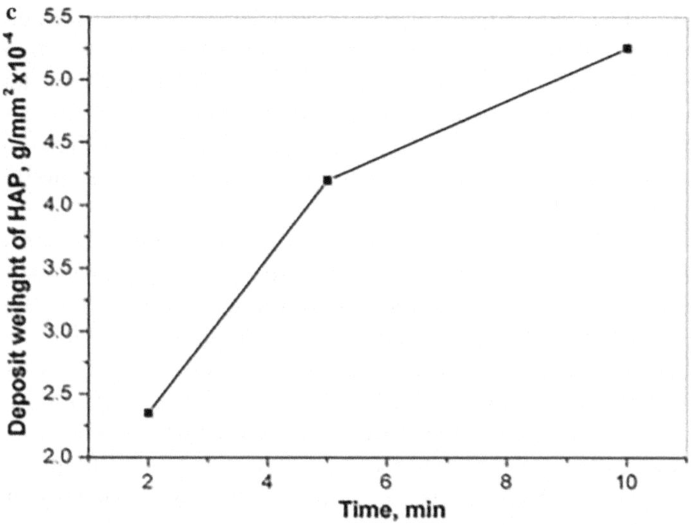

Fig. 4 (continued)

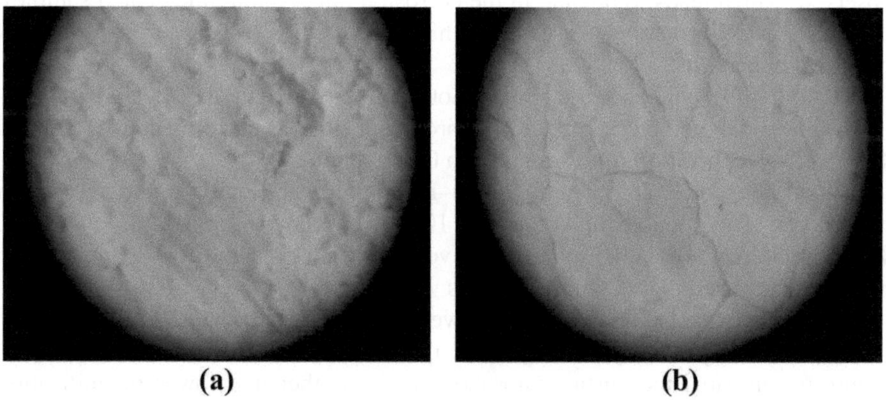

Fig. 5 **a** Coating obtained by EPD at 75 V and 10 min, and **b** coating obtained by EPD at 30 V and 5 min. Reprinted after Stojanovic et al. [44], with permission from Elsevier. Copyright (2007)

respectively. However, the E_{corr} of BG-HA-coated samples at 30 V, 50 V, 60 V, and 90 V persisted at (0.171 $\pm$ 0.5), (0.264 $\pm$ 0.5) V, (0.410 $\pm$ 0.5) V, and (0.388 $\pm$ 0.5) V, implying that the optimum coating potential varies from 50 to 70 V. The E_{corr} for all the coated samples was observed to be more noble relative to the Ti-6Al-4V, thus demonstrating improved corrosion resistance.

Zhitomirsky et al. [46] investigated the fabrication of BG-HA-CS and BG-HA-alginate coatings by EPD. EPD was performed at (10–30 V) in an electrolyte containing 0–2 g/L BG and 0–1 g/L HA. Coatings performed at 10 V were uniform and crack free, however porous structure observed at 30 V. Due to the use of polymers like chitosan (CS) and alginate, problems related to the sintering of BG and HA coatings on the metallic substrate are avoided. In addition, ambient temperature processing allows the integration of other functional materials as reinforcements into the coatings. The examples are signaling of cell biomolecules or antibiotic agents may be applied to the polymer and thereby homogeneously integrated into the polymer matrix.

Kwok et al. [47] fabricated HA coatings with and without carbon nanotubes (CNTs) on Ti-6Al-4V. The corrosion and in vitro bioactivity properties of the HA-coated specimens were evaluated. The coating was crack free having 10 μm thickness, and deposited at 200 V for 30 min. The coated samples were more resistant to corrosion than the substrate. This enhancement was endorsed by a noble shift in the corrosion potential and a lower I_{corr}. Besides, after soaked in Hanks' solution for 4 weeks apatite was formed on the HA coatings, showing high bone bioactivity.

Zhang et al. [48] developed HA coating with different concentrations of CNTs (4–25%) by EPD on Ti-6Al-4V substrate to increase the bone bioactivity and osteointegration with metallic implants. Cathodic EPD was performed under a current of 2 mA and deposition time of 8 s. No cracks were observed on the surface of coated samples with 10% of CNTs. As the content of CNTs increased, cracks and pores were observed, which leads to decline in the corrosion rate. Among the coated samples, the HA-10% CNTs coating possessed the highest corrosion resistance and prominent bone bonding ability.

Bai et al. [49] synthesized carbon nanotube-hydroxyapatite (CNT-HA) composites by chemical method before an EPD process. EPD was performed at 10, 20, and 40 V, and the deposition time varied from 0.5 to 8 min.

The surface morphologies of the CNTs–HA coatings appeared similar and contain micropores. The coating thickness was 10, 18, and 34 μm at 10, 20, and 40 V. The morphology was alike at different voltages, however, the low magnification micrographs reveal that the morphologies were actually different. At 40 V, the coatings have more cracks, relative to the lower voltages (i.e., 10 and 20 V). The crack formation at a high voltage occurs due to the evolution of hydrogen at the cathode. Figure 6 compares the surface morphologies at higher and lower magnification obtained at different voltages. Therefore, it was concluded that crack free coatings were attained at a voltage of 20 V, which provide effective corrosion protection and good biocompatibility.

Bai et al. [50] fabricated HA-TiO_2 coating on Ti by micro-arc oxidation (MAO) combined with EPD with various amounts of HA particles. The EPD was conducted at 165 V for 5 min. Without the incorporation of HA particles to the electrolyte, the samples exhibited a rough and porous TiO_2 layer developed by MAO on the surface of Ti. Figure 7 shows the surface morphology at different concentrations of HA particles. It was found that the samples produced at lower HA concentration contain a number of pores.

The coating surface was less porous as the concentration of HA raised to 20 g/L. The sample developed at 5 g/L HA concentration revealed the highest corrosion resistance and better formation of apatite layer over the entire surface.

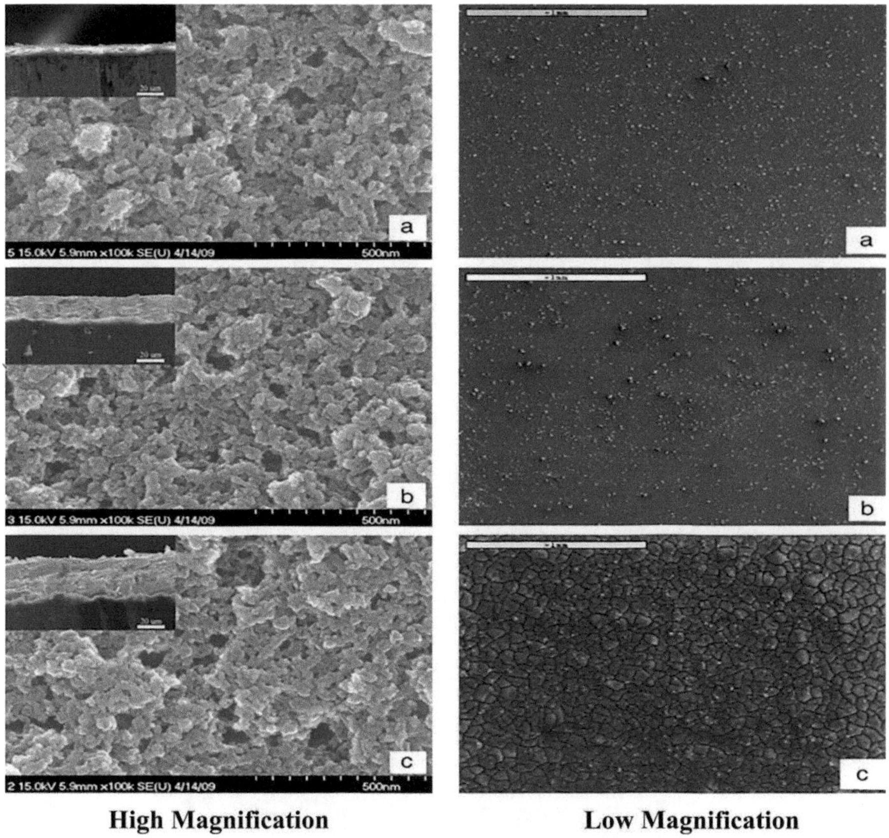

Fig. 6 Scanning electron microscope (SEM) images of the CNTs-HA coatings produced at 8 min duration with voltages: **a** 10 V; **b** 20 V, and **c** 40 V. Reprinted from Bai et al. [49], with permission from Elsevier. Copyright (2010)

Abdeltawab et al. [51] produced HA coating on Ti substrate by EPD. The effect of applied voltage (20–140 V) and time on deposition rate, thickness, and structure of coating were studied. From the analysis of results, authors observed that, at higher applied field, turbulence takes place in the suspension and the deposition process may be affected by the flow disturbance in the surrounding medium. The reason may be the fast movement of particles in the suspension as they cannot get adequate time to find the appropriate positions to develop a close-packed structure. Figure 8 depicts the weight of the deposit, and the thickness as a function of applied voltage.

Although powders may be deposited rapidly if larger applied fields are used, the performance of the deposit can be affected. The particle size increases with more applied voltage as observed from the SEM images of Fig. 9. No cracks were detected at 20 and 30 V voltage, and the coating was found to be smooth. Nonetheless, marginally larger particles have been found in coatings processed at 40 and 60 V

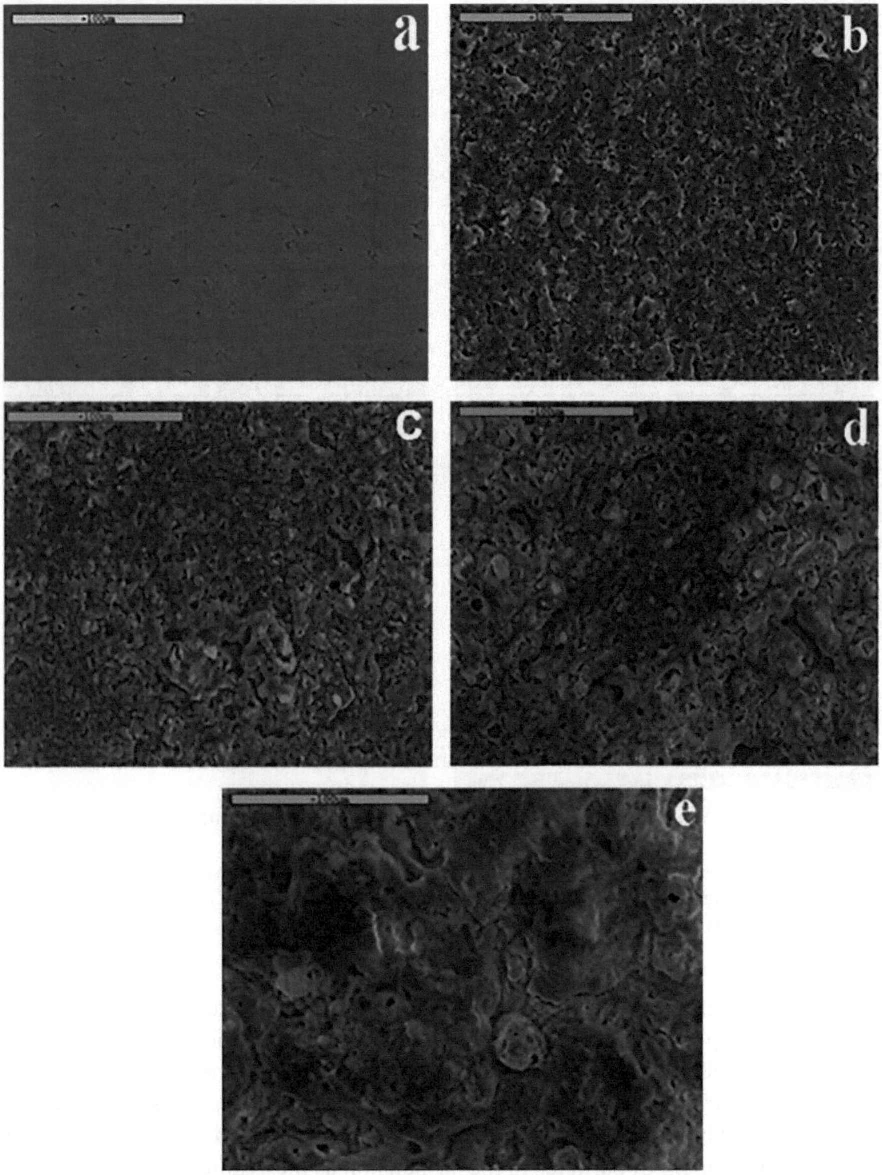

Fig. 7 Surface morphology of Ti (**a**), and coated samples with HA concentration of **b** 0 g/L, **c** 5 g/L, **d** 10 g/L, and **e** 20 g/L. Reprinted from Bai et al. [50], with permission from Elsevier. Copyright (2011)

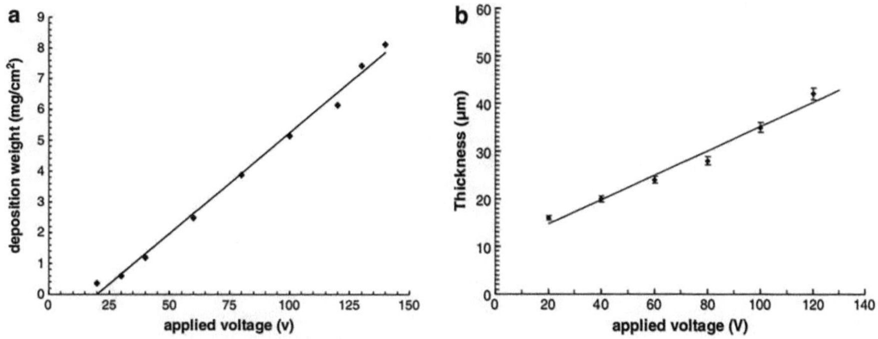

Fig. 8 **a** Weight of HA deposited, and **b** HA coating thickness versus voltage at constant time duration 5 min. Reprinted from Abdeltawab et al. [51], with permission from Elsevier. Copyright (2011)

voltages. Besides, small cracks (40 V) and some larger cracks (60 V) were detected on the surface of coatings.

Mohan et al. [52] investigated the corrosion behavior of TiO_2 + 50% HA-coated Ti-13Nb-13Zr alloy. A thin coating was attained by EPD at 30 V followed by sintering at 850 °C. The microstructures of the sintered and unsintered samples were evaluated by optical microscopy. The SEM-EDAX results exhibited that the sample sintered at 850 °C was densely packed. The unsintered coating exhibited a large number of partially bonded composite particles. The HA-TiO_2 sample depicted a lower corrosion rate and nobler passive potential relative to the Ti-13Nb-13Zr sample.

Sreekanth and Rameshbabu [53] developed magnesium oxide and hydroxyapatite (MgO/HA) coating on AZ31 Mg alloy to control the biodegradability rate. The coating was produced via hybrid method coupled with plasma electrolytic oxidation (PEO) and EPD. Degradation of HA coating was examined by performing electrochemical tests (pH 4.5), replicating the osteoblast resorption condition in vitro and also in the SBF environment (pH 7.4). The corrosion rate of composite coating has demonstrated a significant increase at pH 4.5. However, smaller improvement was observed at a pH of 7.4. This can be ascribed to the existence of Cl^- ions in SBF which may be accountable for corrosion at a pH of 7.4. There was also no major difference in the electrolyte pH during electrochemical testing in both 7.4 pH and 4.5 pH solutions. The analysis of results revealed that the MgO/HA coating substantially enhanced the corrosion protection of Mg alloy under both environments.

Maleki-Ghaleh et al. [54] developed HA coating on NiTi alloy via EPD. A stable suspension was made by the addition of 80 g/L HA powder to butanol. EPD was conducted at a different voltage (40, 60, and 80 V) in 120 s. It has been shown that the specimen coated at 60 V has a dense coating. Figure 10 displays the SEM morphology of HA coating at three different voltages.

As shown in Fig. 10, the coating at 60 V consists of dense and smooth coating whereas the coating at 80 V results in a coating with large pores. The findings of the SBF immersion test showed that the sample coated at 60 V performed most effectively

Fig. 9 SEM images of HA-coated samples at **a** 20 V, **b** 30 V, **c** 40 V, and **d** 60 V. Reprinted from Abdeltawab et al. [51], with permission from Elsevier. Copyright (2011)

and acted as a shield against the penetration of Ni ions. The uniform apatite formation on the coated surface at 60 V after one week of immersion exhibited the HA coating biocompatibility after EPD and sintering process.

Rojaee et al. [55] developed the MAO and MgF_2 as pre-surface treatments of AZ91 Mg alloy to produce a nanostructured n-HA coating. The EPD was done at 30 V for 5 min. The electrochemical tests were performed to evaluate the corrosion performance of the specimens, and in vitro bioactivity evaluation was carried out in SBF. Results have shown that the MAO/n-HA-coated samples with rough structure and reduced corrosion current density resulted in a declined degradation rate of Mg followed by high bioactivity. The presence of MgF_2 and MAO coatings could offer a strong barrier to rapid degradation of the substrate. In addition, this interlayer plays a significant role in improving the biocompatibility of Mg alloy.

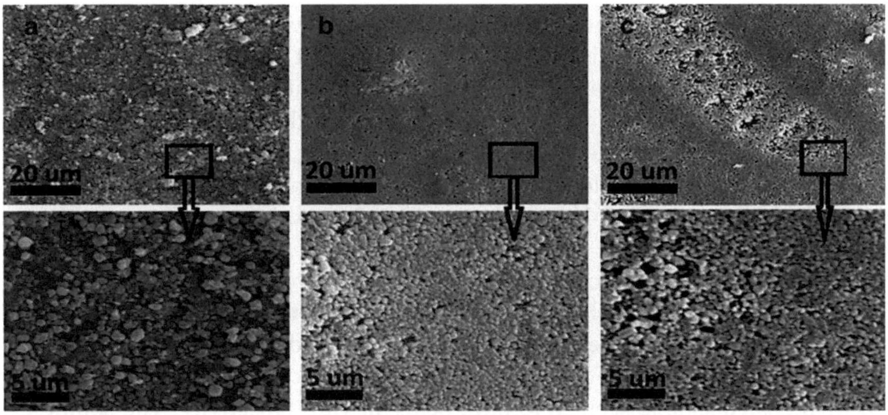

Fig. 10 SEM images at voltages of **a** 40 V, **b** 60 V, and **c** 80 V with a time duration of 120 s. Reprinted from Maleki-Ghaleh et al. [54], with permission from Elsevier. Copyright (2012)

Rojaee et al. [56] fabricated the nanostructured HA coating via sol–gel and EPD (50 V for 15 min) method on AZ91 Mg alloy. The E_{corr} of the uncoated sample was (-1448 ± 51) V. The E_{corr} values of the sol–gel and EPD produced samples were observed to be (-1343 ± 35) V and (-1406 ± 21) V, respectively. The I_{corr} of AZ91, sol–gel, and EPD coated specimens were (22.14 ± 3.21), (2.83 ± 0.04), and (2.21 ± 0.02) $\mu A/cm^2$, respectively. Generally, lower corrosion current density depicts a smaller corrosion rate. With respect to the E_{corr} and I_{corr} values, all coated specimens moved in noble directions in corrosive environments. The better corrosion protection not only monitors the Mg consumption for bone restoration, but also improves the structural integrity and cell adhesion, at the same time.

Kaabi et al. [57] synthesized hydroxyapatite (HA) coating on AZ31 substrate to alter the corrosion rate of magnesium and improve its potential applicability as implant. Coatings were deposited electrophoretically at different potentials from 10 to 30 V. The surface morphologies at different voltages are shown in Fig. 11.

(a) (b) (c)

Fig. 11 SEM micrographs of (HA) coatings on AZ31 substrates at **a** 10 V, **b** 20 V, and **c** 30 V deposition potential. In all cases, the deposition time was 20 min. Reprinted from Kaabi et al. [57], with permission from Elsevier. Copyright (2014)

When the applied voltage increases, the HA particles can travel faster and accumulate quickly, thus increasing the amount of the deposit with more applied potential. There were few cracks on the HA-coated coating produced at 20 and 30 V relative to the 10 V. It was found that by enhancing voltage, the corrosion resistance was improved due to the formation of thick layer of coating.

Huang et al. [58] illustrated the corrosion behavior and bioactivity of manganese-doped hydroxyapatite (MnHAp) film on Ti. The coating was deposited at the following conditions: pH = 4.3, $I_{corr} = 0.85$ mA cm^{-2}, and 20–30 min deposition time. The MnHAp crystals consist of needle-like shape with few Mn incorporations. The inclusion of Mn^{2+} into HA substantially decreased the porosity, and the coating surface became denser. The MnHAp coating displayed more corrosion protection in SBF as compared to the HA coating. SBF immersion experiments showed that MnHAp-coated sample seemed to have better in vitro bioactivity. The addition of Mn and deficiency of Ca in the coating were the primary reasons for the superior bioactivity. The MnHAp film is a potential material to produce advanced biomedical implants.

Farnoush et al. [59] fabricated nano-HA coatings on the Ti-CaP layer by the combination of friction stir processing (FSP) and EPD. The corrosion behavior was evaluated by potentiodynamic polarization tests. By implementation of FSP, the Ti-CaP inner layer serves as a barrier to the penetration of Cl^- ions, and inhibits the electrochemical reactions between the coating and substrate. The findings referring to the HA-Ti-CaP sample suggest a substantial decline in the I_{corr}, and an improvement in the E_{corr}, and corrosion rate.

Li et al. [60] prepared homogeneous graphene oxide/hydroxyapatite (GO/HA) coating on Ti by EPD. Ethanol was used as electrolyte medium to mitigate the gases evolution at the electrodes which may have detrimental effects on the performance of the coatings. Three kinds of suspension were formulated and represented as HA, 2GO/HA, and 5GO/HA, having 0, 2, and 5 (wt. %) GO contents, respectively. EPD was executed at 30 V for 1–5 min. The rate of deposition was calculated by weight test. The deposit mass indicates a liner dependency on time, and the potential on the deposit is constant. Figure 12(a) and (b) shows the weight versus deposit time relationship at different concentrations of GO content.

As examined in Fig. 12(a), the rate of deposition declines along with the extended time of deposition, due to the development of GO/HA composite layer, which would contribute to the decrement of potential drops in the suspension. Figure 12(b) shows the inclusion of GO will decrease the HA particles deposition rate. As contrast with pure HA coating, this new coating can inhibit the crack formation and enhance the corrosion protection of Ti. The in vitro bioactivity of the samples reveals that the presence of GO fillers promotes the bioactivity and cell proliferation.

Rad et al. [61] investigated the effects of the EPD at dynamic voltage on the biological and physical properties of HA coatings on pure Ti substrates. The voltages of 10 V, 20 V, 40 V, and 80 V were applied, and the corresponding damping currents were attained as 0.07 mA, 0.15 mA, 0.25 mA, and 0.35 mA. Figure 13 exhibits the surface morphology of coated samples obtained at different damping currents, and views of their cross sections.

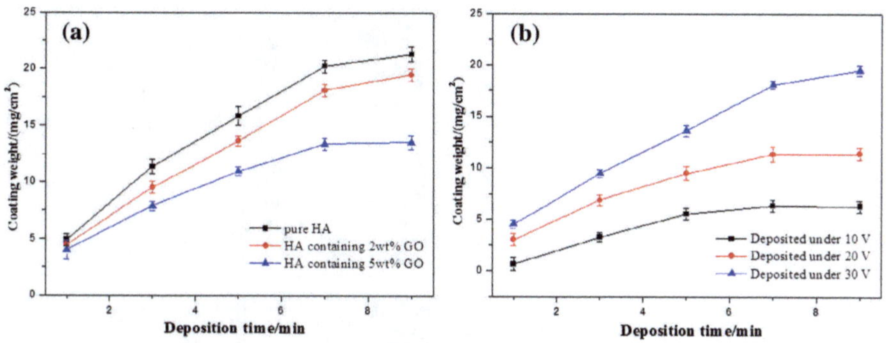

Fig. 12 The coating weight versus deposition time with various GO concentrations at voltage of 30 V (**a**), and with 2 wt.% GO at different voltage (**b**). Reprinted from Li et al. [60], with permission from Elsevier. Copyright (2014)

At higher voltages, the HA particles begin to accumulate in a more porous structure. The samples attained at 40 and 80 V showed highly porous, cracked, and non-stable coating. The optimum results were found at 20 V. The corrosion studies were done in Ringer's solution. The coated samples showed more resistance to corrosion relative to the uncoated sample. The biological experiments were performed by MG63 cells cultured on HA-coated samples. The number of cells attached to the sample increased with longer incubation time. The MG63 cells enlarge on the HA-coated and uncoated specimens after 72 h. The cells on the HA sample proliferate over the coating material and grow more efficiently on the substrate. The cells on the uncoated sample depict a flattened and osteoblast-like morphology.

Mehboob et al. [62] investigated polyethylene glycol (PEG) coated HA coatings on 316L SS via EPD. DC power supply was regulated at 30 V for 2 min. The amount of PEG in HA suspension was in different HA:PEG ratios; 1:0.1, 1:0.2, 1:0.5, and 1:1. Uniform coverage and increased adhesion to substrate was achieved in the case of 1:1 HA:PEG ratio. PEG was used to enhance the binding strength and packing density between coating and substrate. It was found that as the potential difference across the electrodes was constant, the electric field affecting electrophoresis reduced with time due to deposition of ceramic particles on the electrode, which acts as an insulating layer for the incoming particles. The incorporation of PEG improved the coating adhesion, and exhibited greater degree of packing density, which will potentially induce osteointegration of the implants with the living tissues of human.

Janković et al. [63] evaluated the HA and graphene (HA-Gr) composite coating done by EPD on Ti to attain a uniform coating with enhanced corrosion stability in SBF. The corrosion resistance of HA-Gr and HA coatings was determined by electrochemical impedance spectroscopy (EIS) measurements. On the basis of EIS data, it was apparent that the HA-Gr composite coating revealed biomimetic mineralization preeminent to its pure HA counterpart, and increased precipitation of a newly developed apatite layer.

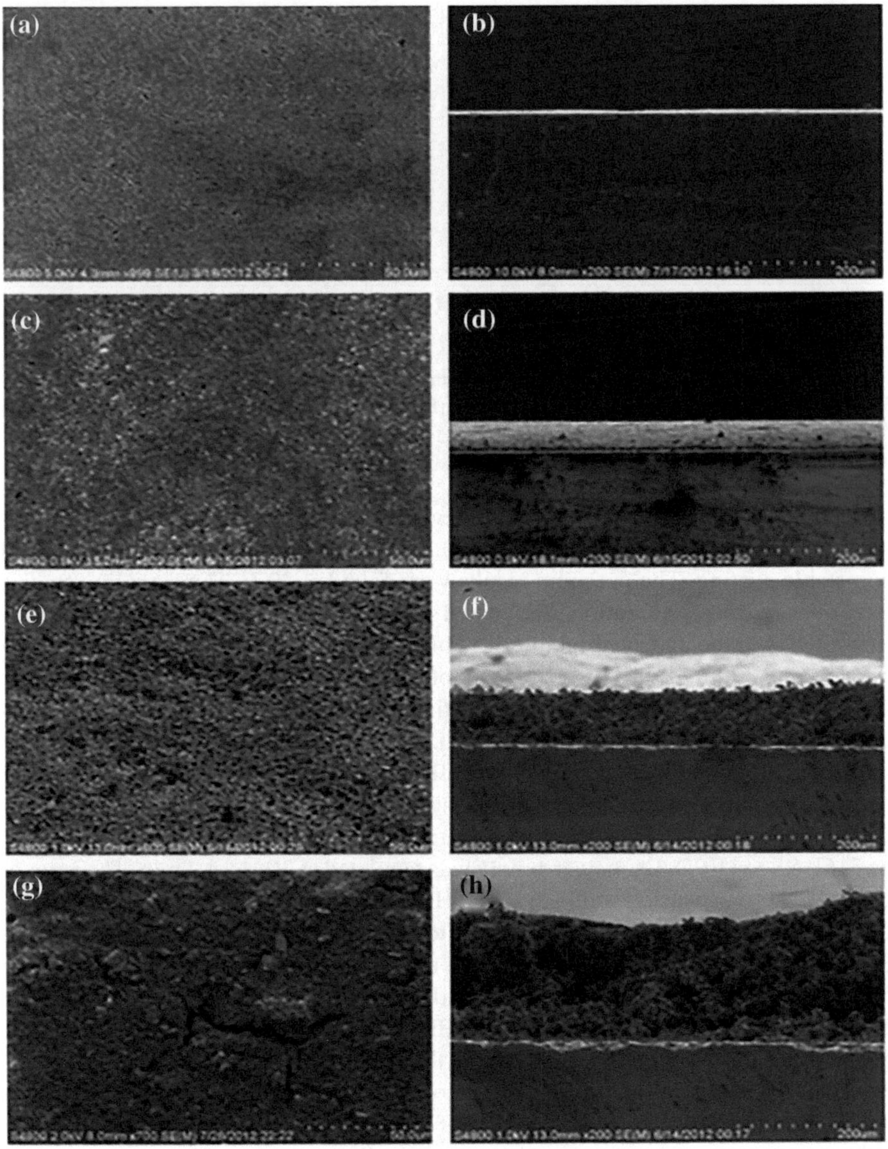

Fig. 13 SEM and the corresponding cross-sectional images of HA coatings on Ti substrates obtained using EPD process at dynamic voltage-constant current, **a**, **b** 0.07 mA, **c**, **d** 0.15 mA, **e**, **f** 0.25 mA, and **g**, **h** 0.35 mA. Reprinted from Rad et al. [61], with permission from Elsevier. Copyright (2014)

Molaei et al. [64] illustrated EPD of chitosan-bioglass-hydroxyapatite (CS-BG-HA) coatings on the Ti substrate by examining the parameters such as pH, voltage, and deposition time. Figure 14 shows the relationship between deposited weight and thickness of coating versus pH values. The deposition was performed at various pH values of 3.3, 4.5, and 5.

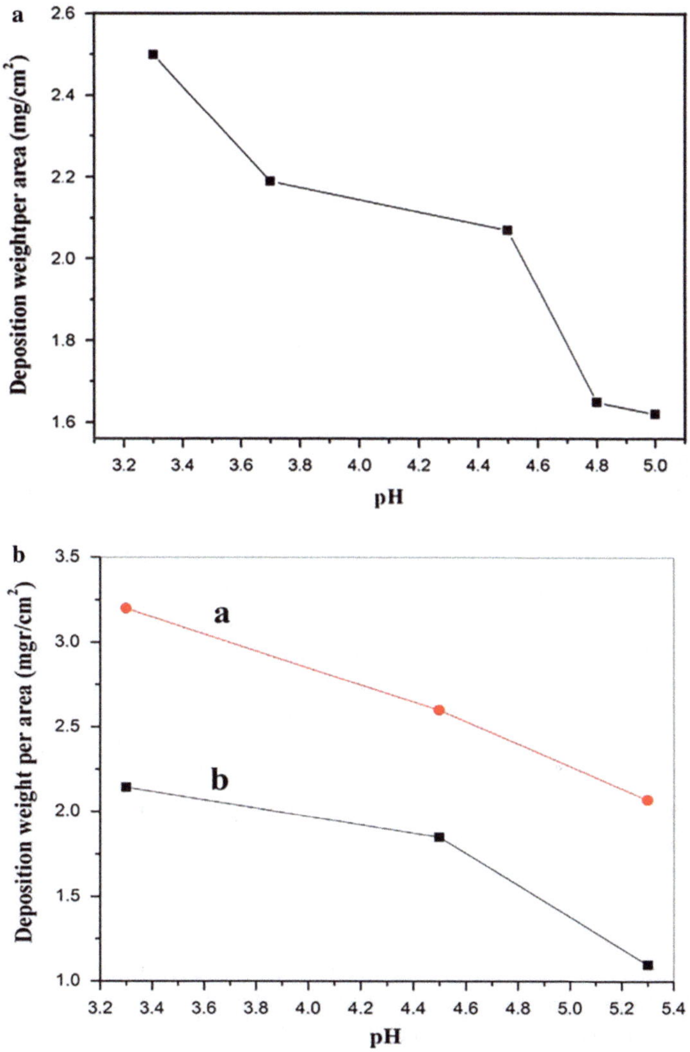

Fig. 14 **a** Deposition weight of CS-BG-HA coating at 30 V for 5 min, **b** variation of weight deposit of CS-BG-HA fabricated on (a) isolated and (b) non-isolated Ti at 20 V for 15 min. and **c** thickness of the CS-BG-HA fabricated on isolated Ti at 20 V for 15 min. All against pH of suspension used. Reprinted from Molaei et al. [64], with permission from Elsevier. Copyright (2015)

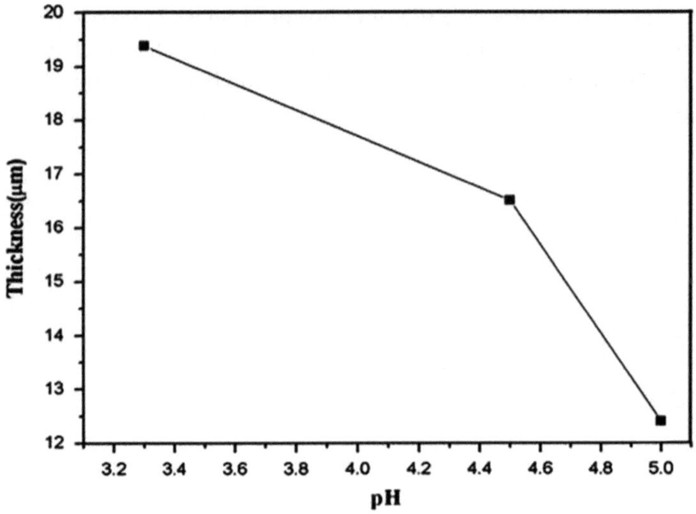

Fig. 14 (continued)

It was observed that deposited weight decrease at higher pH value due to its effect on zeta potential. In ethanol-based suspension, BG particles have higher zeta potential than particles of HA. Particles with larger zeta potential have higher mobility, so they travel quickly to the electrode of the opposite charge. Therefore, by decreasing pH, BG particles have higher rate of deposition, and corrosion resistance was improved. The sample formed at pH = 3.3 displayed the superior bioactivity and corrosion rate.

Razavi et al. [65] investigated the fluoridated hydroxyapatite (FHA) coating on AZ91 Mg by MAO and EPD to improve their corrosion through surface modifications. The electrochemical test was carried out with the help of the Potentiostat apparatus. The values of E_{corr} in samples were observed as follows: FHA coating (−1.39 V) > MAO (−1.56 V) > AZ91 (1.6 V) whereas the position of the I_{corr} value of FHA coating (12.5 nA/cm^2) was declined as compared to the MAO (53,700 nA/cm^2) and AZ91 (63,100 nA/cm^2). Therefore, the findings of E_{corr} and I_{corr} values indicated that the FHA-coated sample has a noticeably more corrosion ability than that of the MAO-coated samples.

Farnoush et al. [66] performed MAO on Ti to produce porous TiO_2 coatings retaining CaP. The MAO was performed at 300 V, 330 V, and 360 V for 5 min. The MAO treatment was done for providing an interlayer that improves the mismatch between titanium substrate and final coating. The electrochemical behavior and bonding strength were also enhanced. The EPD was done at 30 V for 120 s subsequently on the MAO substrate to deposit HA-BG composite coatings. The SEM images of samples after MAO at various voltages is displayed in Fig. 15. It can be seen that as the voltage rose to 360 V, the size of the HA clusters became larger and they connected to each other.

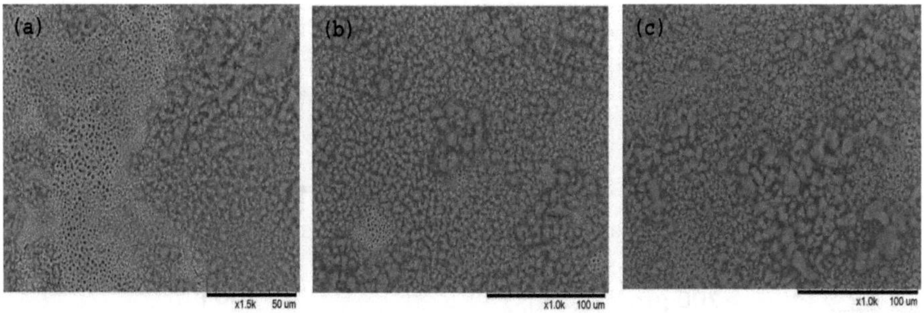

Fig. 15 SEM images of the MAO samples at **a** 300 V, **b** 330 V and **c** 360 V. Reprinted from Farnoush et al. [66], with permission from Elsevier. Copyright (2015)

The porous TiO_2 layer was completely encased with enlarged HA clusters due to the higher sparks and current density on the samples surface. Based on the electrochemical corrosion behavior of samples, the result of samples denoted by HA-BG-MAO 360 Ti and HA-MAO 360 Ti (where 360 refers to the voltage and Ti to the substrate) implies a significant decline in the current density, and increase in corrosion potential. The higher bonding strength was achieved with MAO. It helps in reducing the thermal expansion mismatch between substrate and coating, which results in higher densification.

Drevet et al. [67] studied the synthesis of HA coating by EPD of a nanosized powder. The influence of potential and deposition time variation on the corrosion behavior was explored. EPD was done in a suspension formulated by adding 2 g of HA nanopowder in 25 ml ethanol. The voltage was varying from 5 to 20 V and time from 5 to 20 min. The SEM micrographs obtained at different voltages are shown in Fig. 16.

At the higher voltage, and with the increase in time, the samples revealed cracks, and uncovered areas. These observations can be ascribed to particles agglomeration that progressively enhanced the surface electrical resistance, which can increase the corrosion rate in SBF. The best EPD conditions were obtained at 10 V for 10 min, which leads to compact and uniform coating.

Bakin et al. [68] produced Mg-substituted Cap coating on Ti-6Al-4V substrate for improving surface bioactivity and corrosion performance of coating. Four distinct solutions were made with various Mg contents of 0, 10, 20, and 40 (in wt.%). All coatings showed a needle-like morphology. The coating developed with 10% Mg content acquired the superior corrosion resistance relative to 20 and 40% addition of Mg. The samples were covered with apatite layer after immersion of 7 days. The surface morphology of coating transformed to spheres, and these spheres enclosed the whole surface completely. The surface of coatings becomes compact after 14 days of immersion and the formation of apatite distributed on the entire surface.

Molaei et al. [69] evaluated the effect on CS-BG-HA-halloysite (Hal) nanotube films by varying the concentration of halloysite on pure Ti. EPD was done in a solution containing 0.7 g/L HA, 0.6 g/L Hal, and 0.7 g/L BG in (17%) water–ethanol mixture,

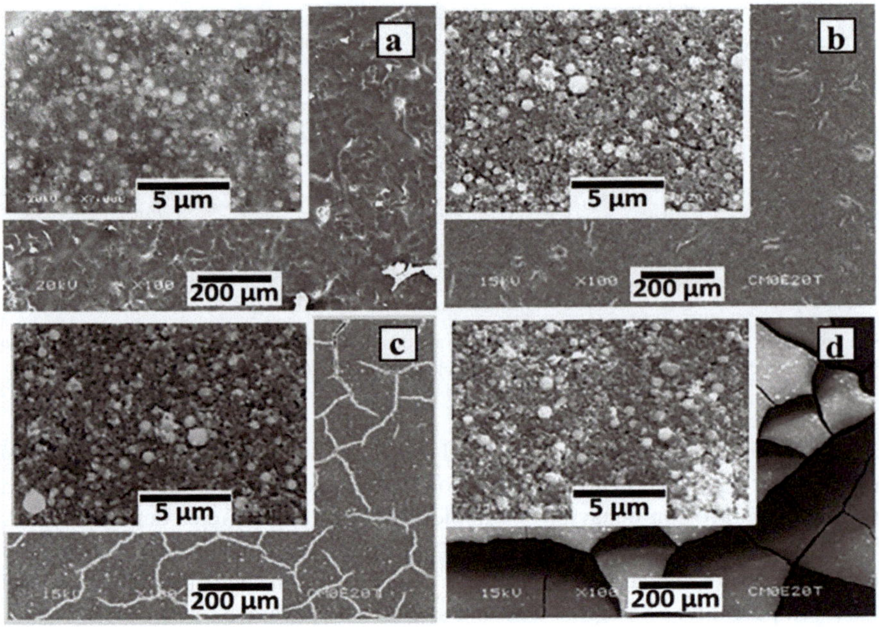

Fig. 16 SEM micrographs at low magnification and high magnification images (insets) for HA coatings attained in 10 min at **a** 5 V, **b** 10 V, **c** 15 V, and **d** 20 V. Reprinted from Drevet [67], with permissions from Elsevier. Copyright (2016)

because of fast sedimentation, whereas the dispersion was stable with (30%) water–ethanol solvent during 30 min. Hence, the desirable mixture for CS-BG-HAwas chose 30% water–ethanol solvent as shown in Fig. 17.

The optimum pH chosen was in the range of $2.5 < pH < 3$. From potentiodynamic polarization curves it was observed that a better corrosion resistance of CS-BG-HA-Hal coated substrate $E_{corr} = -0.507$ V and $I_{corr} = 0.56$ μA/cm^2) in SBF at 37 °C with respect to bare Ti ($E_{corr} = -587$ V and $I_{corr} = 91$ μA/cm^2).

Jugowiec et al. [70] reported EPD on Ti-13Nb-13Zr with chitosan (CS) coatings. The effect of the composition of colloidal solution and EPD variables on the performance of CS coatings was evaluated. The solution of CS (2 g/l) in a solution of distilled water and 0.5 vol % of acetic acid and 50 vol % of ethanol was employed for deposition. The voltage was varying from 5 to 30 V and the time (4 min) remained constant. It was observed that the coating uniformity is dependent on the composition of colloidal solution and stirring before EPD. The most homogenous coatings were deposited at a potential of 10 V and 4 min as shown in Fig. 18.

Kumar et al. [71] presented HA-coated Mg alloy via EPD. Corrosion behavior of the produced sample in the SBF solution was analyzed to recognize the potential of EPD deposited coating for orthopedic applications. The EPD was performed at 20 V for 10 min. It was observed that EPD deposited HA enhanced the corrosion protection of Mg alloy up to a magnificent 25 times. The penetration rate of Mg^{2+}

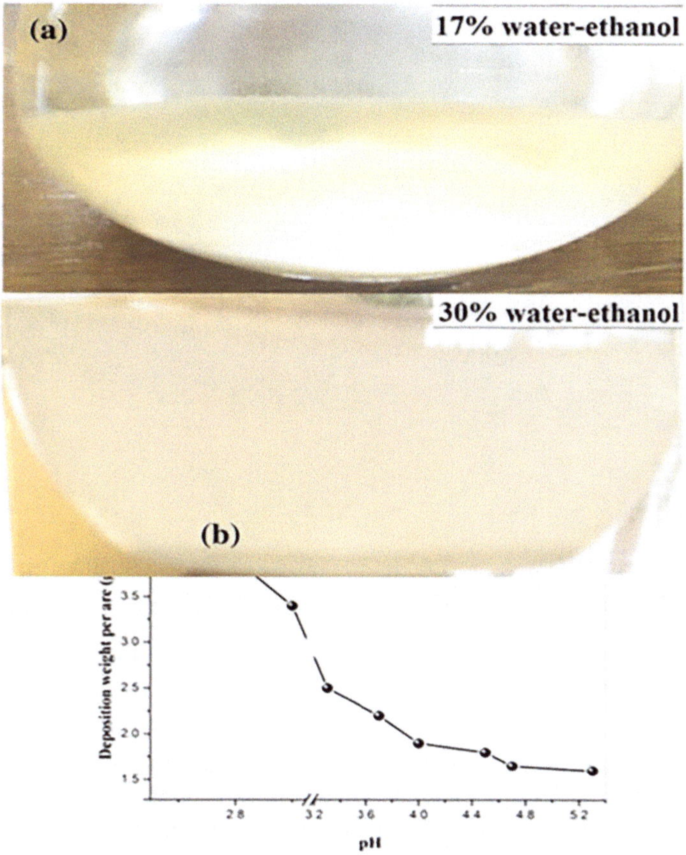

Fig. 17 **a** Digital image of dispersion having 17% water-ethanol (5 min) (**a**), and 30% water–ethanol (30 min). **b** The deposition weight as a function of pH produced from dispersion having 0.6 g/L Hal at 30 V. Reprinted from Molaei et al. [69], with permission from Elsevier. Copyright (2017)

ions in SBF for the coating with less roughness and more annealing temperature was small, implying the lowest degradation rate. The substantial increase in osteogenic cell adhesion was observed in the HA-coated Mg surface. Therefore, it has shown its potential to use for orthopedic implant applications.

Sankar et al. [72] compared the corrosion performance of HA-coated Mg, by employing two methods (a) EPD and (b) Pulsed laser deposition (PLD). By PLD, thin films were formed with the energetic condensation of atomic species. It efficiently forms highly adherent films. The layer of EPD coating was thick which leads to unstable behavior. The HA porous layer leads to more variation in the potential. Due to this, the corrosion rate of PLD-HA was 0.073 mm/yr, but of EPD-HA it was 0.194 mm/yr, and of the substrate it was 0.97 mm/yr. The adhesion of PLD-HA coating was observed to be 30% more as compared to that of the EPD-HA.

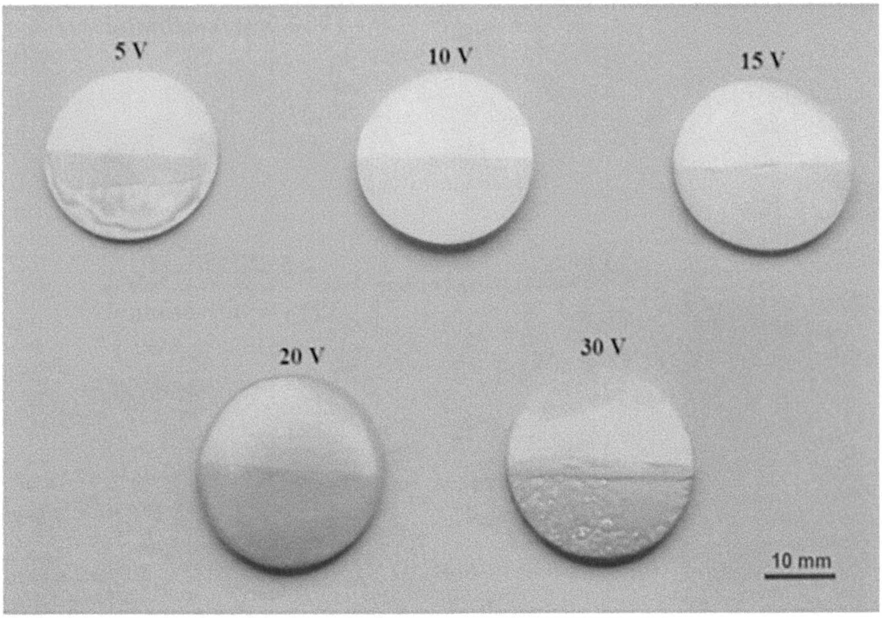

Fig. 18 Uncoated samples with 5–30 V and constant time of 4 min. Reprinted from Jugowiec et al. [70], with permission from Elsevier. Copyright (2017)

Huang et al. [73] investigated magnesium phosphate (MgP) and zinc phosphate (ZnP) doped HA composite coating deposited on AZ31 alloy by EPD. The corrosion behavior was observed through electrochemical tests and degradation tests. Pure HA, HA and MgP, HA and ZnP. HA, MgP, and ZnP coatings were plated with an electrochemical work station at room temperature. Compared with other samples, the electrochemical data showed that HA, MgP, and ZnP have the lowest I_{corr} value. The combination of HA, MgP, and ZnP proves great improvement in corrosion resistance for magnesium alloy.

Jugowiec et al. [74] deposited nanocomposite nc-CS/HA-p and nc-CS/HA-s coatings on Ti-13Nb-13Zr by EPD. The influence of HA particles in two different forms as nanopowder (p) and nanoparticles (s) on coating microstructure, adhesion and corrosion was investigated. Different suspensions of nc-HA-p (1, 2, 3, 4, 5 g/l) and nc-HA-s (4, 10, 15, 30 g/l) were formulated. EPD was carried out at 8–30 V for the time duration of 1, 2, 3, 4, 5, and 6 min. The homogeneous nc-CS/HA-s coatings were developed from 2 g/l of CS and 4 g/l of nc-HA-s suspension, at the potential of 10 V and 4 min as shown in Fig. 19. The formation of microcracks occurs at higher HA concentration in the nc-CS/HA-s coating. Moreover, the nc-CS/HA-s coating provides better adhesion and corrosion protection than the nc-CS/HA-p coating due to the absence of HA-s agglomerates.

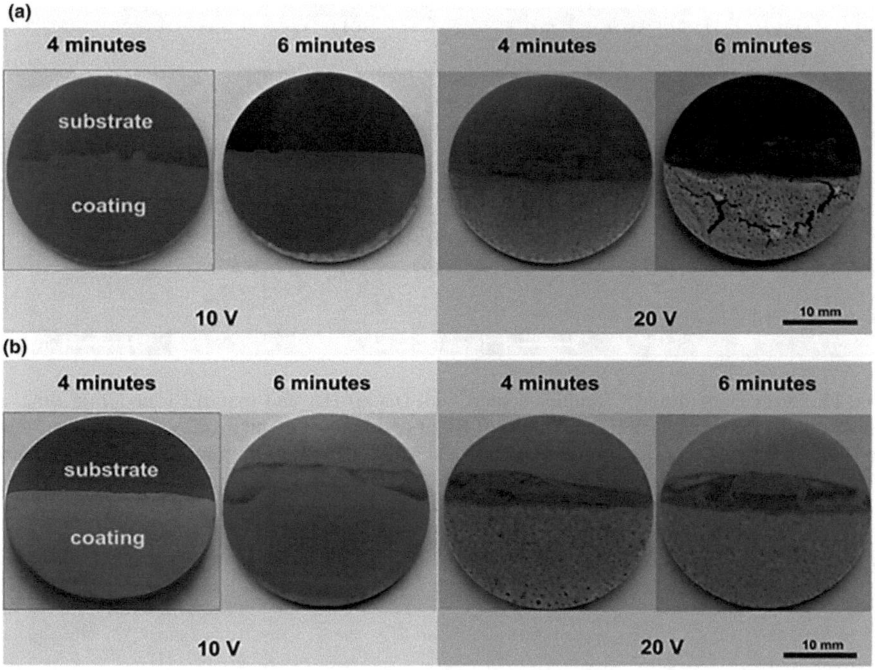

Fig. 19 Ti-13Nb-13Zr samples with **a** nc-HA-p/chitosan, and **b** nc-HA-s/chitosan. Reprinted from Jugowiec et al. [74], with permission from Elsevier. Copyright (2017)

Zhong and Ma [75] developed zinc substituted HA/silk fibroin (SF) coatings (HA-Zn-SF) on Ti by EPD. The SF has been widely explored to develop HA nanocomposite, which has similar structure relative to the natural bone minerals. The voltage and time duration were set as 30 V and 120 s. It was observed that the attained coatings were crack free and there was the excellent formation of apatite in SBF. The in vitro cell tests exhibited that the produced coatings offered good cytocompatibility.

Bartmanski et al. [76] examined the influence of EPD time on the coating homogeneity and the corrosion resistance. The electrochemical tests were conducted for all nano-HA coatings with nano-copper and reference Ti-13Zr-13Nb uncoated specimen. The Ti-13Zr-13Nb alloy was deposited with HA in a solution having 0.1, 0.2, or 0.5 g nano-HA in 100 mL of suspension and at 30 V for 1–2 min. Figure 20 illustrates the coatings derived at 1 min of EPD.

The coatings were inhomogeneous, and several HA agglomerates formed. The individual nano-HA layer, spheroidal nano Cu particles, passing through the HA layer were observed. Figure 21 presents the deposits obtained at 2 min. The HA agglomerates were increased in number due to the increase of time. The coating was more uniform relative to that deposited at 1 min. All nano-HA coatings had more noble corrosion potential than the uncoated sample.

Bakhsheshi-Rad et al. [77] prepared the TiO_2 incorporated MAO coatings (TM) on Mg-0.7Ca alloy. Afterward, zinc-doped hydroxyapatite (ZH) coating was formed

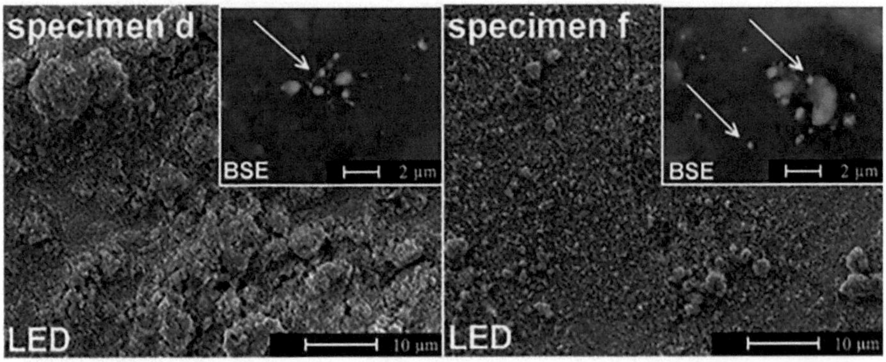

Fig. 20 Surfaces of d and f samples coated with (nano) HA and (nano) Cu at 1 min time on non-oxidized (left specimen d) and oxidized (right specimen f) titanium alloy. Nano Cu particles are indicated by arrows. Reprinted from Bartmanski et al. [76], with permission from Elsevier. Copyright (2019)

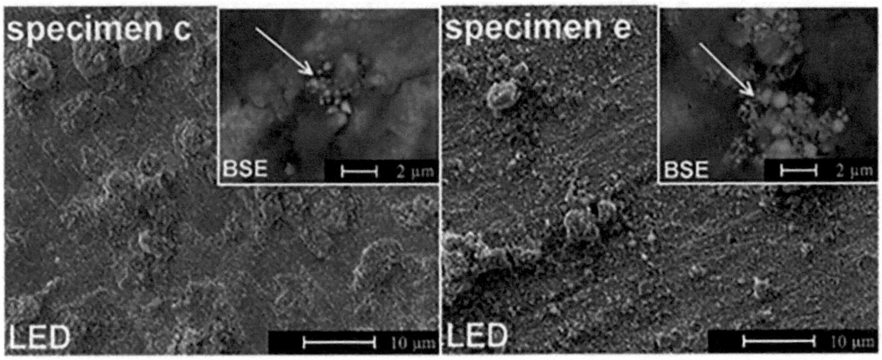

Fig. 21 Surfaces of c and e samples coated with (nano) HA and (nano) Cu coatings at 2 min time on non-oxidized (left specimen c) and oxidized (right specimen e) titanium alloy. Nano Cu particles are indicated by arrows. Reprinted from Bartmanski et al. [76], with permission from Elsevier. Copyright (2019)

by EPD on the MAO coating to reduce the Mg alloy degradation rate. The electrochemical test was conducted in SBF. The polarization plots revealed that cathodic and anodic branches transfer to lessen currents after coating. This may be ascribed to decreased solution diffusion into the substrate. The I_{corr} of Mg sample was detected to be (197.2 ± 8.1) $\mu A/cm^2$, while, after coating, the I_{corr} was declined to (9.8 ± 3.4) $\mu A/cm^2$. The ZH coating exhibited less I_{corr} (1.2 ± 0.3) $\mu A/cm^2$ than the TM coating, which was probably associated to the careful action of the top layer, by covering the TM film open pores. The E_{corr} of the TM coating was (−1600 ± 17) mV, which is about 65 mV more relative to the Mg substrate. After deposition of ZH,

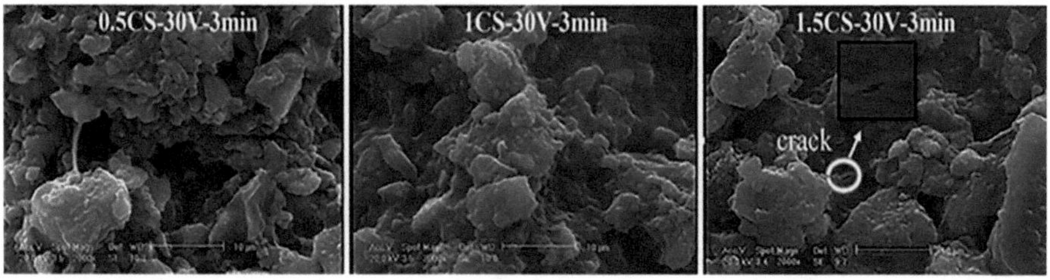

Fig. 22 SEM images of the CS-GO-HA coatings with different contents of chitosan, Reprinted from Karimi et al. [34], with permission from Elsevier. Copyright (2019)

the E_{corr} was moved to the positive position (-1520 ± 16) mV, and the corrosion rate was substantially inhibited.

Tozar and Karahan [78] investigated the single and integrated impacts of collagen (COL) and hexagonal boron nitride (h-BN) matrix on the corrosion performance and mechanical properties of HA-CS composite coatings. Collagen which is the primary human tissues ligament was employed as organic doped material owing to its perfect biodegradability and biocompatibility. The h-BN is a popular lubricant used as inorganic material for their excellent mechanical and corrosion properties. The EPD process was carried out with 15 V and 5 min duration. It was found that HA/CTS/COL/h-BN coatings can improve the mechanical properties and enhance the corrosion protection of Ti-6Al-4V alloys. The synergetic influence of collagen and h-BN in composite coating results in the formation of nucleation sites, which improves the bioactivity to a significant extent.

Karimi et al. [34] fabricated chitosan (CS) reinforced HA-graphene oxide (CS-GO-HA) nanocomposite coatings with different concentrations of CS (0.5, 1, and 1.5 mg/ml) by the EPD process to examine the effect of CS on the coating morphology. The EPD was performed at 30 V for 3 min. Figure 22 presents the SEM micrographs of CS-GO-HA coatings of different concentrations of chitosan.

The HA-GO nanoparticles were uniformly distributed over the entire surface of coatings. It was observed that the CS fills the pores among the interspaces of nanoparticles. However, an increment in the concentration of CS led to lower HA-GO nanoparticles deposition rate which may be due to the more electrolyte viscosity resulting in the lower mobility of particles. Mahmoodi et al. [79] documented an optimized viscosity concentration for obtaining the particles maximum deposition rate. In fact, an increase in the concentration of CS contributes to the microcracks formation, which can be seen in the sample denoted by (1.5CS-30V-3 min). It can be attributed to more absorption of water by enhancing the content of CS that could contribute to its evaporation process and resulting in coating crack formation. Of these coatings, the coating having 0.5 mg/ml CS showed I_{corr} of about 21 nA/cm^2 which was substantially less than the Ti with I_{corr} of 200 nA/cm^2.

Tayyaba et al. [80] synthesized HA coating using EPD on ZK60 Mg alloy. The degradation behavior of coatings has been compared with the Mg alloy in the Ringer's

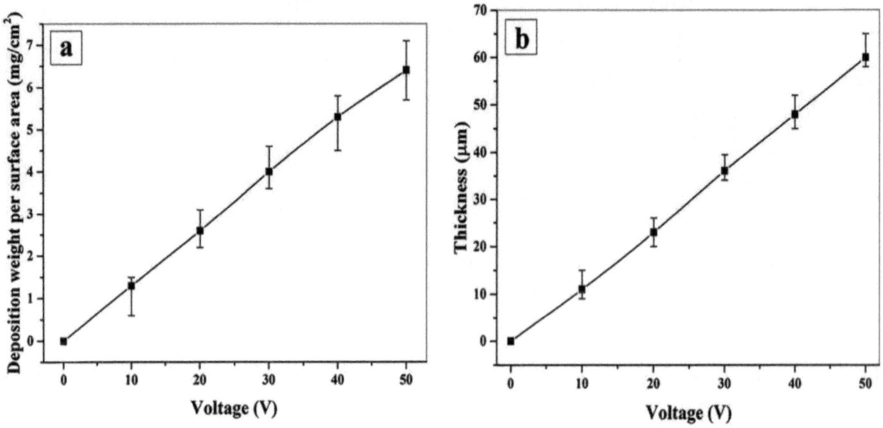

Fig. 23 **a** Deposition weight, and **b** coating thickness as functions of voltage in the range (10–50) V for 5 min. Reprinted from Molaei et al. [81], with permission from Elsevier. Copyright (2008)

solution (24–150) h. The DC power source was set at 100 V for 10 min. The electrochemical test revealed that the HA coatings were not degraded till 48 h and began to degrade at −1.8 V after immersion time of 72 h. The coatings were completely degraded in 128 h. The degradation rate of coated and uncoated samples was not significantly different after 150 h of immersion due to the $Mg(OH)_2$ precipitate formation on the coating surface. The 15 μm thick coating on ZK60 alloy declined the I_{corr} by one fold of magnitude and resulted in more than five times lower corrosion rate.

Molaei et al. [81] carried out EPD of CS-based nanocomposite coatings having BG, HA, and halloysite nanotubes (HNTs) on the Ti to study the effect of process variables such as voltage and deposition kinetics. The deposition was done at the (10–50) V with the duration of (5–25) min. Figure 23(a) and (b) shows the weight and thickness as functions of voltage.

As shown in Fig. 23, the deposition increases by increment in voltage. The lowest and highest deposition thickness and weight are developed from the suspension at 10 and 50 V. Such variations in the deposit thickness and weight are outcomes of the gap in the electric field strength. SEM images at various voltages are shown in Fig. 24. The result of Fig. 24(a) shows a thin film at 10 V. Owing to the lack of adequate force, there are no micro particles of BG in the coating. Figure 24(b) exhibits the coatings prepared at 30 V. The morphology of the composite shows that nano and micro particles were dispersed compactly and uniformly distributed throughout the CS matrix. In fact, increasing the voltage to 30 V ensures a uniform deposition with small cracks and porosity. When the voltage exceeds 50 V, the coating starts to deteriorate [Fig. 24(c)].

The explanation for this behavior is the creation of larger agglomerations and water electrolysis, because of the low voltage threshold. Coated sample produced at

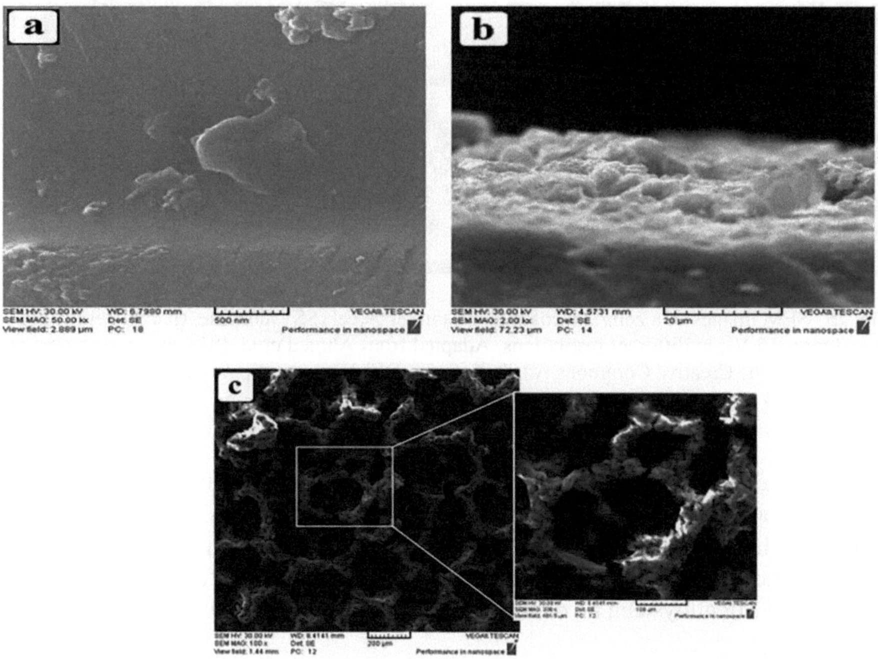

Fig. 24 SEM micrographs of CS-based coatings at **a** 10 V, **b** 30 V, and **c** 50 V. Reprinted from Molaei et al. [81], with permission from Elsevier. Copyright (2008)

30 V acquires excellent corrosion ability relative to the coated samples deposited at 10 and 50 V.

Ahmed et al. [82] obtained zein/HA coatings on Stainless steel (316L SS) via EPD. Zein/HA coatings need to be produced because they improve biocompatibility and corrosion resistance properties. EPD was conducted at voltages (6–15) V, and time duration (3–9) min. In an earlier study [83] it was demonstrated that the suspension becomes unstable, the HA particles concentration in suspension rises to 7.5 g/L and more voltages result in gas bubbles formation. Reportedly, the zein coatings developed at (3–5) V are very thin [84]. A better deposition of zein was obtained at 10 V. Figure 25 shows the SEM images of the zein/HA coatings (obtained at 15 V for 3 min duration, and HA concentration of 5 g/L). These parameters were found optimal during the deposition process.

The homogenous porous film on zein coating was observed. The HA particles are distributed homogeneously in the pores of the zein coating. The pores are blocked by the inclusion of agglomerated HA, which enhances the mechanical stability of the coating. Zein/HA coatings effectively increase the corrosion resistance of bare 316L SS. The coatings were bioactive upon immersion in SBF and the formation of HA crystals occurred after 3 days of immersion.

Antoniac et al. [85] examined the biodegradation rate of ZMX410 and ZM21 Mg alloys through HA coating via EPD. The corrosion rate (CR) of the Mg alloys are

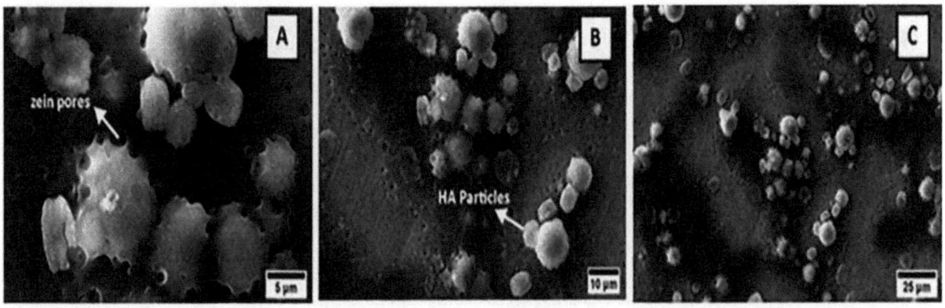

Fig. 25 SEM images of a zein/HA coating on Stainless steel (SS) substrate, (parameters: 5 g/L of HA, 3 min, 15 V); at different resolutions. Adapted from Ahmed et al. [82], under the terms and conditions of the Creative Commons Attribution (CC BY) license. (http://creativecommons.org/licenses/by/4.0/)

approximately equal, with the difference is 0.51 mm/year, in favor of the ZMX410 alloy. On the basis of efficiency toward corrosive attack, it was established that the coating with HA on ZM21 contributes by 40.6% to corrosion protection, and only 4.23% to the corrosion resistance in the ZMX410 alloy. Therefore, the degradation rate of both Mg alloys can be controlled by the EPD of HA coating.

4 Conclusion

The reviewed literature exhibited that EPD is an effective method for the processing of superior quality coatings. It was seen that a number of materials such as inorganic and organic–inorganic composite coatings can be developed in well-controlled manner. The HA-reinforced deposits showed improved characteristics including crack free films, oriented microstructures, enhanced corrosion resistance, and bioactivity properties. It was observed from the literature that the amount of deposit enhances with the increase in applied potential, but it leads to wrinkling, film roughness, and cracks in the coating. In addition, more deposition time results in insufficient attachment of coating to the base metal, and development of non-uniform and thick coatings. At high temperature, there were chances of cracking due to the difference in the sintered coating thermal expansion coefficients and the substrates. Therefore, a number of researchers employed different polymers as an effective binder to avoid the inconvenience of elevated temperature. Based on these findings, it may be concluded that a pilot study must be performed to optimize the parameters of EPD to obtain high quality, smooth, and homogeneous coatings. Furthermore, new experiments are to be conducted with the great potential of EPD for manipulation and assembly of HA composite coatings in which the effect of several other deposition parameters such as particle size, the dispersion medium, substrate, anode material, and temperature can be studied.

References

1. B. Ferrari, R. Moreno, J. Eur. Ceram. Soc. **30**(5), 1069 (2010)
2. O.O. Van der Biest, L.J. Vandeperre, Annu. Rev. Mater. Sci. **29**(1), 327 (1999)
3. L. Corni, M.P. Ryan, A.R. Boccaccini, J. Eur. Ceram. Soc. **28**(7), 1353 (2008)
4. Y. Ma, J. Han, M. Wang, X. Chen, S. Jia, J. Materiomics **4**(2), 108 (2018)
5. J. Kim, S.A. Guelcher, S. Garoff, J.L. Anderson, Adv. Coll. Interface. Sci. **96**(1–3), 131 (2002)
6. J.R. Woodard, A.J. Hilldore, S.K. Lan, C.J. Park, A.W. Morgan, J.A.C. Eurell, S.G. Clark, M.B. Wheeler, R.D. Jamison, A.J.W. Johnson, Biomaterials **28**(1), 45 (2007)
7. R. Ma, Y. Li, I. Zhitomirsky, JOM **62**(6), 72 (2010)
8. C.P. Klein, J.M.A. de Blieck-Hogemrst, J.G.C. Wolket, K. De Groot, Biomaterials **11**(7), 509 (1990)
9. R. Palanivelu, S. Kalainathan, A.R. Kumar, Ceram. Int. **40**(6), 7745 (2014)
10. P. Amaravathy, S. Sathyanarayanan, S. Sowndarya, N. Rajendran, Ceram. Int. **40**(5), 6617 (2014)
11. A. El-Ghannam, J. Biomed. Mater. Res. Part B: Appl. Biomater. **99**(2), 369 (2011)
12. S. Torgbo, P. Sukyai, Mater. Chem. Phys. **237**, 121868 (2019)
13. N. Saha, A.K. Dubey, B. Basu, J. Biomed. Mater. Res. Part B: Appl. Biomater. **100**(1), 256 (2012)
14. V.A. Thampi, B. Subramanian, J. Alloy. Compd. **649**, 1210 (2015)
15. D. Mishra, B. Bhunia, I. Banerjee, P. Datta, S. Dhara, T.K. Maiti, Mater. Sci. Eng. C **31**(7), 1295 (2011)
16. H. Sawalha, K. Schroën, R. Boom, J. Appl. Polym. Sci. **107**(1), 82 (2008)
17. J.G.P. Biner, *Advanced Ceramic Processing and Technology*, vol. 1, 1st ed. (Noyes Publications, 1990), pp. 255–283.
18. B. Ferrari, R. Moreno, J. Eur. Ceram. Soc. **17**(4), 549 (1997)
19. B. Ferrari, R. Moreno, Mater. Lett. **28**(4–6), 353 (1996)
20. M. Zarbov, I. Schuster, L. Gal-Or, J. Mater. Sci. **39**(3), 813 (2004)
21. J. Ervina, Z.A. Ghaleb, S. Hamdan, M. Mariatti, Compos. A Appl. Sci. Manuf. **117**, 1 (2019)
22. D.R. Brown, F.W. Salt, J. Appl. Chem. **15**(1), 40 (1965)
23. T. Mori, H. Nagashima, Y. Ito, Y. Era, J. Tsubaki, Miner. Eng. **133**, 119 (2019)
24. K. Hasegawa, S. Kunugi, M. Tatsumisago, T. Minami, J. Sol-Gel Sci. Technol. **15**(3), 243 (1999)
25. I. Zhitomirsky, L. Gal-Or, J. Mater. Sci. Mater. Med. **8**(4), 213 (1997)
26. R.N. Basu, C.A. Randall, M.J. Mayo, J. Am. Ceram. Soc. **84**(1), 33 (2001)
27. L. Sansone, V. Malachovska, P. La Manna, P. Musto, A. Borriello, G. De Luca, M. Giordano, Sens. Actuators, B Chem. **202**, 523 (2014)
28. V. Lee, L. Whittaker, C. Jaye, K.M. Baroudi, D.A. Fischer, S. Banerjee, Chem. Mater. **21**(16), 3905 (2009)
29. L. Besra, M. Liu, Prog. Mater. Sci. **52**(1), 1 (2007)
30. E. Tabesh, H.R. Salimijazi, M. Kharaziha, M. Mahmoudi, M. Hejazi, Surf. Coat. Technol. **364**, 239 (2019)
31. Y. Su, I. Zhitomirsky, Colloids Surf. A **436**, 97 (2013)
32. A.T. Koh, T. Chen, L. Pan, Z. Sun, D.H. Chua, J. Appl. Phys. **113**(17), 174909 (2013)
33. J. Du, Y. Zhang, S. Deng, N. Xu, Z. Xiao, J. Shi, Z. Wu, H. Cheng, Carbon **61**, 507 (2013)
34. N. Karimi, M. Kharaziha, K. Raeissi, Mater. Sci. Eng. C **98**, 140 (2019)
35. Y. Liu, D. Zhang, S. Pang, Y. Liu, Y. Shang, J. Sep. Sci. **38**(1), 157 (2015)
36. M. Diba, A. Garcia-Gallastegui, R.N.K. Taylor, F. Pishbin, M.P. Ryan, M.S. Shaffer, A.R. Boccaccini, Carbon **67**, 656 (2014)
37. P. Sarkar, P.S. Nicholson, J. Am. Ceram. Soc. **79**(8), 1987 (1996)
38. A. Aziz, S.A. Binti, S.H. Amirnordin, A. Hamimah, H.Z. Abdullah, H. Taib, Adv. Mater. Res. **488**, 1358 (2012)
39. W. He, L. Zhu, H. Chen, H. Nan, W. Li, H. Liu, Y. Wang, Appl. Surf. Sci. **279**, 416 (2013)

40. P. Subramanian, J. Niedziolka-Jonsson, A. Lesniewski, Q. Wang, M. Li, R. Boukherroub, S. Szunerits, J. Mater. Chem. A **2**(15), 5525 (2014)
41. M.S. Wu, Y.P. Lin, C.H. Lin, J.T. Lee, J. Mater. Chem. **22**(6), 2442 (2012)
42. R. Drevet, J. Fauré, H. Benhayoune, Adv. Eng. Mater. **14**(6), 377 (2012)
43. M. Dash, F. Chiellini, R.M. Ottenbrite, E. Chiellini, Prog. Polym. Sci. **36**(8), 981 (2011)
44. D. Stojanovic, B. Jokic, D. Veljovic, R. Petrovic, P.S. Uskokovic, D. Janackovic, J. Eur. Ceram. Soc. **27**(2–3), 1595 (2007)
45. A. Balamurugan, G. Balossier, G.J. Michel, J.M.F. Ferreira, Electrochim. Acta **54**(4), 1192 (2009)
46. D. Zhitomirsky, J.A. Roether, A.R. Boccaccini, I. Zhitomirsky, J. Mater. Process. Technol. **209**(4), 1853 (2009)
47. C.T. Kwok, P.K. Wong, F.T. Cheng, H.C. Man, Appl. Surf. Sci. **255**(13–14), 6736 (2009)
48. B. Zhang, C.T. Kwok, F.T. Cheng, H.C. Man, J. Nanosci. Nanotechnol. **11**(12), 10740 (2011)
49. Y. Bai, M.P. Neupane, I.S. Park, M.H. Lee, T.S. Bae, F. Watari, M. Uo, Mater. Sci. Eng. C **30**(7), 1043 (2010)
50. Y. Bai, K.A. Kim, I.S. Park, S.J. Lee, T.S. Bae, M.H. Lee, Mater. Sci. Eng. B **176**(15), 1213 (2011)
51. A.A. Abdeltawab, M.A. Shoeib, S.G. Mohamed, Surf. Coat. Technol. **206**(1), 43 (2011)
52. L. Mohan, D. Durgalakshmi, M. Geetha, T.S. Narayanan, R. Asokamani, Ceram. Int. **38**(4), 3435 (2012)
53. D. Sreekanth, N. Rameshbabu, Mater. Lett. **68**, 439 (2012)
54. H. Maleki-Ghaleh, V. Khalili, J. Khalil-Allafi, M. Javidi, Surf. Coat. Technol. **208**, 57 (2012)
55. R. Rojaee, M. Fathi, K. Raeissi, Appl. Surf. Sci. **285**, 664 (2013)
56. R. Rojaee, M. Fathi, K. Raeissi, IEEE Trans. Nanobiosci. **13**(4), 409 (2014)
57. S. Kaabi Falahieh Asl, S. Nemeth, M.J.T. Tan, *Paper presented at the 7th International Conference on Materials for Advanced Technologies (ICMAT 2013)*, Suntec City, Singapore, 30 June 2013 to 5 July 2013
58. Y. Huang, Q. Ding, S. Han, Y. Yan, X. Pang, J. Mater. Sci. Mater. Med. **24**(8), 1853 (2013)
59. H. Farnoush, A.A. Bastami, A. Sadeghi, J.A. Mohandesi, F. Moztarzadeh, J. Mech. Behav. Biomed. Mater. **20**, 90 (2013)
60. M. Li, Q. Liu, Z. Jia, X. Xu, Y. Cheng, Y. Zheng, T. Xi, S. Wei, Carbon **67**, 185 (2014)
61. A.T. Rad, M. Solati-Hashjin, N.A.A. Osman, S. Faghihi, Ceram. Int. **40**(8), 12681 (2014)
62. H. Mehboob, M. Awais, H. Khalid, S.A. Siddiqi, I. Rehman, Biomed. Eng. Appl. Basis Commun. **26**(6), 1450073 (2014)
63. A. Janković, S. Eraković, M. Mitrić, I.Z. Matić, Z.D. Juranić, G.C. Tsui, C.Y. Tang, V. Mišković-Stanković, K.Y. Rhee, S.J. Park, J. Alloy. Compd. **624**, 148 (2015)
64. A. Molaei, M. Yari, M.R. Afshar, Ceram. Int. **41**(10), 14537 (2015)
65. M. Razavi, M. Fathi, O. Savabi, D. Vashaee, L. Tayebi, Metall. Mater. Trans. A **46**(3), 1394 (2015)
66. H. Farnoush, F. Muhaffel, H. Cimenoglu, Appl. Surf. Sci. **324**, 765 (2015)
67. R. Drevet, N.B. Jaber, J. Fauré, A. Tara, A.B.C. Larbi, H. Benhayoune, Surf. Coat. Technol. **301**, 94 (2016)
68. B. Bakin, T.K. Delice, U. Tiric, I. Birlik, F.A. Azem, Surf. Coat. Technol. **301**, 29 (2016)
69. A. Molaei, M. Yari, M.R. Afshar, Appl. Clay Sci. **135**, 75 (2017)
70. D. Jugowiec, M. Kot, T. Moskalewicz, Arch. Metall. Mater. **61**(2), 657 (2016)
71. R.M. Kumar, K.K. Kuntal, S. Singh, P. Gupta, B. Bhushan, P. Gopinath, D. Lahiri, Surf. Coat. Technol. **287**, 82 (2016)
72. M. Sankar, S. Suwas, S. Balasubramanian, G. Manivasagam, Surf. Coat. Technol. **309**, 840 (2017)
73. W. Huang, B. Xu, W. Yang, K. Zhang, Y. Chen, X. Yin, Y. Liu, Z. Ni, F. Pei, Surf. Coat. Technol. **326**, 270 (2017)
74. D. Jugowiec, A. Łukaszczyk, Ł Cieniek, K. Kowalski, Ł Rumian, K. Pietryga, M. Kot, E. Pamuła, T. Moskalewicz, Surf. Coat. Technol. **324**, 64 (2017)
75. Z. Zhong, J. Ma, J. Biomater. Appl. **32**(3), 399 (2017)

76. M. Bartmanski, A. Zielinski, M. Jazdzewska, J. Głodowska, P. Kalka, Ceram. Int. **45**(16), 20002 (2019)
77. H.R. Bakhsheshi-Rad, E. Hamzah, A.F. Ismail, M. Aziz, M. Daroonparvar, E. Saebnoori, A. Chami, Surf. Coat. Technol. **334**, 450 (2018)
78. A. Tozar, İH. Karahan, Surf. Coat. Technol. **340**, 167 (2018)
79. S. Mahmoodi, L. Sorkhi, M. Farrokhi-Rad, T. Shahrabi, Surf. Coat. Technol. **216**, 106 (2013)
80. Q. Tayyaba, M. Shahzad, A.Q. Butt, M. Khan, A.H. Qureshi, Surf. Coat. Technol. **375**, 197 (2019)
81. A. Molaei, M. Lashgaroo, M. Yousefpour, J. Aust. Ceram. Soc. **56**(1), 1 (2020)
82. Y. Ahmed, M. Yasir, M.A. Ur Rehman, Surfaces **3**(2), 237 (2020)
83. H. Piri-Moghadam, M.N. Alam, J. Pawliszyn, Anal. Chim. Acta **984**, 42 (2017)
84. S. Kaya, A.R. Boccaccini, J. Coat. Technol. Res. **14**, 683 (2017)
85. I. Antoniac, F. Miculescu, C. Cotrut, A. Ficai, J.V. Rau, E. Grosu, A. Antoniac, C. Tecu, I. Cristescu, Materials **13**(2), 263 (2020)

Geopolymers Prepared by Microwave Treatments

Yuta Watanabe and Takaomi Kobayashi

Abstract Geopolymers occur easily in a polycondensed polymer framework in high-alkali conditions. They are known as sustainable materials with a simple process that uses waste for raw materials like fly ash, slag, and rice husk etc. This chapter highlights the use of microwaves as efficient auxiliary processing for geopolymer production. Compared to conventional heating, the microwave cure reduced the curing time as an advantage. In addition, depending on the microwave irradiation conditions, the resulting geopolymers were changed in material properties such as high porosity, and low mechanical strength using a high-water foaming agent content due to water evaporation in the geopolymer. Also, the microwave curing had the advantage over the conventional oven heating, of densely structuring geopolymers when the water foaming agent content was low by combining the pre-heating and accelerating the polymerization. In contrast, the alkali activator concentration caused enhancement of the morphology of geopolymers in the case of microwave cure. Thus, microwave curing is expected to be an effective method for the fabrication of geopolymers. This chapter describes the feasibility of synthesizing the porous morphology and densely structured geopolymer without micro-sized pores using microwave energy.

Keywords Geopolymer · Microwave · Alkali activator · Silica-aluminum linkage · Replacement concrete

1 Overviewing Geopolymers

Geopolymers are usually activated in highly alkaline solutions and alumino-silicate geopolymers are inorganic materials consisting of tetrahedral AlO_4 and SiO_4 units [1, 2]. Essentially a mineral chemical compound or mixture of compounds consist of repeating units, for example, silico-aluminate (polysialate) (–Si–O–Al–O–) and is created through a process of polymerization. The polymerization process involves

Y. Watanabe · T. Kobayashi (✉)
Department of Science of Technology Innovation, Nagaoka University of Technology, 1603-1 Kamitomioka, Nagaoka, Niigata 940-2188, Japan
e-mail: takaomi@vos.nagaokaut.ac.jp

S. J. Ikhmayies (ed.), *Advanced Ceramics*, Advances in Material Research and Technology, https://doi.org/10.1007/978-3-031-43918-6_8

chemical reactions under highly alkaline solution conditions in Si and Al materials that provide covalent Si–O– and Al–O bonds, with the following formula [3].

$$M_n\{-[SiO_2]_z - AlO_2\}_n \cdot wH_2O$$

where M is an alkali cation such as sodium (Na^+) or potassium (K^+),—refers to bonding, n is the polymerization degree, z is usually a number of 1, 2, 3 and sometimes higher, and w is the number of moles of water. When strong alkalis activate the silicon and aluminum in the raw materials, geopolymerization allows transforming glassy structure partially or totally into a very compact composite [3, 4]. In these processes, as shown in Fig. 1, the alkali activator accelerates the condensation reaction, then leaches to form alumino-silicate precursor in the aqueous solution and finally the condensates to be geopolymer gels and then progressing polymerization. As a result, the mixture becomes hardened geopolymer [3, 5]. The geopolymer network generally proceeds at moderate temperature below 160 °C. For the Si and Al materials, generally alumino-silicates such as slag [6], metakaolin [7, 8], and fly ash [9, 10] were mixed with high concentration of alkaline solution at low temperatures. For example, NaOH and KOH are mainly used as highly concentrated alkalis, and the aforementioned materials containing Si and Al become soluble in alkaline solutions [1, 3, 11]. Here, Si and Al rich materials are the primary requirement for geopolymerization like rice husk [12, 13], clays like metakaolin [14, 15], natural soil [16] and bentonites [17, 18] as seen in Table 1. Geopolymers can be made from several types of alumino-silicate materials, including metakaolin, calcined clay [15], agricultural waste ashes [12, 13], and industrial sludge waste [19]. In the waste, the geopolymers based on industrial waste like slags [20–22], and aluminum waste [23] were helpful in the geopolymer fabrication.

Looking at the content of Al_2O_3 and SiO_2 in Table 1, the wastes contain the necessary components for geopolymer. In the cases of agricultural waste like rice husks, the ashes involve high silica contents depending on sintering temperature. The higher temperature the higher silica contents remain like about 90%, although

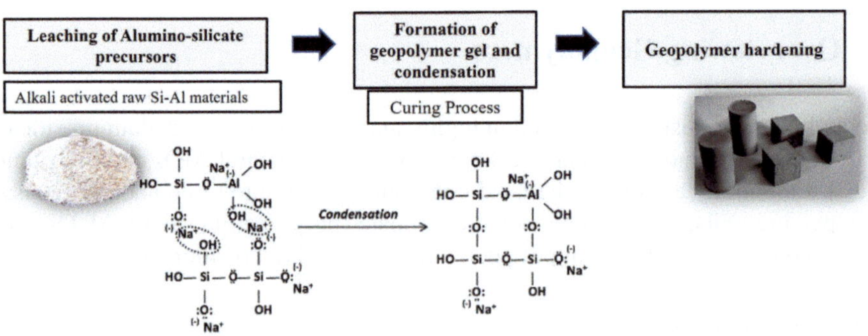

Fig. 1 Scheme of geopolymerization of silica alumina precursor by geopolymerization process

Table 1 Chemical compositions of raw materials for synthesis geopolymers

Compositions (%)	Al_2O_3	SiO_2	Na_2O	MgO	P_2O_5	SO_3	K_2O	CaO	TiO_2	Fe_2O_3	References
Rice husk*	0.09	92.19	1.64	0.41	0.01	0.41	0.05	0.09	0.71	0.1	[12]
Rice husk**	1.4	76.8	–	0.5	0.6	0.7	3.7	0.5	–	0.1	[13]
Metakaolin	42.63	50.00	0.28	0.13	–	0.44	0.34	1.29	1.71	2.11	[14]
Na-bentonite	15.79	68.28	2.57	4.08	0.04	0.02	0.87	5.5	0.24	2.29	[18]
Kaolin	43.98	53.39	–	0.08	–	0.01	0.86	0.04	–	0.63	[19]
Calcined kaolin waste	33.85	56.56	0.23	0.05	–	–	5.15	0.05	0.16	2.16	[15]
Fly ash***	18.37	31.94	1.2	1.86	0.16	5.13	3.05	19.84	0.4	17.39	[19]
Aluminum hydroxide waste	63.28	0.18	–	0.31	0.01	<0.01	0.01	0.01	0.01	–	[10]
GCBS****	33.6	12.9	0.3	7.9	–	2.2	0.4	41.5	0.7	0.6	[22]

* Incinerating at 650–700 °C for 1 h in India
** Calcined at 400 °C for 10 h in Taiwan
*** Supplied from Mae Moh power plant in Lumpang (Thailand)
**** Granulated Blast-Furnace Slag (from volcanic rocks)

alumina contents are less. In this case, metakaolin was used as alumina source, and the mixture with rice husk ash enabled the geopolymerization [8, 12].

Among these starting materials for geopolymer, fly ashes are interesting from the perspective of sustainable materials. Thus, fly ash-based geopolymer is an important material because such ash is the one of bulky wastes resulting from burning coal or lignite in power generation facilities. For using an environmentally friendly method [20, 21], fly ash of chemical composition of silica and alumina typically ranging from 40 to 60%, and from 20 to 30%, respectively is available worldwide for a geopolymer source. Thus, the condensated alumino-silicate is known as an eco-friendly material due to recycling of burnt ash, and another advantage is obtaining materials at low temperature for the polycondensation, which easily occurs under high alkaline conditions. In various types of slags used individually or as a partial replacement of fly ash, lightweight geopolymers have been obtained using palm oil fuel ash [23], ground granulated blast furnace slag [24], low calcium fly ash [25], and class F fly ash, ground granulated blast furnace slag [26] and metakaolin/biomass ash composite [27]. In the geopolymer preparation processes, heating and ambient curing condition were investigated using sodium silicate [28], alkali activator [9, 15, 29], curing temperature as sintering condition [7, 30] and additives [31]. Geopolymer made from industrial waste and by-products is low cost and environmentally friendly from environmental point of view. Therefore, scientists are still researching the many possible uses of geopolymer processes. In addition, geopolymers have various applications, and their properties can be engineered. Some papers reported that geopolymers in general have material structure with high apparent bulk density in the range of over 1 g/cm^3 causing high mechanical strength [10]. On the other hand, when geopolymer bulk has pores using foaming agents in the preparation, in general, the density and mechanical strength decrease. In the latter cases with pore structure, the bulk had a high specific surface area and high porosity, therefore, such geopolymers were expected to apply to adsorbents [8, 32–34].

There are disadvantages of geopolymers as compared to traditional cement in certain aspects like long maintenance time and slow curing till enough strength development, becoming the limiting factor for practical applications of geopolymers. However, geopolymers are interesting in technologies that enable the rapid development of strength in the fields. The common geopolymer products use conventional thermal curing under water, air, solar, or oven curing conditions to accelerate geopolymerization process. Table 2 summarizes advantages and disadvantages of each curing method of geopolymers. As seen in Table 2, water-cured geopolymer was found to be of lower compressive strength than air-cured, solar- and oven-cured. Water curing had disadvantages of reducing strength due to leaching reactant of geopolymerization and alkali activator solution [35]. Air and solar-cured geopolymer could be of higher compressive strength than the one cured with other conventional curing methods, these curing methods had the advantages that curing with both of them needs low energy source, and decreases manufacturing cost and carbon footprint. However, there were still limitations on the application of these curing methods due to the long curing time (air: 7–90 days, and solar: 3–28 days), even though these changes depend upon condition and geological location [36, 37]. Traditionally, oven heating was used as the most effective method for producing geopolymers with higher compressive strength, because of easy adjustment of the curing temperature and producing at 60–120 °C for 6–24 h in relatively a short time and low temperature [4, 10, 22]. However, among the conventional curing methods, oven curing especially has high heat loss due to the distribution of heat in the geopolymer from the exterior to the interior leading to the non-uniform [38]. Moreover, it was necessary for long heating within 6–24 h and high energy consumption for the curing. In addition, a limitation was the use of a foaming agent for a porous geopolymer. On the other hand, microwave curing could be attractive method for geopolymerization in recent years. Compared with the conventional heating methods, microwave curing by standardized volumetric measurement has some advantages of heating as shown below.

(1) Shorter curing time for geopolymerization (1–60 min) [39–42].
(2) Easier operation for fabrication of geopolymers [39, 43].
(3) No requirement of foaming agents for porous morphology geopolymers [16, 39].

2 Effect of Microwave on Geopolymerization

Recently, from the perspective of reducing energy consumption and humbleness, microwave energy has been utilized as an innovative tool for heating to influence chemical reactions of geopolymerization [39–41, 43, 44]. This method includes the benefits of microwave heating; higher heating rates in less times, no direct contact between the reactants and energy source, and clean, selective, and remote heating of the reactants in the desired atmosphere [45]. Microwave method utilized the bottom ash to produce thermal insulating bricks for foaming of the mortar by geopolymerization [44].

Using microwave radiation, the conventional heating was helpful for obtaining geopolymers attaining thermal stability as the dense matrices. Nevertheless, reports are few in the literature about microwave cured geopolymers. But, with these benefits, microwave-assisted synthesis of concrete [46, 47] and zeolite [48–50] were known to be the effectiveness, because the activation time in hydrothermal synthesis was reduced automatically. Table 3 lists researches of microwave cure for geopolymerization. There are two types of geopolymers in the dense and porous conformation.

It is known that the conventional heating has been used as the popular way to accelerate the geopolymerization, but heat is distributed in the sample from the exterior to interior leading to the non-uniform and long curing time [38, 40]. Then, microwave consisting of an electric and a magnetic field component has the ability to penetrate deep into the materials in nature within a short time. Samples containing inorganic ions that absorb microwave energy can heat volumetrically [51]. The electric component of an electromagnetic field causes heating by two main mechanisms: dipolar polarization and ionic conduction [45, 52, 53]. Under microwave condition

Table 2 Advantages and disadvantages of each curing method for geopolymer fabrication

Curing method	Water [42, 35]	Air [37]	Solar [36]	Oven [4, 10, 22]	Microwave [39, 39–42]
Raw materials	Fly Ash + Slag	Fly Ash	Fly Ash + Ground Granulated Blast-Furnace Slag (GGBS)	Fly Ash, GCBS	Fly Ash, Natural Soil
Reaction temperature	23 °C	Room temperature (22- 25 °C)	20–60 °C	60–120 °C	40–200 °C
Curing time	Long (1–28 days)	Very long (7–90 days)	Very long (3–28 days)	Short (6–24 h)	Very short (1–60 min)
Morphology	Dense	Dense	Dense	Dense	Dense, Porous
Compressive strength (MPa)	10–48 MPa	30–65 MPa	25–90 MPa	20–70 MPa	Dense: 30–90 MPa
					Porous: 3.1–8.8 MPa
Advantages	Small temperature changes	Low energy force	Decreasing the manufacturing cost and carbon footprint	Simple operation	Short time curing
				Easy controlling the curing temperature	Easy operation
					No requirement of foaming agents for porous morphology
Disadvantages	Leaching of alkaline activator	Long curing time	Depending on weather condition and geographical location	High energy requirements	Difficulty in reaction controlling
		Depending on the environment temperature		High heat loss	

Table 3 Researches of microwave cure for geopolymerization

Raw materials	Morphology	Microwave curing (MWC) conditions	Geopolymerization conditions	Chemical composition (XRF)	Compressive strength (CS)	Density	Reference
Fly ash (less than 10% Ca component)	Dense morphology	Pre-curing at ambient temperature + MWC 240,360 and 600W MW for 120 min after ambient curing for 12 h	NaOH: 4, 6, 8 and 10 M River sand	SiO_2: 50.2%, Al_2O_3: 28.7%, Fe_2O_3: 5.72%, CaO: 5.9%, MgO: 1.7%, Na_2O: 1.0%, SO_3: 1.0%	56 MPa (MWC at 240W for 120 min, with 10 M-NaOH)	–	[43]
Fly ash (Higher than 10% Ca component)		MWC + Oven cure 90W MWC for 5 min + oven cure at 65 °C for 3,6 and 12 h	NaOH: 10 M Na_2SiO_3 Sand (The mass ratio of Na_2SiO_3/NaOH was 1.5)	SiO_2: 39.2%, Al_2O_3: 19.7%, Fe_2O_3: 12.1%, CaO: 16.9%, MgO: 1.7%, Na_2O: 1.9%, SO_3: 2.9%	42.5 MPa (MWC at 90W for 5 min + Oven cure at 65 °C for 12 h)	–	[41]
Granulated blast-furnace slag (GCBFS)		Pre-curing at ambient temperature + MWC 350, 540 and 750W Pre-curing for 3 days + MWC for 5 min	NaOH Na_2SiO_3 (Alkali modulus of Na_2O/SiO_2 was 1.25)	SiO_2: 33.6%, Al_2O_3: 12.9%, Fe_2O_3: 0.6%, CaO: 41.5%, MgO: 7.9%, Na_2O: 0.3%, SO_3: 2.2%	43 MPa (pre-curing + MWC at 350W)	–	[22]

(continued)

Table 3 (continued)

Raw materials	Morphology	Microwave curing (MWC) conditions	Geopolymerization conditions	Chemical composition (XRF)	Compressive strength (CS)	Density	Reference
Metakaolin		Pre-curing at ambient temperature + MWC 3 kW MWC for 26 min used the second day at ambient temperature	$Ca(OH)_2$ Potassium water glass (Si/Al = 1.4–1.85)	SiO_2: 52.0%, Al_2O_3: 41.5%, Fe_2O_3: 1.1%, CaO: 0.2%, MgO: 0.8%, Na_2O: –, SO_3: 0.2%	62 MPa (MWC + pre-curing)	–	[56]
Fly ash (Higher than 10% Ca component)	Porous morphology	MWC 200, 500, 700 and 850W MWC for 1 min	NaOH: 2, 5, 10 and 15 M Sodium Silicate (the weight ratio of Sodium silicate/NaOH was 2.5)	SiO_2: 29.8%, Al_2O_3: 18.5%, Fe_2O_3: 16.8%, CaO: 19.8%, MgO: 1.9%, Na_2O: 1.1%, SO_3: 4.9%	–	0.9 g/cm^3 (MWC at 850 W with 15 M NaOH)	[39]

(continued)

Table 3 (continued)

Raw materials	Morphology	Microwave curing (MWC) conditions	Geopolymerization conditions	Chemical composition (XRF)	Compressive strength (CS)	Density	Reference
Natural soil		MWC 575W MWC for 4 min + 1 min to eliminate the moisture	NaOH Na_2SiO_3 (Alkali modulus of Na_2O/SiO_2 was 1.25)	SiO_2: 62.8%, Al_2O_3: 22.4%, $Fe(OH)_3$: 11.5%, Calcite: 0.92%, MgO: –, Na_2O: 0.8%, SO_3: –	3.1 MPa (MWC with 50% natural soil)	0.44 g/cm^3 (MWC with 50% natural soil)	[16]
Bottom ash		MWC 900W MWC for 4 min	NaOH: 14 M Sodium Silicate	SiO_2: 52.0%, Al_2O_3: 22%, Fe_2O_3: 17%, CaO: 6%, MgO: –, Na_2O: –, SO_3: –	3.4 MPa (MWC with 7% NaOH)	0.69 g/cm^3 (MWC at with 7% NaOH)	[57]

at 2.45 GHz, dipolar polarization generates heat in polar molecules such as water. Due to the electric fields associated with the microwaves the dipoles are aligned by rotation [45], providing more uniform heating [38, 45, 53]. In addition, in the high-alkali activator concentration, the access temperature was high [39], due to that in ionic conduction, the ions oscillate back and forth through the specimens by the electric field component of microwave irradiation. At the same time, heat generation by friction and collisions occurring between ions and molecules also supports heating [45, 53, 54]. Therefore, microwave devices are extremely convenient for influencing material reactions related to water at a frequency of 2.45 GHz, where the entire body quickly reaches a higher temperature by volumetric heating [41]. Because of microwave dielectric heating, water containing ions more efficiently heats in comparison to deionized water [55] due to ionic conduction and dipolar re-orientation variation with microwave frequency. In geopolymers, Na^+ and OH^- ions play the ion conduction role, and heat is generated when the cations and anions collide with molecules and others (Fig. 2). By utilizing such advantages of microwave heating like instantaneous, very rapid heat-up time, and easy control, fast microwave synthesis of fly ash based porous morphology geopolymers was investigated in the presence of high-alkali concentration [39].

In the feasibility of synthesis of porous fly ash-based geopolymer using microwave energy, the geopolymer pastes were cured for approximately 1 min when a household microwave oven was operated at 200–850 W with high-alkali concentration. Figure 3 shows the temperature profile of geopolymer pastes containing different sodium hydroxide (NaOH) concentrations in different power of the household microwave for 1 min [39]. For the geopolymerization, NaOH solutions and sodium silicate solutions were used as activators, and microwaves oven was operated at a frequency of 2.45 GHz under identical conditions. The solid fraction in the geopolymer mixture

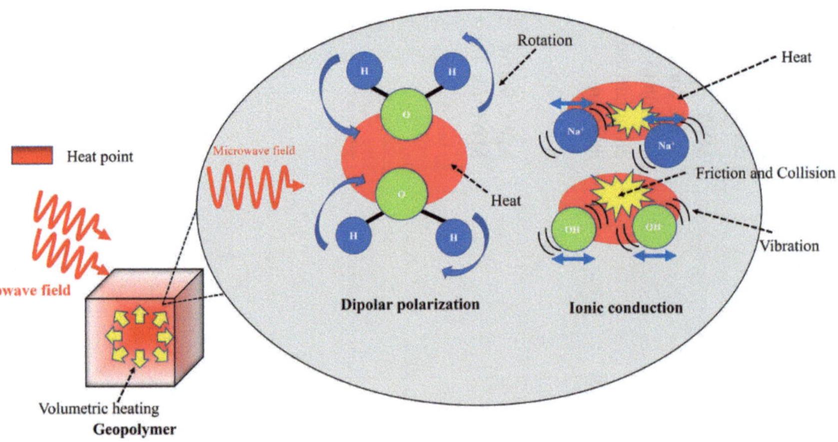

Fig. 2 Illustration of microwave heating of geopolymer by ionic conduction and dipolar re-orientation variation

Fig. 3 The plot of temperature against NaOH concentration of geopolymerized slurry solutions at different microwave oven powers (left), and pictures of geopolymers cured from components containing different NaOH concentration (upper right) and X-ray micro CT for the cross-section of geopolymers (lower right), where the microwave radiation was carried out at power 850 W for 1 min [39]

was 65 wt.%. The ratio of Na_2SiO_3/NaOH solution and SiO_2/Al_2O_3 were fixed at 2.5 and 2.69, respectively, at NaOH solution concentration of 2–15 M.

Actually, as the ion concentration increased in the fly ash aqueous phase for geopolymers, heating occurred effectively by dipolar polarization. It was known about dielectric properties of geopolymer mortar constituents that fly ash, sodium silicate, NaOH and water had relative dielectric constants of 2.9, 2.78, 8.94, 14.32 and 72.18, and relative dielectric loss factor of 0.21, 0.20, 1.70, 1.17 and 11.75, respectively [58]. When NaOH concentration increased in the mortar, microwave heating easily occurs. In the results of relative dielectric constants and relative dielectric loss factor, water strongly contributes to the temperature increase in the slurry. In addition, at high NaOH concentration, the contribution influences the temperature increase. Therefore, as seen in Fig. 3 (left), high NaOH concentration enhances temperature increase. Also as shown in Fig. 3 (right) for the morphological observation of the resulting geopolymers, porosity developed in the matrix when the NaOH concentration became high. On the other hand, it is worth noting that geopolymerization is possible even in a low concentration NaOH environment. This is because microwave dielectric heating efficiently acts on geopolymerization at the molecular level. Namely, microwave generates a phase difference in the absorbed species between the orientation of the field and that of the dipole, causing energy to be lost from the dipole by molecular friction and collisions and thus heat, which leads to efficient geopolymerization [45, 46].

As the results show, the effect of microwave heating was summarized in the geopolymer structure with or without such porous morphology. The properties of geopolymer cured using microwave were explained with the aid of previous works.

First, as comparing the different properties of the porous and dense structure, in the cases of geopolymer of dense structure without micro-sized pores having the density of over 1 g/cm^3, reference [41] reported that compressive strength and bulk density of fly ash-based geopolymers were 12.9–42.5 MPa, and 2.0–2.1 g/cm^3. On the other hand, bottom ash-based geopolymers such as porous morphology geopolymer obtained compressive strength and density of 3.6–6.2 MPa and 0.61–1.3 g/cm^3 [57]. Porous morphology geopolymer is very low compared to the compressive strength and bulk density of geopolymer using construction material [4, 10, 25, 26]. Therefore, porous morphology geopolymer is expected to be applied in lightweight concrete [44, 59] and adsorbent [8, 34] due to the low density.

Second, the parameters affecting the properties of the geopolymer using microwave curing were microwave power, curing time, and concentration of alkali solutions. According to reference [42], the fly ash-based geopolymers were synthesized at different microwave powers of 200–600 W for 15 min. The compressive strength of geopolymers cured at 300 W increased (72 MPa), but at higher microwave power of 400 and 600 W it decreased (67 and 54 MPa). It was seen that higher microwave power led to surface cracks in the geopolymer bulk, because of highly evaporated water vapor. Thus, rapid dry curing is not useful for the high compressive strength of geopolymers at higher microwave power of 400 and 600 W. The compressive strength of geopolymer cured for different curing times of 10–60 min increased with increasing curing time, especially, geopolymer cured for 60 min recorded 90 MPa in short time. Reference [43] reported the effect of concentration of alkali solution on microwave curing. Fly ash-based geopolymers with NaOH concentration of 4–10 M were cured at 240 W for 120 min. The compressive strength of geopolymers increased with increasing NaOH concentrations. When NaOH concentration was 10 M, the compressive strength recorded the maximum value of 60 MPa. This is because increased dissolution rates of fly ash particles under higher NaOH concentrations may have formed higher amounts of alumino-silicate gel even at shorter curing time. Therefore, these factors affected the mechanical properties of geopolymers.

Finally, when comparing the effect of curing with and without microwaves, Chindaprasirt et al. [41] reported that fly ash-based geopolymers cured by oven heating at 65 °C for 24 h obtained compressive strength of 32.7 MPa and bulk density of 1.9 g/cm^3, while geopolymers cured using microwave at power of 90 W for 5 min + additional oven heating at 65 °C for 12 h had higher compressive strength and bulk density of 42.5 MPa and 2.1 g/cm^3. Therefore, microwave curing provided significantly reduced curing time shorter within only 5 min.

Consequently, microwave irradiation affects geopolymer properties of their structure in dense or porosity, depending on the microwave conditions of irradiation power and time. In addition, the resultant compressive strength of geopolymers had a tendency to increase when increasing microwave curing time. This is related to the longer cure times under microwave power, the increase of the internal temperature of the geopolymer matrix, and the promotion of geopolymerization [38, 42, 60]. However, much longer irradiation of microwave conversely and significantly affects the properties of the geopolymer, for example, decreasing the compressive strength [38]. As a result, the longer microwave treatment led to thermal shrinkage and provided thermal stress on the internal structure. Then, these reached the large cracks, pores, burns, and formed glassy bubbles in the geopolymer matrix [39, 61, 62]. Therefore, appropriate parameters need to be chosen for each geopolymer fabrication system by microwave curing.

3 Microwave Cured Geopolymer Having Water Foaming Agent

In the presented geopolymers, various properties were obtained depending on pore size in the cured bulk. When the pore size increased, the weight loss occurred, and surface area was increased in the case of high porosity structure [32, 33, 39, 63]. These geopolymers find applications for adsorbent [8, 34] and lightweight concrete [59], etc., where pore-forming agents like cooking oil and starch create voids within the mixture [64], surfactants [65], and H_2O_2 when added to alkaline environment [66]. In the case of microwave heating to form the porous structure, the forming agent was water, and the porosity was caused by the modified temperature treatment, and the amount of water evaporation. Although it is difficult to control the microstructure of pore size, pore number, and pore distributions, there is an advantage in forming a porous matrix in the absence of any other foaming agents. Actually, the microstructure of the resulting geopolymers depends on the microwave condition with pre-curing as shown in Table 3. For the preparation of the geopolymer matrix, the SEM micrographs indicated a dense structure, having compressive strength in the range of 42–62 MPa [22, 41, 43, 56]. In these cases, it was convenient to operate microwave curing and conventional heating. In contrast, porous geopolymers were fabricated by microwave curing only, especially with high microwave power. This was due to microwave heating as water absorbed microwaves and generated vaporization as a result of polarization caused by external oscillation of polar water molecules [67, 68]. In the microwave curing, the water in the geopolymer paste was heated and then vaporized easily at temperatures higher than 100 °C, due to the generation of free water created bubble [68]. Then, the vaporization process may influence the morphology of the geopolymers, since bubbles caused by evaporation of water in geopolymer paste were fixed inside the geopolymer sample along with the heating curing process, resulting a porous morphology geopolymer structure [39]. Reference [39] also reported the investigation of the effect of the ratio of sodium silicate (Na_2SiO_3)/NaOH during the fabrication of fly ash-based geopolymer using the household microwave. When the water weight ratio in slurry paste increased, numerous pores formed in the geopolymer matrix, indicating that a lot of water evaporation occurred to cause pore voids during curing. In addition, the presence of NaOH in the geopolymer paste increased temperature, because dielectric heating was effective in higher NaOH. Thus, the microwave heating enhanced evaporation of water acting as a forming agent. This could be controlled to tailor the porosity and density of the geopolymers, depending upon microwave heating condition, but few studies on geopolymer foams using microwave heating. There were examples for microwave geopolymerization for porous fly ash-based geopolymer [40]. Although the fly ash-based geopolymers were cured by traditional oven heating at 80 °C for 24 h, they obtained an apparent bulk density of 1.5 g/cm^3. The microwave cured geopolymers with 15 M-NaOH concentration decreased the apparent bulk density to 0.9 g/cm^3. The effect of microwave powers and concentration of NaOH activator is investigated in Fig. 4a. In the higher microwave power with over 500–800 W, the

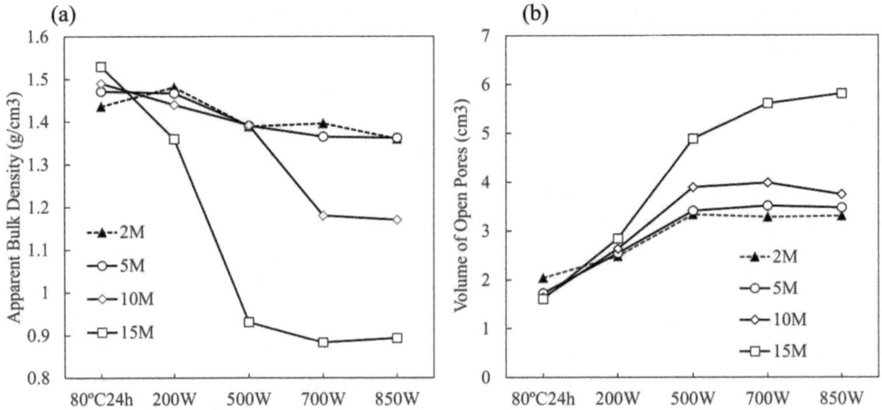

Fig. 4 Apparent bulk density (**a**), and volume of open pores (**b**) of geopolymer cured at different microwave power 200–850W having different NaOH concentration and cured by oven at 80 °C for 24 h [39]

addition of NaOH concentration increased the volume of pores in the geopolymers (Fig. 4b) [39].

Actually, this section summarizes microwave curing for geopolymer to the effective to porous forming without forming agents. Especially, higher microwave power promoted to the more porous structure in the geopolymer.

4 Pre-curing and Microwave Cured Geopolymer Concretes

The most practical application of geopolymers is as a replacement material for existing concrete [9, 69–72]. This is because of the fact that cement manufacture is highly energy intensive and emits CO_2 and other greenhouse gases responsible for global warming [69, 73]. In this way, the use of ceramics as structural materials requires a dense structure and high mechanical properties. In these cases, the geopolymer is densely packed, providing less pores. But, the microstructures of geopolymer mortars were influenced by the use of conditions under different temperatures [74], crystalline compound [75], alkali activator [76, 77], and silica to alumina ratio in the geopolymer components [78, 79].

Due to dipolar polarization and ionic conduction mechanisms, microwave heating was effective for geopolymerization in the advantage of short time preparation. If anything is ideal to decrease the evaporation of water during microwave heating, resulting in less evaporation of water even at over 100 °C, microwave radiation effectively enhanced the geopolymerization. Actually, microwave heating led to high compressive strength of fly ash geopolymers [40, 41]. But, common geopolymerization can take a long curing period to prepare the bulk ceramic body from a few days to as long as 90 days in geopolymer mortars [78, 80]. But, if pre-curing and microwaved curing were combined, the microwave method diversified to prepare densely structured geopolymers (Fig. 5). In the case of microwave heating, the same formulation composition can be changed into porous morphology geopolymer and dense structure geopolymer without micro-sized pores by changing the adjustment conditions. In the geopolymer paste made from fly ash [39], microwave heating can easily produce porous morphology geopolymer without pore forming agents. For example, when the geopolymer paste was formed with the weight ratio of (Solid/Liquid) = 1.86 and (Na_2SiO_3/10 M-NaOH) = 2.5, porous morphology geopolymer was produced within 1 min by simply exposing microwave as shown in the SEM picture inserted in Fig. 5. On the other hand, pre-curing at room temperature after preparing the geopolymer paste could prepare the geopolymer with dense structure. Actually, after the microwave curing for 1 min at both 200 W and 500 W out-put power, the value of the compressive strength (MPa) without and with a 1 h pre-curing process at room temperature (RT) increased from 5.8 MPa to 6.2 MPa, and 1.5 MPa to 3.9 MPa in the case of the 200 W and 500 W of microwave out-put power respectively as shown in Fig. 6. When the geopolymer was without pre-curing, compressive strength decreased due to the formation of pores. This means that pre-curing was possible to easily control the fabrication of the geopolymer dense structure without micro-sized pores with high compressive strength. In addition, the compressive strength of geopolymers cured by conventional oven curing at 25 °C for 7 days was 4.4 MPa, whereas geopolymer with 1 h pre-curing and microwave cured at 200 W for only 1 min was 6.2 MPa.

So, in summary, this section summarized the effect of microwave curing and pre-curing for densely structured geopolymers. By combining both curing methods, it would be possible to have the practical application of the resulting geopolymers, because of shortening the preparation period.

5 Future Prospects of Geopolymers

Given the environmental footprint of geopolymer production and operation, the benefits of concrete mainly lie in the sustainability of geopolymer production with the manufacturing process including lower resource and energy requirements and lower CO_2 emissions [23, 81, 82]. In addition, there are other benefits of geopolymers such as reuses of industrial waste and production of much less fuel, which leads to contribute in the reduction of global warming. From this point of view, geopolymers can fully play the role of alternative concrete. Recently world's first building with geopolymer concrete has been constructed at the University of Queensland's Global Change Institute and also at the Brisbane West Well camp airport in 2014, as applied by the world's largest geopolymer concrete project [82].

As for the future prospects of geopolymers, their industrial applications are of great interest. Many review articles recently summarized these applications [21, 66, 69, 71, 72]. To summarize in terms of geopolymer material form, we can consider bulk body, form body, composite, and coating as shown in Table 4. Among their

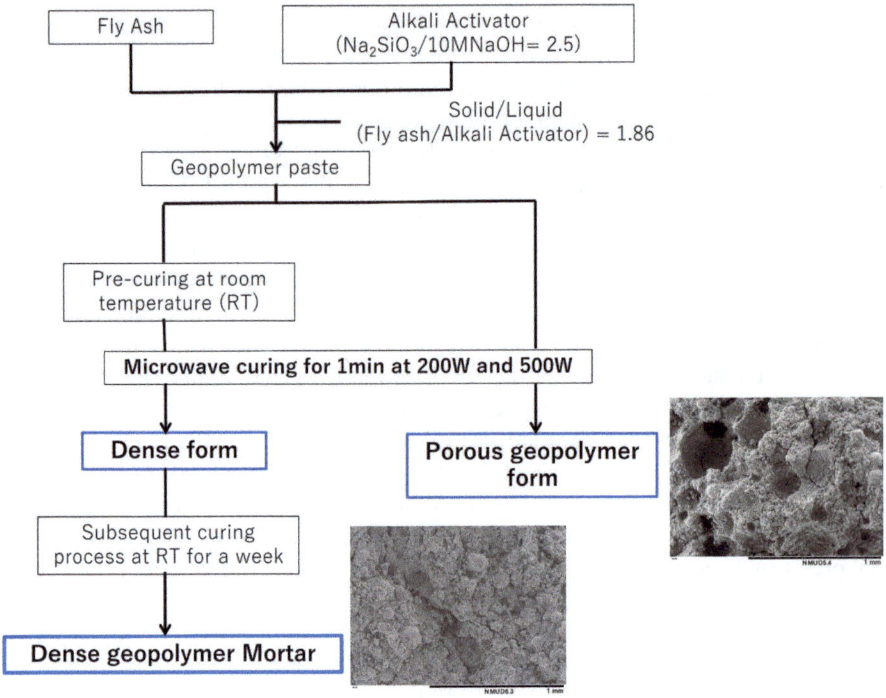

Fig. 5 Flowchart of microwave heating for preparation of porous and dense geopolymers sourced from fly ash and sodium silicate (Na_2SiO_3) in NaOH activator

Fig. 6 Compressive strength of geopolymer without pre-curing and with various pre-curing periods of time. Geopolymers were synthesized at microwave power 200 and 500 W, cured for 1 min, and at 25 °C for 7 days

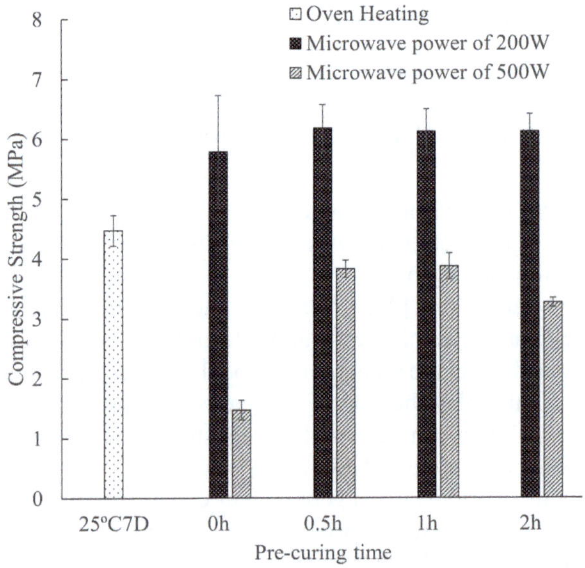

applications, the most common research case is the application of concrete, mortar, and cement for building materials. The application fields of these geopolymers have the potential to contribute significantly to the construction of a sustainable society (Fig. 7).

In particular, apart from the sustainability aspects of the geopolymer binder, the synthesis conditions become important factors affecting the environmental footprint

Table 4 Applications of geopolymer materials having different forms

Form of geopolymer material	Applications	References
Bulk body	Cement and concrete	[4, 25, 26, 29, 47, 69, 71]
	Ceramics	[83–85]
	Storage toxic, and radioactive waste	[1, 86–88]
Foam body	Adsorption of heavy metal	[8, 34]
	Light weight cement	[44, 59]
Coating	Anti-corrosion and acid and base resistant	[80, 89–91]
	Reinforcing	[92]
	Temperature resistance	[89, 93]
Composite	Fire and heat resistant matrix materials	[44, 81, 94]
	Reinforcing	[95, 96]

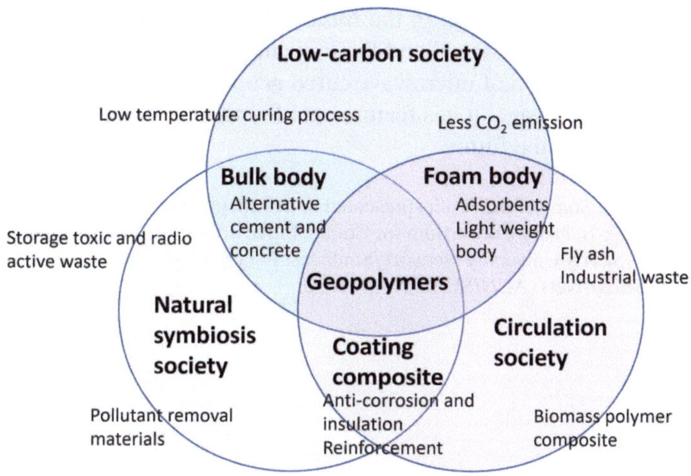

Fig. 7 Geopolymers eco-friendly cooperated with the integrated improvement of economy and environment for a sustainable society having symbiosis with human and nature. Low carbon, circulation and natural symbiosis are reached by environmental life and civilization under SDGs-ship [97]

of the production. In addition, immobilization and adsorption of pollutant metals are interesting technologies of geopolymers using bulk and foam geopolymers. Geopolymer layer coating technology on the substrates could also show reinforcement of the substrate materials and provide insulation against acid or base attack and corrosion of the body materials. Similar effects were found in geopolymer composite with organic polymers and other additives. In the future when dealing with these technologies of geopolymers, it is necessary to consider the concept containing a low-carbon society, circulation society, and natural symbiosis society. As discussed before, if geopolymer is synthesized by a common sintering process, the environmental footprint of the product may be not comparable to concrete materials, due to the consumption of high energy associated with sintering, becoming less "green" process. In other words, the low carbon and low power consumption of geopolymer processing are attractive and consistent with the goals of a sustainable society [98].

6 Final Remarks

This chapter highlights geopolymers prepared by microwave treatments, specifically how their properties can be tuned using microwaves. The incorporation of microwave heating in the geopolymer production process has the advantages of being able to adjust the porous morphology geopolymer without additives, and to make dense structure geopolymer without micro-sized pores concreted under certain conditional settings. Due to the incorporation of microwave heating in increasing efficiency and improving characteristics with less energy consumption, it is expected to be used as an eco-friendly ceramic material in the future. However, several problems such as difficulties in temperature control and the low energy conversion efficiency remain in the practical application of microwave cured geopolymers. Although this method is still in a transitional stage, it has features not found in conventional methods, and thus holds promise for the future.

Acknowledgements Some of the results presented in this chapter are obtained with the support of the Global Academia-Industry Consortium for Collaborative Education (GAICCE) program (Grant No.: CEP 1901), ASEAN University Network/Southeast Asia Engineering Education Development Network (AUN/SEED-Net) (AUN/SEED-Net), JICA.

References

1. J. Davidovits, *Geopolymer Chemistry and Applications*, 3rd edn. (Institut Géopolymère 16 rue Galilée F-02100 Saint-Quentin France, France, 2011). pp. 21–200
2. J. Davidovits, Geopolymers – inorganic polymeric new materials. J. Therm. Anal. **37**(8), 1633–1656 (1991). https://doi.org/10.1007/BF01912193
3. D. Khale, R. Chaudhary, J. Mater. Sci. **42**, 729 (2007)
4. A. Palomo, M.W. Grutzeck, M.T. Blanco, Cement Concrete Res. **29**(8), 1323 (1999)

5. J. Davidovits, J.L. Sawyer, U.S. Patent, 4,509,985, 9 Apr 1985.
6. M.B. Karakoç, I. Türkmen, M.M. Maraş, F. Kantarci, R. Demirboğa, M. Uğur Toprak, Mechanical properties and setting time of ferrochrome slag based geopolymer paste and mortar. Constr. Build. Mater. **72**, 283–292 (2014). https://doi.org/10.1016/j.conbuildmat.2014.09.021
7. P. Rovnaník, Effect of curing temperature on the development of hard structure of metakaolin-based geopolymer. Constr. Build. Mater. **24**(7), 1176–1183 (2010). https://doi.org/10.1016/j.conbuild-mat.2009.12.023
8. F.J. López, S. Sugita, M. Tagaya, T. Kobayash, J. Mater. Sci. Chem. Eng. **2**, 16 (2014)
9. G.S. Ryu, Y.B. Lee, K.T. Koh, Y.S. Chung, The mechanical properties of fly ash-based geopolymer concrete with alkaline activators. Constr. Build. Mater. **4**, 409–418 (2013). https://doi.org/10.1016/j.conbuildmat.2013.05.069
10. O. Sujitra, J. Sirrithan, T. Kobayashi, Constr. Build. Mater. **101**, 298 (2015)
11. J.L. Francisco, S. Sugita, M. Tagaya, T. Kobayashi, Geopolymers using rice husk silica and metakaolin derivatives; preparation and their characteristics. J. Mater. Sci. Chem. Eng. **2**(5), 35–43 (2014). https://doi.org/10.4236/msce.2014.25006
12. N.S. Singh, S. Thokchom, R. Debbarma, Properties of fly ash and rice husk ash blended geopolymer with sodium aluminate as activator solution. Eng. Appl. Sci. Res. **48**(1), 92–101 (2021). https://ph01.tci-thaijo.org/index.php/easr/article/view/240316
13. C.T. Chen, C.H. Chang, AJCE-Special Issue **37**(2), 97 (2019)
14. A. Albidah, M. Alghannam, H. Abbas, T. Almusallam, Y. Al-Salloum, Characteristics of metakaolin-based geopolymer concrete for different mix design parameters. J. Mater. Res. Tech. **10**, 84–98 (2021). https://doi.org/10.1016/j.jmrt.2020.11.104
15. S. Prasanphan, A. Wannagon, T. Kobayashi, J. Sirithan, Constr. Build. Mater. **221**, 409 (2019)
16. M. Nadeem, E.U. Haq, F. Ahmed, M.A. Rafiq, G.H. Awan, M.Z. Abdein, Constr. Build. Mater. **230**(10), 117074 (2020)
17. J.E. Jaimes, A.M. Montañol, C.P. González, Geopolymer derived from bentonite: Structural characterization and evaluation as a potential sorbent of ammonium in waters. J. Phys.: Conf. Ser. **1587**(1), 012008(2020). https://doi.org/10.1088/1742-6596/1587/1/012008
18. Y. Yanga, J. Jianga, L. Hou, Z. Lu, L. Jun, J. Li, J. Wang, Pore structure and properties of porous morphology geopolymer based on pre-swelled bentonite. Constr. Build. Mater. **254**(6), 119226 (2020). https://doi.org/10.1016/j.conbuildmat.2020.119226
19. C. Tippayasam, P. Keawpapasson, P. Thavorniti, T. Panyathanmaporn, C. Leonelli, D. Chaysuwan, Constr. Build. Mater. **53**, 455 (2014)
20. A. Nazari, J.G. Sanjayan, J. Clean. Prod. **99**, 297 (2015)
21. N. Shehata, E.T. Sayed, M.A. Abdelkareem, Sci. Total Envir. **762**, 143166 (2021)
22. G. Kastiukas, S. Ruan, S. Liang, X. Zhou, J. Clean. Prod. **225**(10), 120290 (2020)
23. S. Kumar, R. Kumar, Ceram. Int. **37**(2), 533 (2011)
24. N. Böke, G.D. Birch, S.M. Nyale, L.F. Petrik, Constr. Build. Mater. **75**, 189 (2015)
25. M.O. Yusuf, M.A.M. Johari, Z.A. Ahmad, M. Maslehuddin, Mater. Des. **55**, 387 (2014)
26. A. Islam, U.J. Alengaram, M.Z. Jumaat, I.L. Bashar, Mater. Des. **56**, 833 (2014)
27. A.N. Murri, V. Medri, E. Papa, L. Laghi, C. Mingazzini, E. Landi, Porous morphology geopolymer insulating core from a metakaolin/biomass ash composite. Environments **4**(4), 86 (2017). https://doi.org/10.3390/environments4040086
28. B. Nematollahi, J. Sanjayan, F.U.A. Shaikh, Ceram. Int. **41**, 5696 (2015)
29. M. Torres-Carrasco, F. Puertas, Rev. Ing. de Constr. RIC **32**(2), 5 (2017)
30. M.M.A.B. Abdullah, L. Jamaludin, K. Hussin, M. Bnhussain, C.M.R. Ghazali, M.I. Ahmad, Int. J. Mol. Sci. **13**(4), 4388 (2012)
31. G. Masi, W.D.S. Rickard, L. Vickers, M.C. Bignozzi, A. van Riessen, Ceram. Int. **40**, 13891 (2014)
32. X. Zhang, C. Bai, Y. Qiao, X. Wang, D. Jia, H. Li, P. Colombo, Compos. A Appl. Sci. Manuf. **150**, 106629 (2021)
33. C. Bai, P. Colombo, Ceram. Int. **44**(14), 16103 (2018)
34. J.L. Francisco, S. Sugita, T. Kobayash, Chem. Lett. **43**(1), 128 (2014)
35. H.M. Giasuddin, J.G. Sanjayan, P.G. Ranjith, Fuel **107**, 34 (2013)

36. J. Zhang, J.L. Provis, D. Feng, J.S.J. van Deventer, Ceram. Int. **43**(14), 11233 (2017)
37. P. Nath, P.K. Sarker, V.B. Rangan, Procedia Eng. **125**, 601 (2015)
38. Y. Sun P. Zhang, J. Hu, B. Liu, J. Yang, S. Liang, K. Xiao, H. Houngfei, Constr. Build. Mater. **294**, 123491 (2016)
39. O. Sujitra, J. Sirithan, P. Thavorniti, T. Kobayashi, Ceram. Int. **42**(8), 9866 (2016)
40. P. Chindaprasirt, U. Rattanasak, S. Taebuanhuad, Mater. Struct. **46**, 375 (2013)
41. P. Chindaprasirt, U. Rattanasak, S. Taebuanhuad, Adv. Powder Tech. **24**, 703 (2013)
42. A. Graytee, J.G. Sanjayan, A. Nazaria, Ceram. Int. **44**(7), 8216 (2018)
43. J. Somaratna, D. Ravikumar, N. Neithalath, Cem. Concr. Res. **40**, 1688 (2010)
44. E. Ul. Haq, S.K. Padmanabhan, A. Licciulli, Fuel Process. Tech. **130**, 263 (2015)
45. J. Anwar, U. Shafique, W. Zaman, R. Rehman, M. Salman, A. Dar, J.M. Anzano, U. Ashraf, S. Ashraf, Arabian J. Chem. **8**, 100 (2015)
46. P. Rattanadecho, N. Suwannapum, B. Chatveera, D. Atong, N. Makul, Mater. Sci. Eng. A **472**(1–2), 299 (2008)
47. C.K.Y. Leung, T. Pheeraphan, Cem. Concr. Res. **27**(3), 463 (1997)
48. X. Xu, Y. Bao, C. Song, W. Yang, J. Liu, L. Lin, Microporous Mesoporous Mater. **75**(3), 173 (2004)
49. X. Xu, W. Yang, J. Liu, L. Lin, Sep. Purif. Tech. **25**(1), 241 (2001)
50. H. Tanaka, A. Fujii, S. Fujimoto, Y. Tanaka, Adv. Powder Tech. **19**(1), 83 (2008)
51. T. Gubb, I. Baranova, S.M. Allan, M.L. Fall, H. Shulman, W. Kriven, in *Developments in strategic materials and computational design II*. ed. by A.L. Gyekenyesi, W.M. Kriven, J. Wang, S. Widjaja, D. Singh (Wiley, USA, 2011), p.35
52. C.O. Kappe, in *Comprehensive Medicinal Chemistry II*, ed. by J.B. Taylor and D.J. Triggle (Elsevier Science, Oxford, 2007), p. 837
53. V.G. Gude, P. Patil, E. Martinez-Guerra, S. Deng, N. Nirmalakhandan, Sustain. Chem. Proc. **1**, 5 (2013)
54. K. Nakamura, J. Home Econ. Jpn. **54**(5), 351 (2003)
55. C. Gabriel, S. Gabriel, E. Grant, B. Halstead, D. Michael, P. Mingos, Dielectric parameters relevant to microwave dielectric heating. Chem. Soc. Rev. **27**(3), 213–224 (1998). https://doi.org/10.1039/A827213Z
56. P. Hájková, Minerals **8**(10), 444 (2018)
57. E.U. Haq, S.K. Padmanabhan, A. Licciulli, Fuel Process. Technol. **130**, 263 (2015)
58. S. Jumrat, B. Chatveera, P. Rattanadecho, Int. Commun. Heat Mass Transf. **38**(2), 242 (2011)
59. S. Hanif, M.A. Tahir, K. Rashid, M.U. Rehman, N. Saleem, A. Aslam, G. Naeem, Physico-mechanical performance of lightweight geopolymer foam aggregates developed by geopolymerization through microwave-oven irradiations. J. King Saud Univ. Eng. Sci. (2021). https://doi.org/10.1016/j.jksues.2021.04.002
60. S. Samantasinghar, S. Singh, Constr. Build. Mater. **235**, 117481 (2020)
61. S. Hong, H. Kim, Sci. Rep. **9**, 15694 (2019)
62. S. Hong, H. Kim, Korean J. Chem. Eng. **36**(7), 1164 (2019)
63. E.U. Haq, S.K. Padmanabhan, M. Zubair, L. Ali, A. Licciulli, Constr. Build. Mater. **126**(15), 951 (2016)
64. S.K. Kaliappan, A.A. Siyal, Z. Man, M. Lay, R. Shamsuddin, Effect of pore forming agents on geopolymer porosity and mechanical properties. AIP Conf. Proc. **2016**, 020066 (2018). https://doi.org/10.1063/1.5055468
65. Y. Qiao, X. Li, C. Bai, O. Icon, H. Li, J. Yan, Effects of surfactants/stabilizing agents on the microstructure and properties of porous morphology geopolymers by direct foaming. J. Asian Ceram. Soc. **9**(1), 412–423 (2020). https://doi.org/10.1080/21870764.2021.1873482
66. N.H. Jamil, W.M.A.W Ibrahim, M.M.A.B. Abdullah, A.V. Sandu, M.F.M. Tahir, IOP Conf. Ser.: Mater. Sci. Eng. **209**, 012004 (2017)
67. M. Vollmer, Physics of the microwave oven. Phys. Educ. **39**(1), 74 (2004). https://doi.org/10.1088/0031-9120/39/1/006
68. S. Hanjitsuwan, S. Hunpratub, P. Thongbai, S. Maensiri, V. Sata, P. Chindaprasirt, Cem. Concr. Compos. **45**, 9 (2019)

69. N.B. Singh, M.H. Kumar, S. Rai, Mater. Today: Proc. **29**, 743 (2020)
70. A. Bhutta, M. Farooq, C. Zanotti, N. Banthia, Mater. Struct. **50**(1), 1 (2017)
71. P. Zhang, Y. Zheng, K. Wang, J. Zhan, A review on properties of fresh and hardened geopolymer mortar. Compos. B Eng. **152**, 79–95 (2018). https://doi.org/10.1016/j.compositesb.2018.06.031
72. F. Farooq, J. Xin, M.F. Javed, A. Akbar, M.I. Shah, F. Aslam, R. Alyousef, Geopolymer concrete as sustainable material: a state of the art review. Constr. Build. Mater. **306**, 124762 (2021). https://doi.org/10.1016/j.conbuildmat.2021.124762
73. R. Arellano-Aguilar, O. Burciaga-Diaz, A. Gorokhovsky, J.I. Escalante-Garcia, Constr. Build. Mater. **50**, 642 (2014)
74. D. Adak, M. Sarkar, S. Mandal, Constr. Build. Mater. **70**, 453 (2014)
75. A.R. Kotwal, Y.J. Kim, J. Hu, V. Sriraman, Int J. Concr. Struct. Mater. **9**(1), 35 (2015)
76. A. Bhowmick, S. Ghosh, Int. J. Civ. Struct. Eng. **3**(1), 168 (2012)
77. T.O. Yusuf, M. Ismail, J. Uaman, A.H. Noruzman, Impact of blending on strength distribution of ambient cured metakaolin and palm oil fuel ash based geopolymer mortar. Adv. Civ. Eng. **2014**(24), 1–8 (2014). https://doi.org/10.1155/2014/658067
78. E.J. Guades, Experimental investigation of the compressive and tensile strengths of geopolymer mortar: the effect of sand/fly ash (S/FA) ratio. Constr. Build. Mater. **127**, 484–493 (2016). https://doi.org/10.1016/j.conbuildmat.2016.10.030
79. H.M. Khater, H.A. Abd el Gawaad, Constr. Build. Mater. **102**, 329 (2016)
80. S. Thokchom, P. Ghosh, S. Ghosh, Int. J. Recent Trends Eng. Technol. **1**(6), 36 (2009)
81. D.L.Y. Kong, J.G. Sanjayan, Cem. Concr. Compos. **30**(10), 986 (2008)
82. T.Ch. Madhavi, P.M. Rameshwaran, *Geopolymer Concrete—The Eco Friendly Alternate to Concrete* (NBM & CW Infra Construction & Equipment Magazine, 2020), https://www.nbmcw.com/product-technology/construction-chemicals-waterproofing/concrete-admixtures/geopolymer-concrete-the-eco-friendly-alternate-to-concrete.html. Accessed at 18 May 2020
83. N.I. Zulkifli, M.M.A.B. Abdullah, M.A.A.M. Salleh, R. Ahmad, A.V. Sandu, N.A.M. Mortar, Development of geopolymer ceramic as a potential reinforcing material in solder alloy: short review. IOP Conf. Ser.: Mater. Sci. Eng. **743**, 012023 (2020). https://doi.org/10.1088/1757-899X/743/1/012023
84. L. Yun-Minga, H.C. Yong, L. Li, N.A. Jaya, M.M.A. Bakri, A.T. Soo, J.K. Hussin, Constr. Build. Mater. **156**, 9 (2017)
85. J. Davidovits, Geopolymers: ceramic-like inorganic polymers. J. Ceram. Sci. Technol. **8**(3), 335–350 (2017). https://doi.org/10.4416/JCST2017-00038
86. M. Łach, D. Mierzwiński, K. Korniejenko, J. Mikuła., M. Hebda, J. Air Waste Manag. Assoc. **68**(11), 1190 (2018)
87. J. Davidovits, *Environmental Implications of Geopolymers* (Mater. Today, 2015), https://www.materialstoday.com/polymers-soft-materials/features/environmental-implications-of-geopolymers/. Accessed at 29 Jun 2015
88. T.H. Vu, N. Gowripalan, J. Adv. Concrete Technol. **16**(3), 124 (2018)
89. Q. Tian, S. Wang, Y. Sui, Z. Lv, Int. J. Adhes. Adhes. **110**, 102934 (2021)
90. S. Sikora, E. Gapys, B. Michalowski, T. Horbanowicz, M. Hynowski, E3S Web of Conf. **49**, 00101 (2018)
91. C. Jiang, A. Wang, X. Bao, T. Ni, J. Ling, J. Build. Eng. **32**, 101734 (2020)
92. M.K. Harun, M.Z.A. Yahya, S. Abdullah, C.H. Chan, Adv. Mater. Res. **686**, 227 (2013)
93. X. Rong, Z. Wang, X. Xing, L. Zhao, Review on the Adhesion of geopolymer coatings. ACS Omega **6**, 5108–5112 (2021). https://doi.org/10.1051/e3sconf/20184900101
94. G. Mucsi, R. Szabó, S. Nagy, K. Bohács, I. Gombkötő, Á. Debreczeni, IOP Conf. Ser.: Mater. Sci. Eng. **251**, 012079(2017)
95. J. Davidovits, *Reinforced Geopolymer Composites: A Critical Review* (Mater. Today, 2016), https://www.materialstoday.com/polymers-soft-materials/features/reinforced-geopolymer-composites-a-critical-review/. Accessed at 31 May 2016
96. Y. Huang, L. Gong, Y. Pan, C. Li, T. Zhou, X. Cheng, RSC Adv. **8**, 2350 (2018)

97. Suistainable Development Goals; The 17 Goals. (United Nations: Department of Economic and Social Affairs Sustainable Development, New York). https://sdgs.un.org/goals. Accessed 9 Jan 2023
98. T. Kobayashi, L. Nakajima, Current Opin. Green Sustain. Chem. **28**, 100439 (2021)

Index

S. J. Ikhmayies (ed.), *Advanced Ceramics*, Advances in Material Research and Technology, https://doi.org/10.1007/978-3-031-43918-6

B

C

F

G

N

O

P

T

MIX
Papier aus verantwortungsvollen Quellen
Paper from responsible sources
FSC® C105338

If you have any concerns about our products,
you can contact us on
ProductSafety@springernature.com

In case Publisher is established outside the EU,
the EU authorized representative is:
Springer Nature Customer Service Center GmbH
Europaplatz 3, 69115 Heidelberg, Germany

Printed by Libri Plureos GmbH
in Hamburg, Germany